14th Topical Conference on Gas Utilization 2014

Topical Conference at the 2014 AIChE Spring Meeting and 10th Global Congress on Process Safety

New Orleans, Louisiana, USA
30 March – 3 April 2014

ISBN: 978-1-63439-063-7

14th Topical Conference on Gas Utilization 2014

Topical Conference at the 2014 AIChE Spring Meeting and 10th Global Congress on Process Safety

New Orleans, Louisiana, USA
30 March – 3 April 2014

ISBN: 978-1-63439-063-7

AIChE

Printed from e-media with permission by:

Curran Associates, Inc.
57 Morehouse Lane
Red Hook, NY 12571

Some format issues inherent in the e-media version may also appear in this print version.

Printed by Curran Associates, Inc. (2014)

For permission requests, please contact AIChE
at the address below.

AIChE
3 Park Avenue
New York, NY 10016-5991

Phone: (203) 702-7660
Fax: (203) 775-5177

www.aiche.org

Additional copies of this publication are available from:

Curran Associates, Inc.
57 Morehouse Lane
Red Hook, NY 12571 USA
Phone: 845-758-0400
Fax: 845-758-2634
Email: curran@proceedings.com
Web: www.proceedings.com

TABLE OF CONTENTS

354012 Possible vs. Practical: Engineers Must Lead the Development of Practical Technologies

Abstract:

Chemical Engineering is need today more than ever. There is a strong societal desire for sources of energy and feedstocks that can sustainably support a high standard of living for all. Thermodynamic, material and economic challenges stand in the way. Most of the general public does not have the basic understanding of the energy flows, chemical transformations, and the scale of the systems that support our current way of life. These concepts are fundamental to chemical engineering. Chemical engineers should be on the vanguard of separating hype from truly practical options yet occationally we let society down. This talk will present several examples where the fundamentals of chemical engineering can be used to access the practicality of proposed solutions to energy and water issues, as well as explain why the venture capital model struggles in the materials market.

Key Decisions and Factors Influencing them for a Successful Liquefaction Plant – Perspective From an OEM

Patrice Bardon
Vice President Business Development
Dresser-Rand

14 AIChE Spring Meeting, New Orleans, LA

DRESSER-RAND

Agenda

- Installed LNG Capacity and D-R Experience
- Project Timeline
- Trends and Business Drivers
  - Why Efficiency Matters
- Mechanical Drivers Available
- Selecting a Mechanical Driver – GT options
  - Assumptions
  - CAPEX and OPEX comparisons
  - Other Variables and Benefits
- Conclusions

14 AIChE Spring Meeting, New Orleans, LA

DRESSER-RAND

© Copyright 2014

D-R Experience In LNG

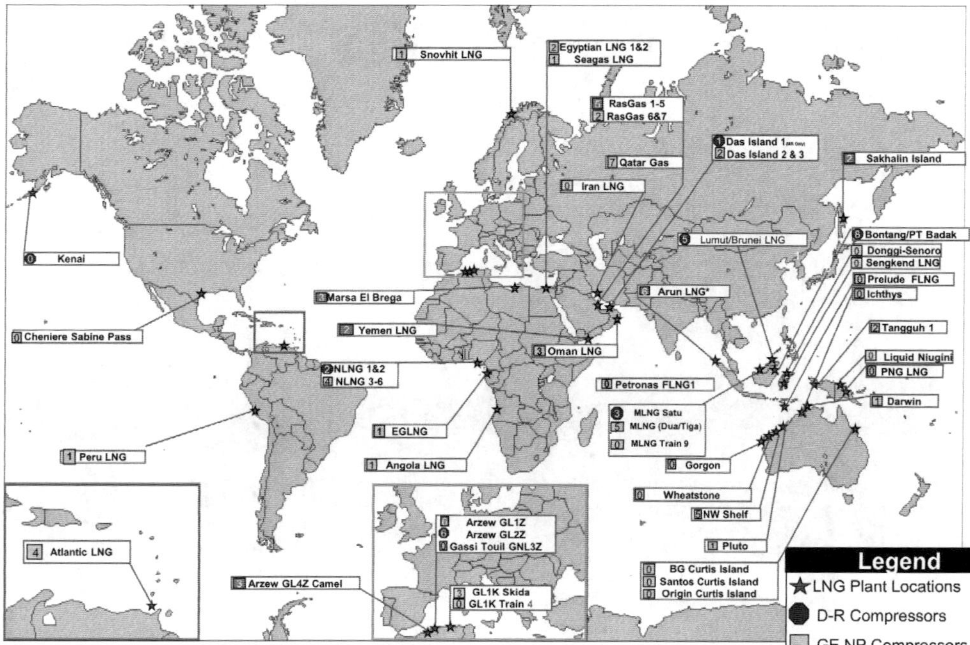

As of January 2014, 25% of the world's line of LNG in operation is achieved using D-R compressors

* LNG Plants shown are greater than 1 MTPA

Source: D-Rdata & industry publications

Full Load String Test Facility

14 AIChE Spring Meeting, New Orleans, LA

DRESSER-RAND

Project Timeline & Key Decisions

- Gas reservoir size
  - Impacts plant size and number of LNG lines
- Gas quality
  - Impacts the gas treatment required to meet the gas quality needed
- Location (onshore / offshore)

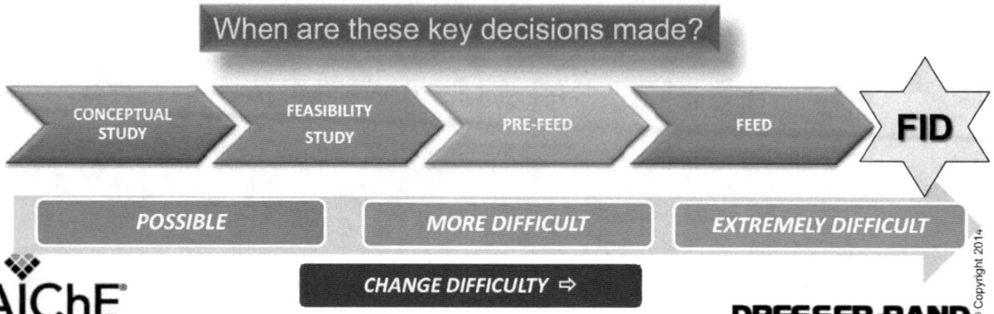

14 AIChE Spring Meeting, New Orleans, LA

DRESSER-RAND

Processes available

If you select the most efficient process, you should select the most efficient driver

	Approximate relative power consumption		MPTA	MW	Relative Fuel consumption
C3 MR	0.24	Kwh/Kg LNG	8	217	100
CoP cascade	0.26	Kwh/Kg LNG	8	239	110
DMR	0.27	Kwh/Kg LNG	8	247	113
SMR	0.29	Kwh/Kg LNG	8	265	122
HFC precooling + N2	0.32	Kwh/Kg LNG	8	290	133
SMR Limum	0.32	Kwh/Kg LNG	8	292	134
Triple N2	0.32	Kwh/Kg LNG	8	292	134
Dual N2	0.34	Kwh/Kg LNG	8	311	143
Niche (CH4 - N2)	0.43	Kwh/Kg LNG	8	388	179
Warm MR Cold MR	0.44	Kwh/Kg LNG	8	402	185
Nitrogen	0.45	Kwh/Kg LNG	8	411	189

More processes exist - not all these processes are practical for FLNG

Sources: various - above are order of magnitude
(no commercial or technical tolerance)

14 AIChE Spring Meeting, New Orleans, LA

DRESSER-RAND

Market Trends

- Historically main refrigeration units were steam turbine driven
- Darwin was the first Cascade process to use aero gas turbines
- Most subsequent Cascade projects have used aero (ex Angola LNG Fr.6):
  - Brass – the 3 Gladstone projects – Wheatstone – Sabine Pass...etc.
- PNG LNG is the first C3 MR process to use aero gas turbines
- Prelude FLNG has selected Steam Turbine but the next 'lean LNG' will consider aero gas turbines
- All other FLNG are going aero gas turbines
- SnØhvit, Cove Point and Freeport LNG are the only projects using electric motors
  - Pars, Persian LNG, Shtokman and Kitimat had considered E-LNG solutions
- Several projects remain Frame 7 driven: Ichthys, Dominion Cove, Hackberry LNG, Yamal.
- Lakes Charles LNG moved from Frame 7 to Trents
- Hitachi is willing to enter the market with the H-80

Aero gas turbines and modern Industrial gas turbines are very fuel efficient
⟹ more gas liquefied

14 AIChE Spring Meeting, New Orleans, LA

DRESSER-RAND

Efficiency

- When the power available from the driver is fixed, one percent of better compressor efficiency equals to one percent of incremental production
  - This is confirmed by all process providers – M$ 181 NPV for a 10 MTPA plant for each percent of compressor efficiency
- Driver fuel burnt is LNG that will not be produced
  - How is the project going to evaluate this?
- If electrical motor are chosen, the losses between the GT and the electrical motor shaft can be assumed to be 9.5%
- Steam turbines provide an efficiency is around 26%
- Gas turbine will have thermal efficiencies between 34 %to 42%

The driver and driven compressor efficiencies will have a direct impact on the fuel consumption and the production rate

14 AIChE Spring Meeting, New Orleans, LA

DRESSER-RAND

Mechanical drivers for LNG refrigeration units

- **Steam Turbines**
  - Poor thermal cycle efficiency, high system footprint and weight
  - Good reliability
  - Boilers can accept nearly any kind of fuel gas
- **Electric Motors**
  - Grid stability issues and high total system CAPEX
  - Very good driver availability
  - No power variation with ambient conditions
- **Industrial Gas Turbines (HDGT)**
  - Good reliability
  - High specific fuel consumption
  - Higher emissions
  - Longer downtime for scheduled or unscheduled maintenance
  - Large foot print and heavy
- **Aero Derivative Gas Turbines**
  - Capex competitive
  - As reliable as HDGTs
  - Highest Efficiency
  - Lightest weight
  - Low production losses due to short downtime

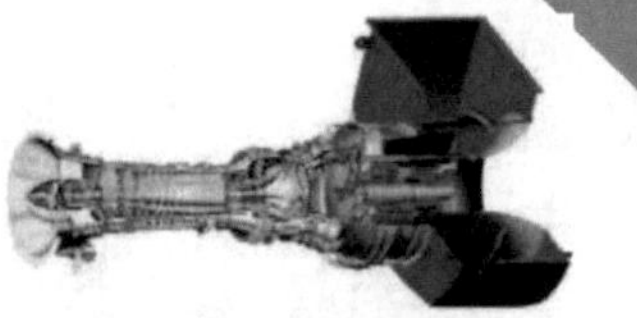

D-R VECTRA (LM2500 Family) HSPT

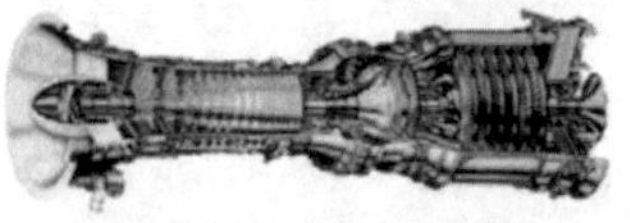

LM2500 six pack LSPT

RRES TRENT

LM6000 LSPT

Entering the market: Hitachi H-80

14AIChE
Spring Meeting, New Orleans, LA

DRESSER-RAND.

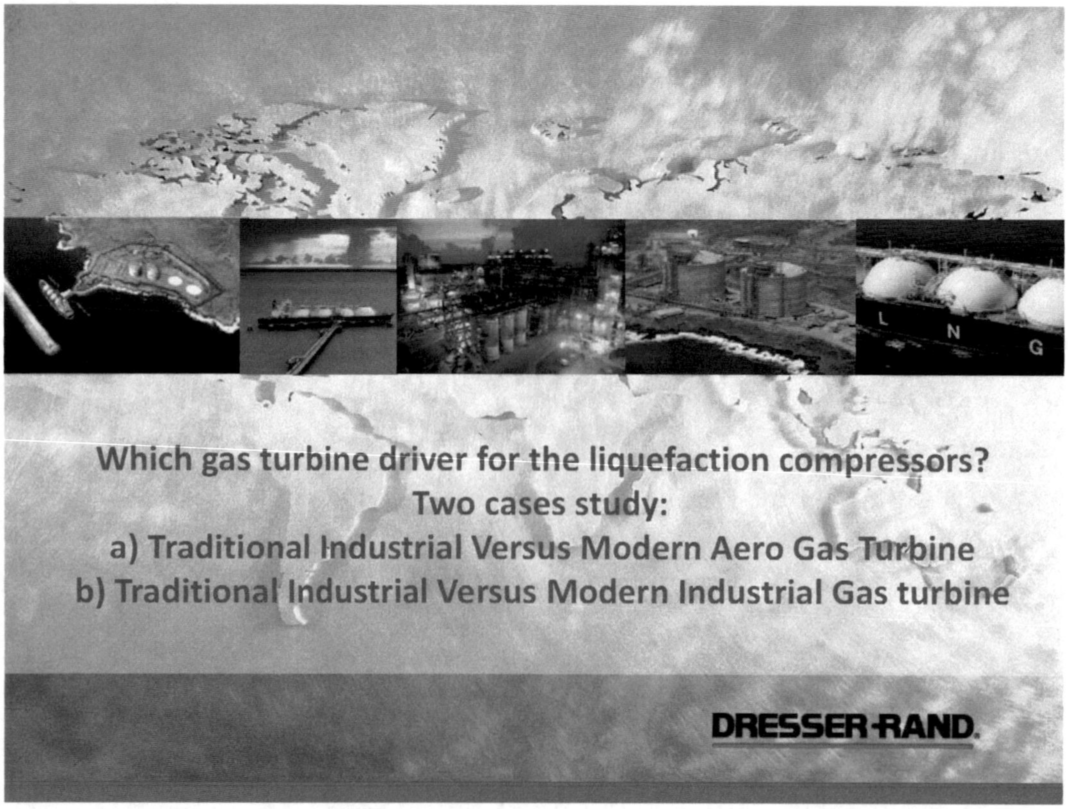

Assumptions made

- Gas price 4 $/Mbtu
- LNG price 11.47 $/Mbtu (600$ / T)
- Average gas 52.32 Mbtu/tonnes of LNG
- Site rating for the turbine 30 Deg C
- 120/260 mm H2O inlet exhaust losses for gas turbine performance
- 10% turbine aging and fouling
- Capex exclude civil work
- Scheduled maintenance based on published information
- Gas quality gives 52.3 Mbtu/tonne
- Fuel gas contains up to 30% of N2
- Average fuel gas for the turbines
- CO2 emission 55 Kg/Gigajoules
- 20.8 Euro per tonne of CO2
- $/Euro exchange rate 1.3
- NPV base on 10% financial rate
- NPV at 10 and 20 years
- One spare aero engine available at site

14AIChE®
Spring Meeting, New Orleans, LA

DRESSER-RAND®

Assumptions Continued

- Capacity 10 MTPA (no safety factor or commercial margin)
- 20 MW starter for industrial gas turbine
- helper VSD motor for the Large Industrial gas turbine
- Various compressor configuration
- 9% Electrical losses
- Capex based on D-R market knowledge and real cost
- Comparison between several drivers:
  - Large industrial with out helper
  - Large industrial with helper
  - 30 MW class aero
  - 40 MW class aero
  - 50 MW class aero
  - VSD with large industrial power plant
- 10 MW of incremental electrical power per MPTA to be produced
- Process piping evaluated as twice the price of the compressors without drivers

APCI C3 MR process with an assumed power requirement of 0.24 KW per Kg of LNG

14AIChE®
Spring Meeting, New Orleans, LA

DRESSER-RAND®

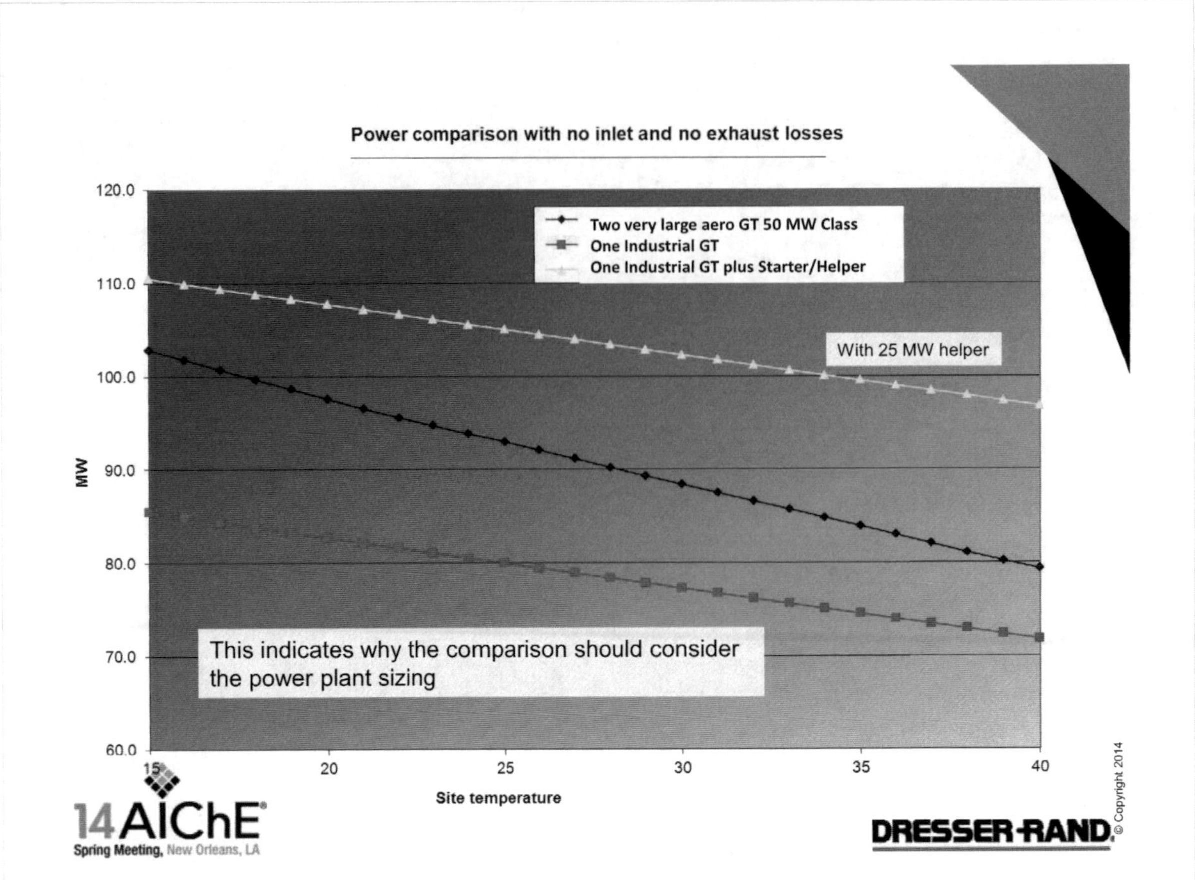

Power comparison with no inlet and no exhaust losses
Two very large aero GT 50 MW Class
One Industrial GT
One Industrial GT plus Starter/Helper
With 25 MW helper
This indicates why the comparison should consider the power plant sizing
MW
120.0
110.0
100.0
90.0
80.0
70.0
60.0
15
20
25
30
35
40
Site temperature
14AIChE
Spring Meeting, New Orleans, LA
DRESSER-RAND
© Copyright 2014

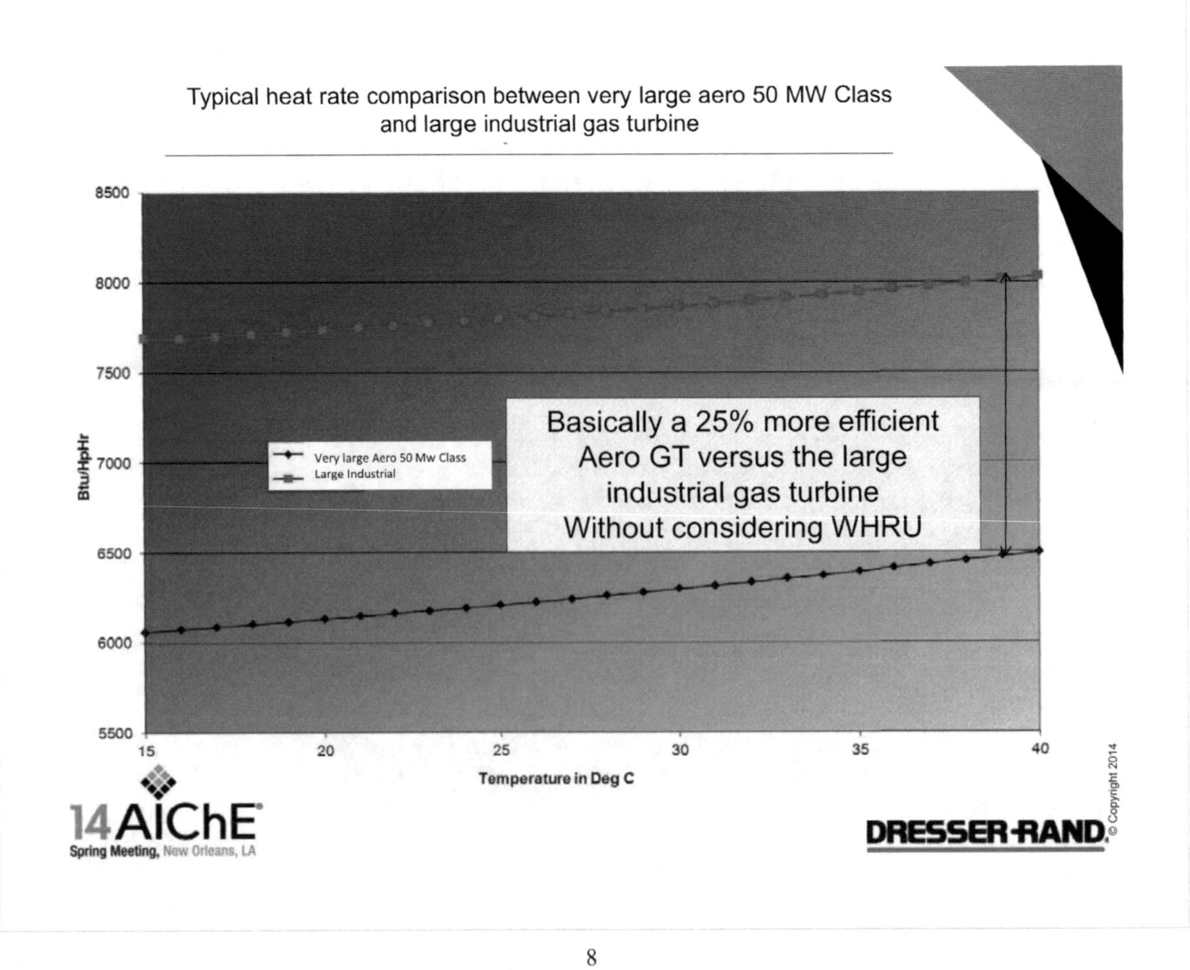

Typical heat rate comparison between very large aero 50 MW Class and large industrial gas turbine
Very large Aero 50 Mw Class
Large Industrial
Basically a 25% more efficient Aero GT versus the large industrial gas turbine Without considering WHRU
Btu/HpHr
8500
8000
7500
7000
6500
6000
5500
15
20
25
30
35
40
Temperature in Deg C
14AIChE
Spring Meeting, New Orleans, LA
DRESSER-RAND
© Copyright 2014

Opex of one case 10 MTPA

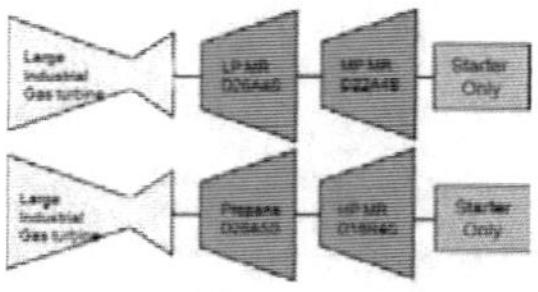

Very Large Industrial GT two lines

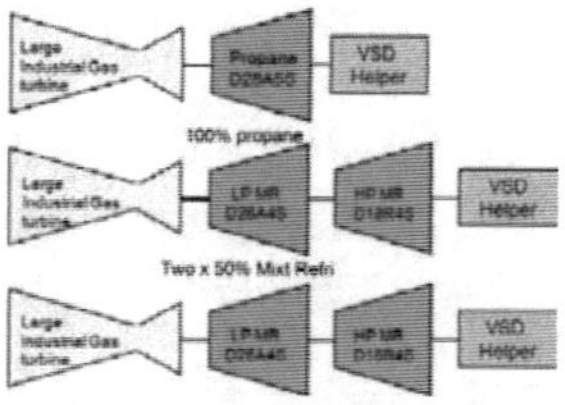

Very Large Industrial GT One line

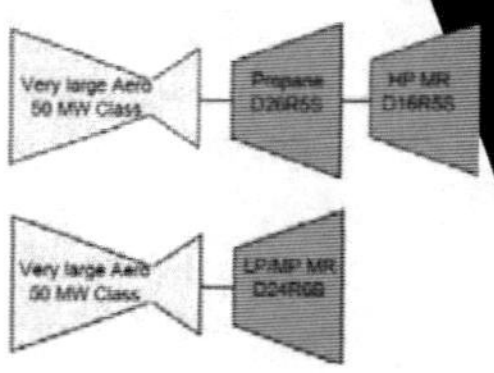

50 MW Class Aero GT four lines

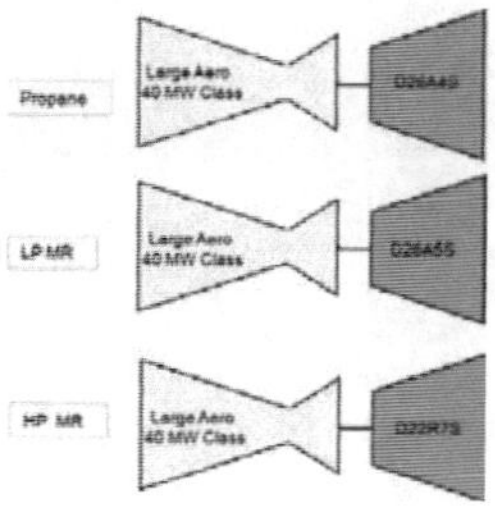

40 MW Class Aero GT three lines

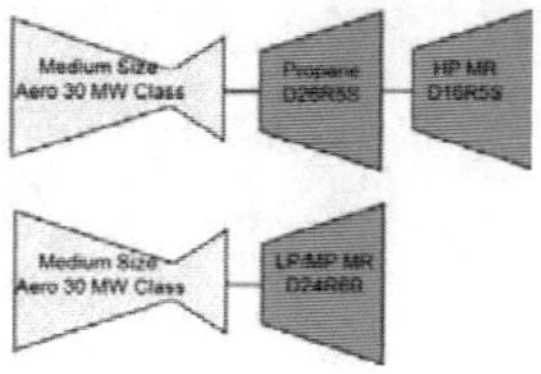

30 MW Class Aero GT six lines

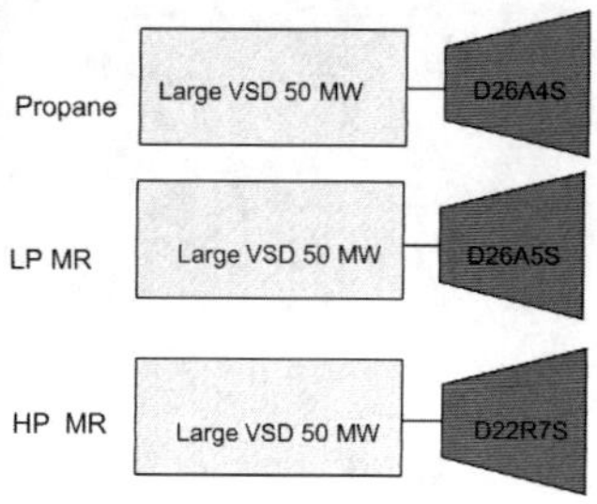

Very large VSD two lines

14 AIChE® Spring Meeting, New Orleans, LA

DRESSER-RAND®

Relative fuel saved based on LNG price

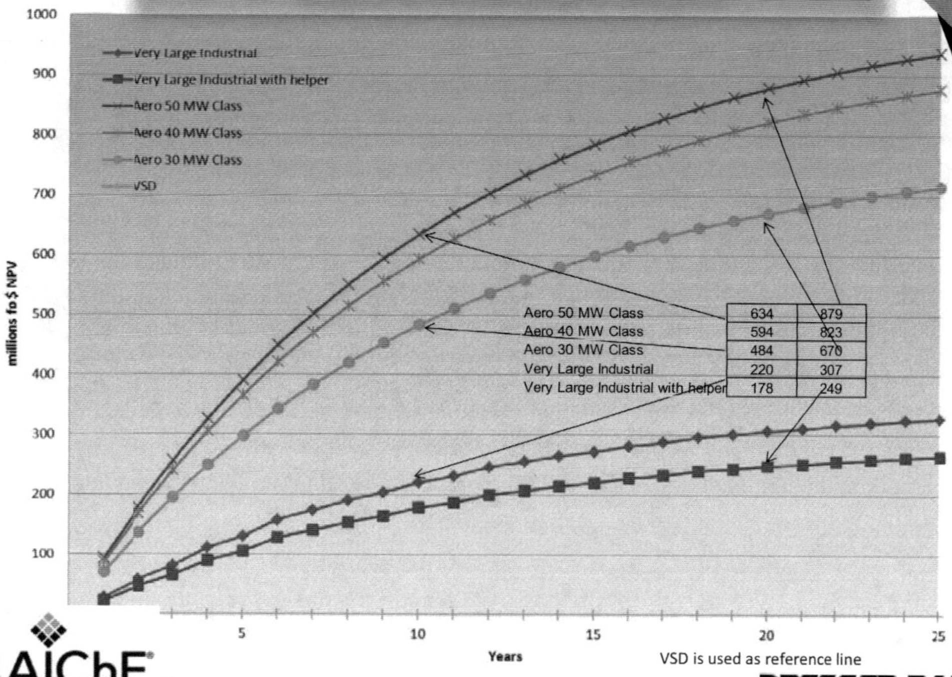

Aero 50 MW Class	634	879
Aero 40 MW Class	594	823
Aero 30 MW Class	484	670
Very Large Industrial	220	307
Very Large Industrial with helper	178	249

VSD is used as reference line

14 AIChE® Spring Meeting, New Orleans, LA

DRESSER-RAND®

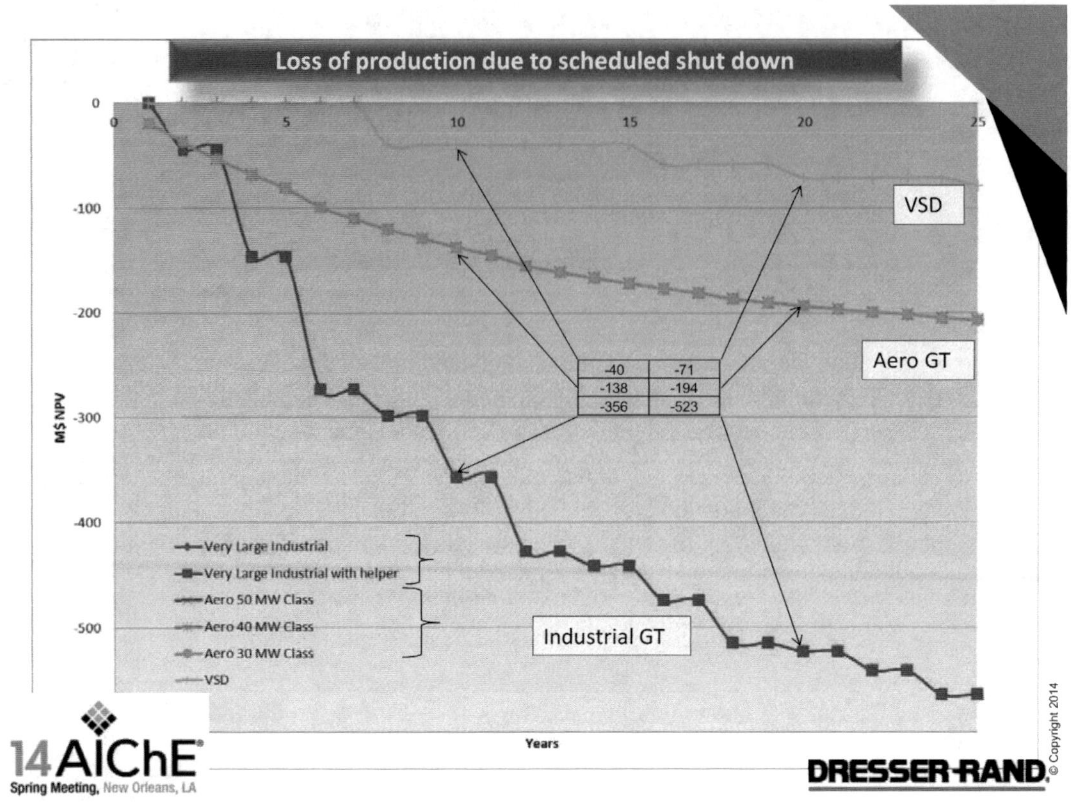
Loss of production due to scheduled shut down
0
-100
-200
-300
-400
-500
M$ NPV
0 5 10 15 20 25
Years
VSD
Aero GT
Industrial GT
-40 -71
-138 -194
-356 -523
Very Large Industrial
Very Large Industrial with helper
Aero 50 MW Class
Aero 40 MW Class
Aero 30 MW Class
VSD
14 AIChE Spring Meeting, New Orleans, LA
DRESSER-RAND
© Copyright 2014

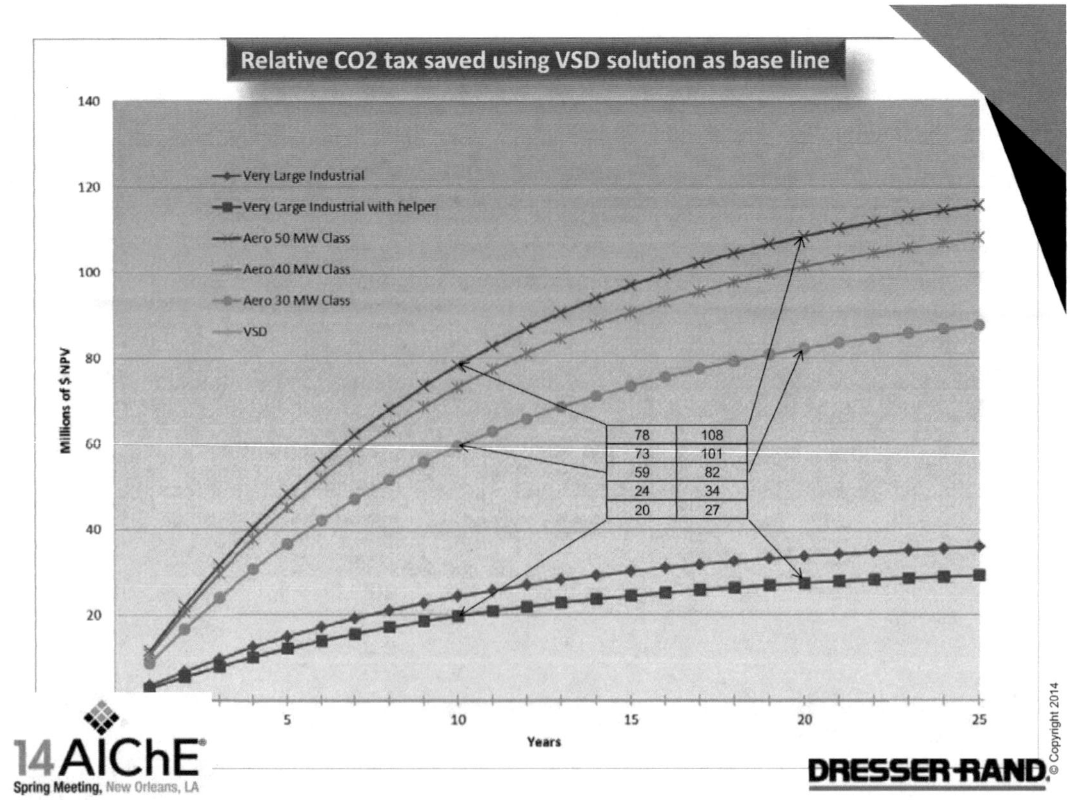
Relative CO2 tax saved using VSD solution as base line
140
120
100
80
60
40
20
Millions of $ NPV
5 10 15 20 25
Years
Very Large Industrial
Very Large Industrial with helper
Aero 50 MW Class
Aero 40 MW Class
Aero 30 MW Class
VSD
78 108
73 101
59 82
24 34
20 27
14 AIChE Spring Meeting, New Orleans, LA
DRESSER-RAND
© Copyright 2014

Operating Income linked to the incremental spare power available

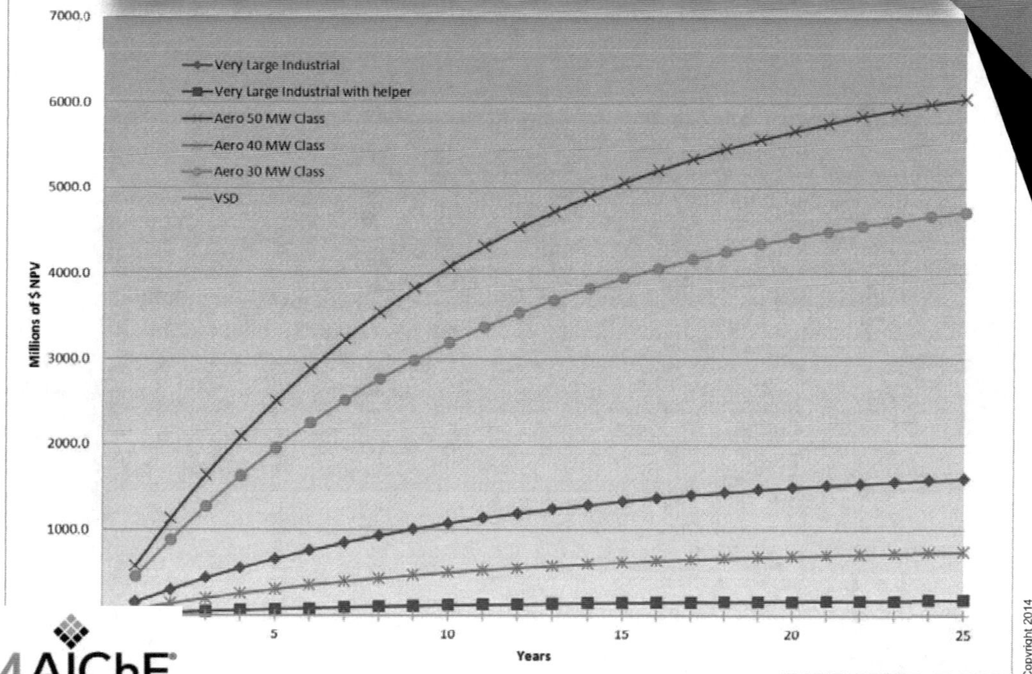

14 AIChE Spring Meeting, New Orleans, LA

DRESSER-RAND

Summary At 10 Years

	10 years					
	Very Large Industrial No Helper	Very Large Industrial With Helper	50 MW Class Aero GT	40 MW Class Aero GT	30 MW Class Aero GT	VSD
Capex	599	535	620	547	573	854
Fuel consumption	2344	2386	1930	1970	2080	2564
Scheduled shut down	356	356	138	138	138	40
Realiability loss of production	963	1475	867	1155	1251	444
CO2 tax	296	301	242	247	261	320
Total	4558	5053	3796	4057	4303	4222
Potential incremental OI due to the spare power	1080	122	4083	502	3187	0

All values are in Millions of Dollars NPV

14 AIChE Spring Meeting, New Orleans, LA

DRESSER-RAND

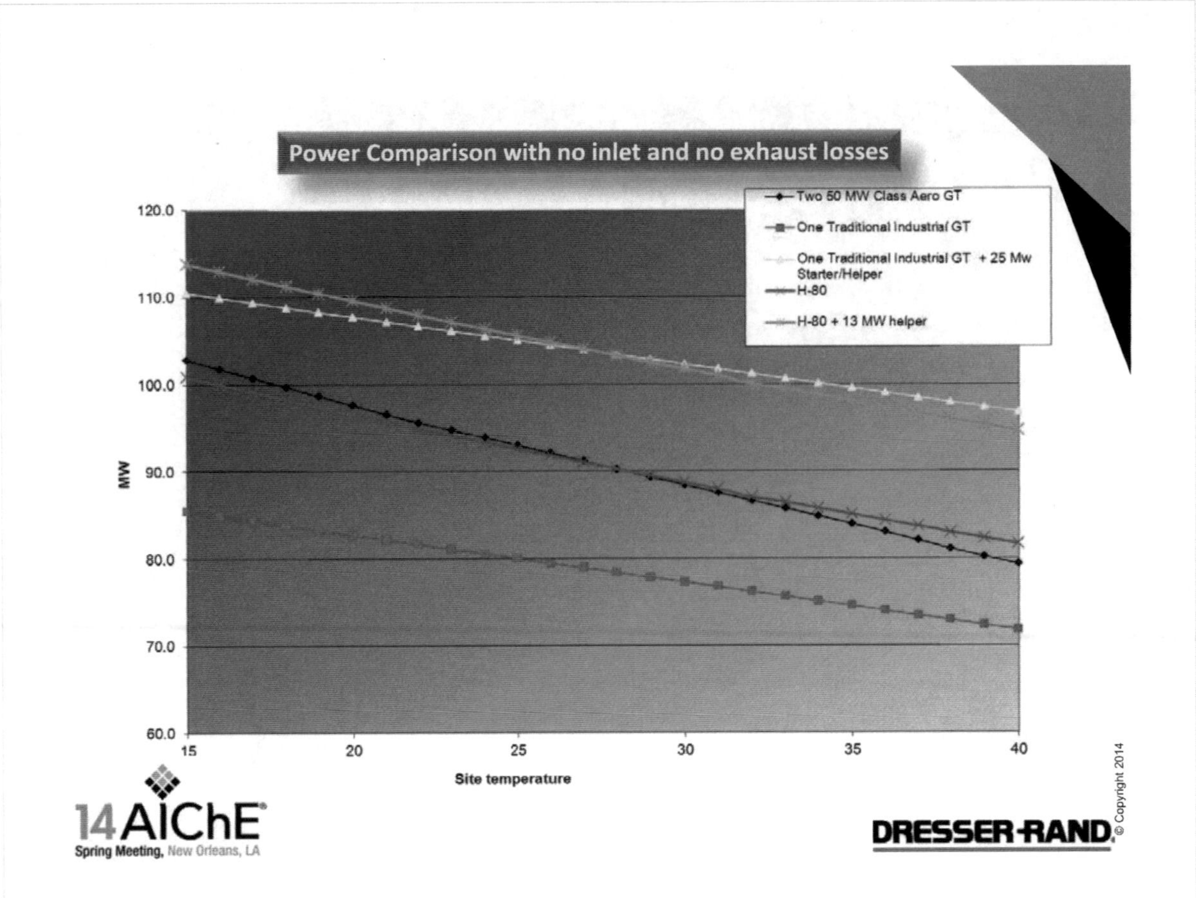
Power Comparison with no inlet and no exhaust losses
Two 50 MW Class Aero GT
One Traditional Industrial GT
One Traditional Industrial GT + 25 Mw Starter/Helper
H-80
H-80 + 13 MW helper
120.0
110.0
100.0
90.0
80.0
70.0
60.0
MW
15
20
25
30
35
40
Site temperature
14AIChE
Spring Meeting, New Orleans, LA
DRESSER-RAND
© Copyright 2014

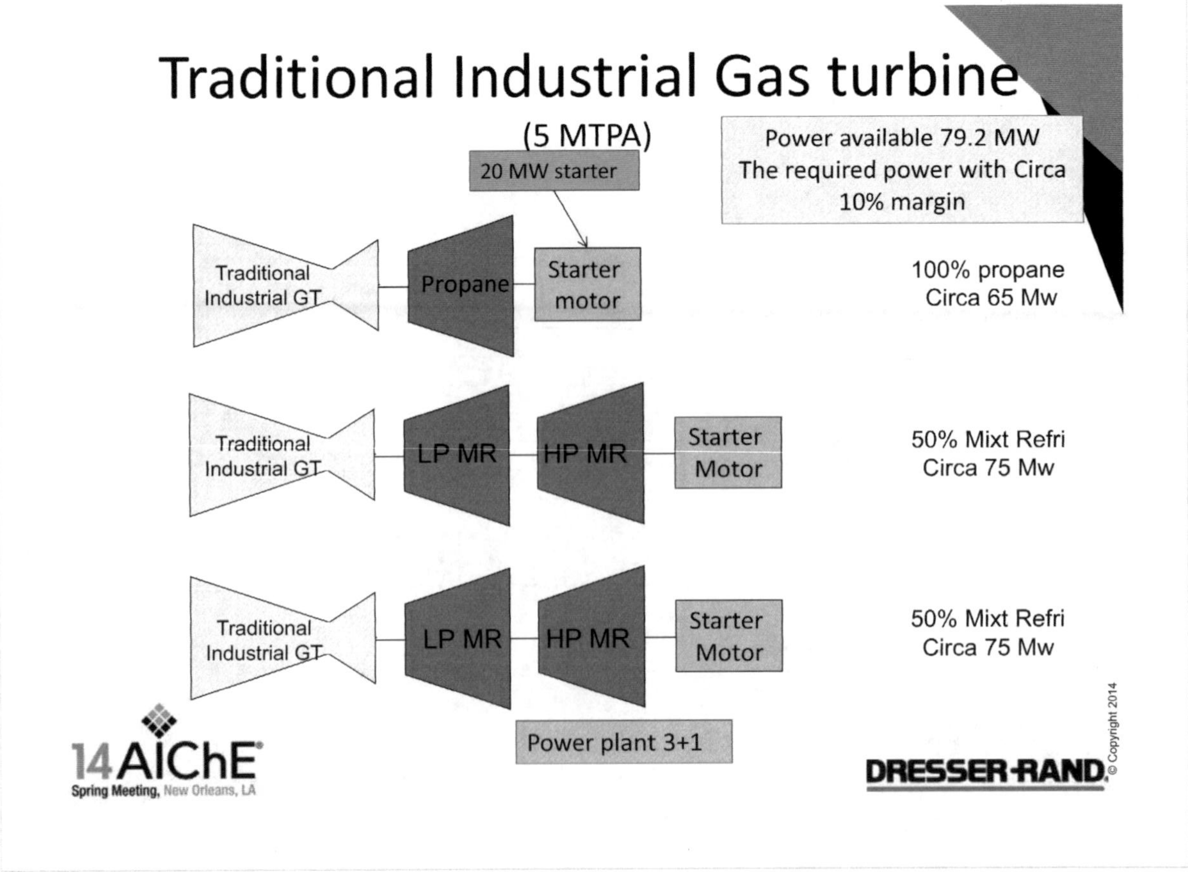
Traditional Industrial Gas turbine
(5 MTPA)
20 MW starter
Power available 79.2 MW
The required power with Circa 10% margin
Traditional Industrial GT
Propane
Starter motor
100% propane
Circa 65 Mw
Traditional Industrial GT
LP MR
HP MR
Starter Motor
50% Mixt Refri
Circa 75 Mw
Traditional Industrial GT
LP MR
HP MR
Starter Motor
50% Mixt Refri
Circa 75 Mw
Power plant 3+1
14AIChE
Spring Meeting, New Orleans, LA
DRESSER-RAND
© Copyright 2014

Modern Industrial Gas turbine H-80

(5 MTPA)

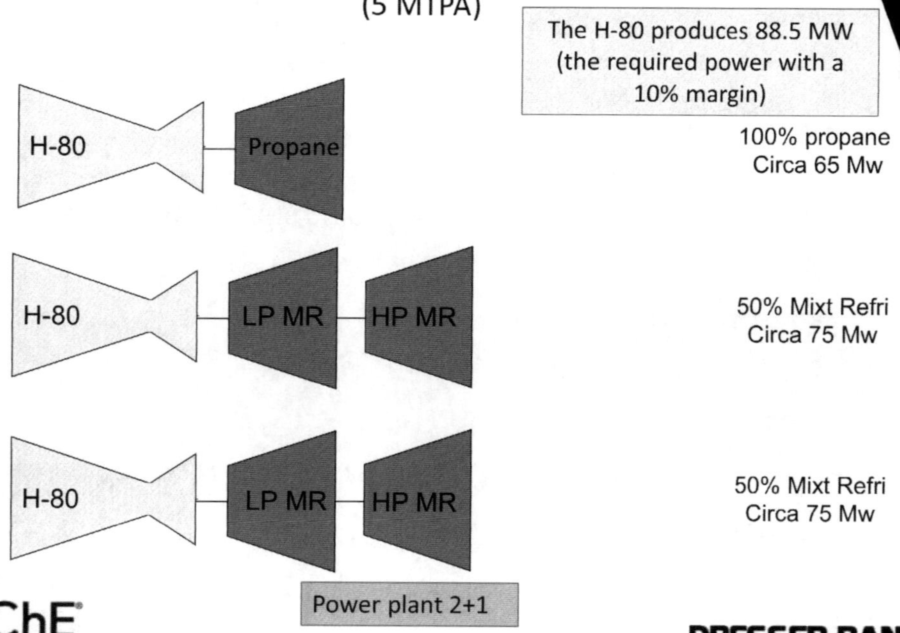

14 AIChE Spring Meeting, New Orleans, LA

DRESSER-RAND

H-80 Versus & Heavy duty gas turbine at 10 years of operation

	Traditional Heavy Duty	H-80	Benefits
Compressors and power generation	601	481	120
Fuel cost based on LNG price	1801	1594	207
Losses due to schedule shutdown	226	163	63
Losses due to reliability factor	908	638	270
Potential CO2 tax	455	400	55
Total in M$			**715**

The H-80 driver is a very attractive driver that provides with the DATUM compressor value to the operator
An incremental production is also available with the H-80

14 AIChE Spring Meeting, New Orleans, LA

DRESSER-RAND

Aero and Modern Industrial Gas Turbine Benefits

- Variable speed (flexibility of operation)
- Re-start from hot condition
- Start from Settle Out Pressure (no flaring of the refrigeration gas)
- Maintainability
- No turbine turning gear
- Reduced power plan (No starter motor)
- Simplicity

The large LNG plants need to select the driver very early in the project and optimize the process for the driver power available

Conclusions

- The liquefaction island of an LNG plant consumes a large quantity of power.
- The most efficient solution should be driving the decision
  - Most efficient process
  - Most efficient driver
- Helpers are reducing the efficiency of the solution
- The production rate per line of LNG shall be calculated based on the driver selected and then multiply the number of line as needed
- Aero gas turbines are always more efficient than traditional industrial GT solution
- The largest aero gas turbine will provide the highest plant efficiency.
- Modern industrial gas turbines provide an excellent compromise between aero gas turbine and traditional industrial

The liquefaction Opex and Capex is always lower when efficient driver is selected

www.dresser-rand.com
info@dresser-rand.com
DRESSER-RAND.

30b

StarLNG™: a Family of Small-to-Mid-Scale LNG Processes

Heinz C. Bauer, Linde Engineering, Pullach, Germany

Abstract

Linde Engineering has successfully developed and marketed a range of processes, which cover the range between 30 t/d and 3,000 t/d LNG product. These processes are part of the StarLNG™ family and are typically based on the efficient LIMUM® process, which is Linde's proprietary SMR technology. However, expander based processes like EcoRel (a trademark of Cryostar, a Linde company) can also be chosen, if the focus lies on ease of operation and safety. Cryogenic heat exchangers can be selected from a large portfolio including brazed aluminum plate-fin heat exchangers (also available in block-in-shell configuration) and coil wound heat exchangers fabricated from aluminum or stainless steel.

Introduction

World class base load LNG plants with a typical capacity between 3 and 8 mtpa LNG are thermodynamically well understood. The impact of operating conditions like feed gas pressure and composition as well as ambient temperature have been tested in operation since decades and the lessons learnt found their way into the design of e.g. propane precooled mixed refrigerant liquefaction systems or various concepts with cascaded refrigeration units.

Small-to-mid-scale LNG plants with a nameplate capacity less than 1.5 mtpa are just starting to get established. For certain applications, however, including floating LNG this train size seems to become more attractive in the future. This paper will discuss various influences on the energy efficiency for a selected choice of typical liquefaction processes.

To focus on the competing liquefaction technologies the natural gas conditions have been selected such that straight forward liquefaction can be done without further processing (e.g. a scrub column for heavies removal), which otherwise might favor one specific technology. The dry feed gas composition, which has been used for process simulation is 0.8 mol% N_2, 98.3 mol% CH_4, 0.8 mol% C_2H_6, 0.1 mol% C_3H_8 plus 50 vppm CO_2. The resulting molecular weight is 16.3 only, which has some impact on the specific energy consumption (see below). Gases with a similar composition can be found widely including coal seam gas. The operating conditions during liquefaction have been set in such a manner that 5 mol% of the LNG leaves the end flash system as vapor. Reheating the cold end flash gas was not considered, as small-to-mid-scale LNG plants may use cold suction end flash compressors. Other parameters like pressure drop in equipment and piping as well as efficiency of rotating equipment have been selected on a fair and equitable basis for all processes.

To narrow the number of liquefaction technologies, which have been used for this comparison, processes with one refrigerant only (MR or N_2) and relevant references in the industry have been picked.

Process A

The first process is based on a single mixed refrigerant (SMR) concept with 4 components (N_2, C_1, C_2, and C_4) and plate-fin heat exchangers (PFHE). Even if Process A may look like the PRICO® process, which is licensed by Black & Veatch, it is a proprietary process (patent pending) by Linde Engineering. With a temperature approach (i.e. minimum temperature difference in the composite heat-temperature curves for all warm and cold streams at any temperature level within the heat exchanger) of min. 2.5 K within the PFHEs about 10 full size cores are required for 1 mtpa capacity.

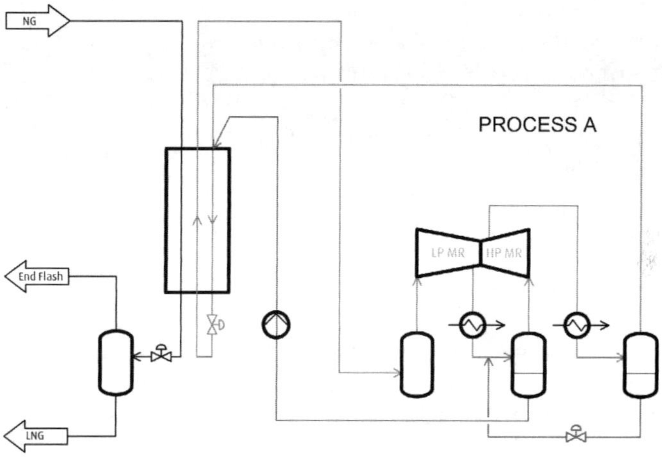

Process B

The second process is similar to Process A as far as the refrigeration compressor set-up and the refrigerant composition are concerned. A coil wound heat exchanger (CWHE) with a more relaxed temperature approach (min. temperature difference 4 K) is used for heat transfer, however. This concept is patented by Linde Engineering (e.g. US patent 6,334,334) and is marketed as LIMUM® process.

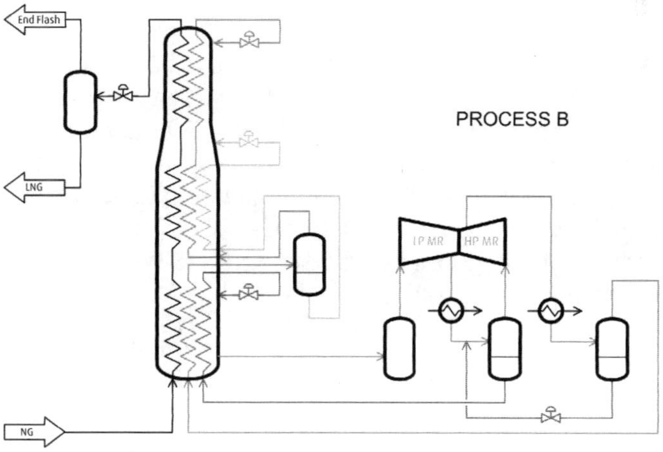

Process C

The third process is a typical representative of the various double N_2 expander processes, which are offered in the market. Again, PFHEs (only 4 cores required for 1 mtpa) with a min. temperature difference of 2.5 K have been assumed for the process calculations. This concept had been patented formerly by BHP Petroleum (expired US patent 5,768,912) and is marketed now by Linde AG with proprietary modifications.

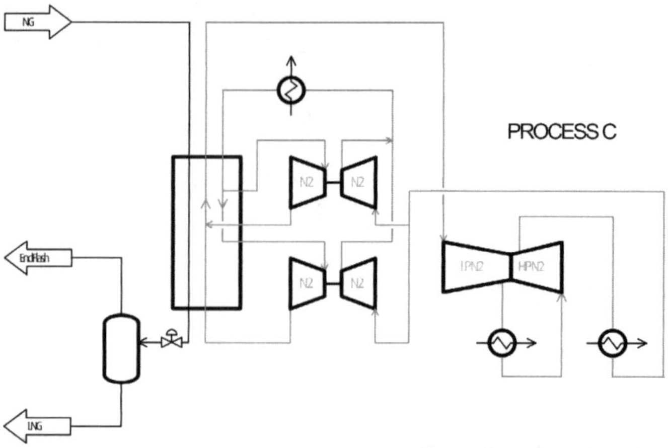

Results and Interpretation

A systematic variation of the parameters 'feed gas pressure' and 'fluid temperature' at the warm end of the cryogenic heat exchangers (PFHE or CWHE) has been executed. A feed gas pressure variation between 20 and 80 bar has been selected to cover most of the realistic conditions. Further, the temperature of all fluids (natural gas, liquid and gaseous refrigerants) entering the cryogenic exchanger has been set to +40°C (warm climate) and ly -40°C (pre-cooled process or extremely cold climate). A check calculation at 0°C and 50 bar has been made to ease interpolation between the temperature extremes.

Other parameters like compressor efficiency and temperature approach have been selected conservatively, but consistent amongst the compared processes. Thus, the derived figures are not as optimistic, as they may be found in other advertisement oriented publications.

Efficiency

The following diagrams for the Processes A, B and C show the results for the specific energy consumption in kWh/t_{LNG}. To convert the figures from kWh/t_{LNG} into $kW/t_{LNG}pd$ divide by 24 (example: 360 kWh/t_{LNG} is equivalent to 15 $kW/t_{LNG}pd$). For mixed refrigerant compression 80% isentropic efficiency had been chosen, for N_2 compression only 75%. The accuracy of the consumption figures can be assumed to be within a margin of ±3% to reflect simplifications in the simulation model.

Process A and Process B have comparable specific energy consumption within a wide range of parameters and are about 10-20% higher compared to a conventional propane precooled MR process. Process B tends to get less efficient at low feed gas pressure as the high tube side pressure drop of a CWHE reduces the feed gas pressure during liquefaction significantly. At low inlet temperature, however, Process B is very competitive. Both SMR Processes A and B have been simulated with propane instead of butane as refrigerant component in case of -40°C inlet temperature to allow for a suitable refrigerant boiling behavior. The choice of either C_2H_4 or C_2H_6 as refrigerant component has little influence on the energy consumption and shall be made on the basis of ready availability.

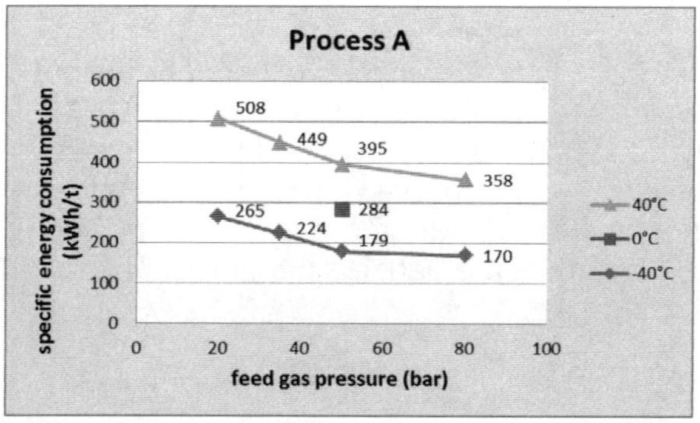

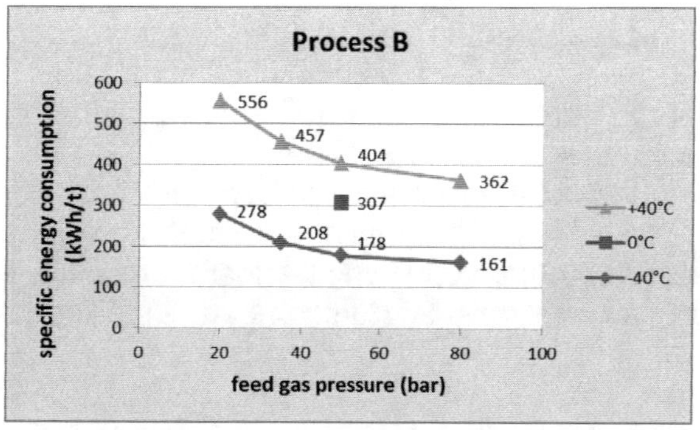

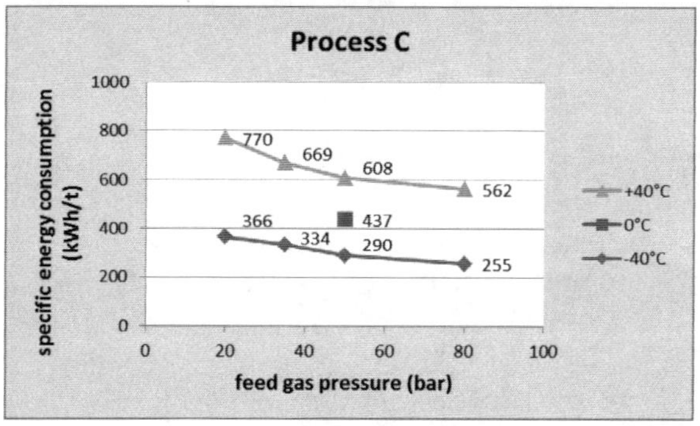

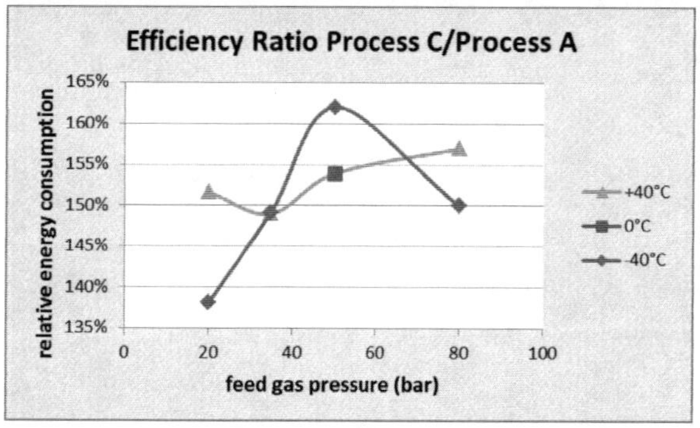

Process C requires significantly more power over the full parameter range compared to both MR based processes. In comparison to Process A about 50% extra compressor duty has to be installed. Under certain conditions, however, like low pressure and low temperature the ratio between Process C and Process A shrinks substantially. It is known in the industry that the LNG sub-cooling cycle in the Qatar mega trains, which is a N_2 expander cycle, is very efficient and on par with MR sub-cooling concepts.

All information given above refers to the lean natural gas composition with only 16.3 g/mol, which has been defined before. In case the hydrocarbon composition of a real feed gas stream differs from the selected one, the energy consumption figures can be converted with the following formula

$$power_b = power_a \times MW_a / MW_b$$

If the actual feed gas translates into a molecular weight of 18 or 19 g/mol, the figures for the specific energy consumption look much better, as they would be 10 to 15% lower now. The correction via the molecular weight (MW) is an estimate only and may fail, if the concentration of light ends (nitrogen, etc.) is significantly higher than assumed in this report.

Heat Exchanger Size

A set of enthalpy-temperature diagrams for 40°C and 50 bar is shown underneath. Process A shows the best temperature approach over most of the temperature range. This fact, however, is due to the high circulation rate of the refrigerant, which causes a high thermal duty of the heat exchanger (the heat duty in the diagrams refers to 3,000 t/h LNG production). The condensation plateau of natural gas between 200 and 190 K is almost hidden by the large contribution of the refrigeration cycle. Process B with its wider temperature approach does not carry the entire refrigerant stream to the cold end, but adds heavier refrigerant fractions subsequently to the vaporizing refrigerant mixture. Thus, the duty of the heat exchanger can be reduced by almost 25% in comparison to Process A.

Process A

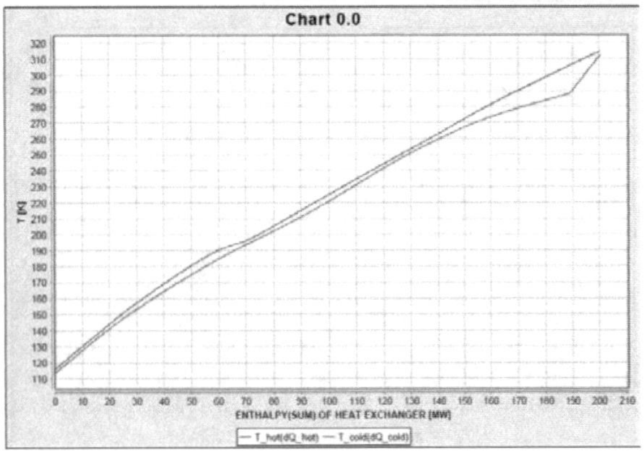

Process B

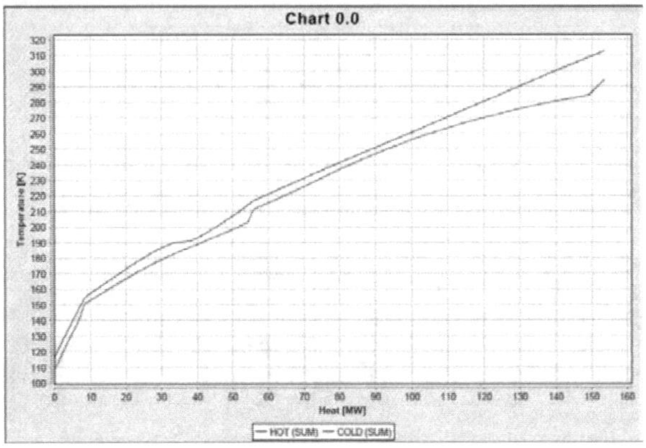

Process C

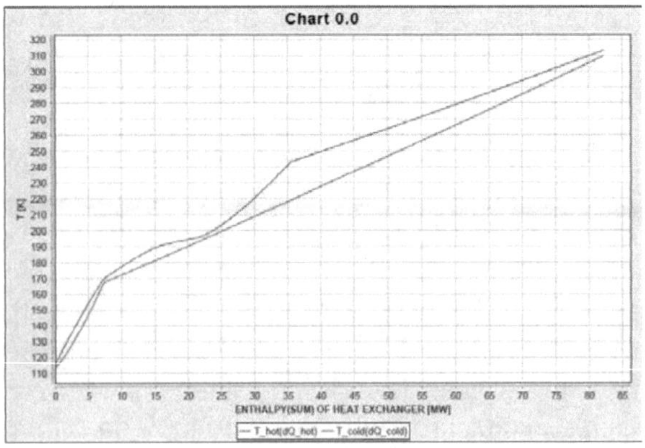

The diagram for Process C may look sub-optimally heat integrated. The cold part starting at 200 K, however, is satisfying and underlines the competitiveness of expander cycles at low temperatures. The large temperature differences in the warmer temperature range could be mitigated by either a third N_2 expander or a pre-cooling cycle. These options have not been investigated in detail. In total the low refrigerant circulation rate of Process C provides a very low overall heat duty, which is only 40% compared to Process A.

A more quantitative comparison of the heat exchangers is shown in the diagrams above. The unit area (UA) for the full range of parameters is shown relative to Process A. Process B offsets the more expensive heat exchanger type (CWHE vs PFHE) by a reduction of the required heating surface of about 50%. Thus, Process B can compete with Process A economically, as soon as the plant size justifies the use of a coil wound heat exchanger. This break-even between Process A and Process B is somewhere between 0.2 and 0.5 mtpa depending on project specific influences.

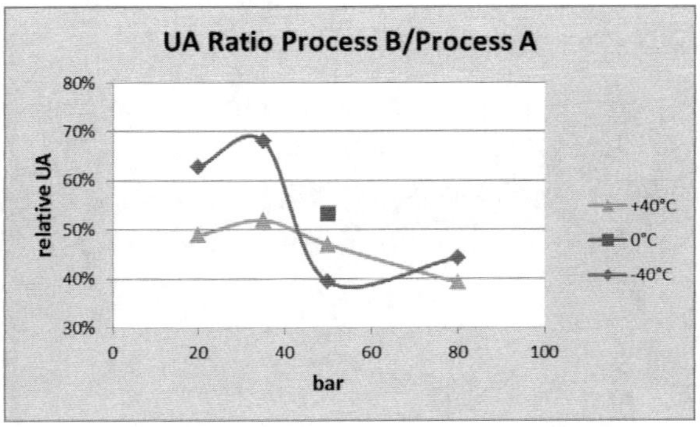

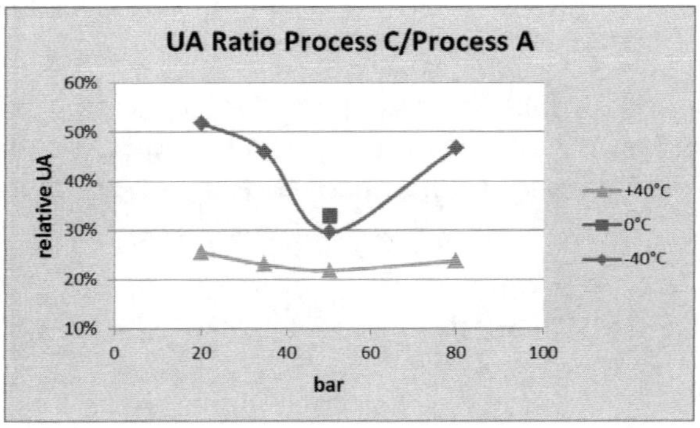

As expected, low heat duty and large mean temperature differences translate into a very low UA number of Process C especially for warm operating temperatures. A difference in the heat transfer coefficients between MR and N_2 refrigerant explains the actual number of PFHE cores. Process C has only 25% UA under warm conditions, but requires 4 cores for 1 mtpa compared to 10 cores for Process A.

Processes A and C are based on relatively inexpensive plate-fin heat exchangers. Thus, narrow temperature differences can be accepted without penalizing the investment cost too much. If the installed heat exchanger surface is reduced by about 25% (e.g. by limiting the min. temperature approach to 4 K instead of 2.5 K), Process A requires about 9% more power in comparison to only 4% extra power for Process C. This underlines that Process A depends on large heat exchangers to realize its superior efficiency.

Cycle Compressor Size (figures for LNG plants with 1 mtpa)

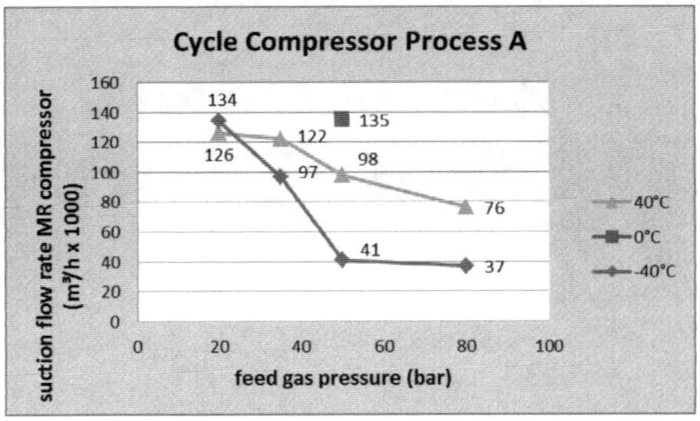

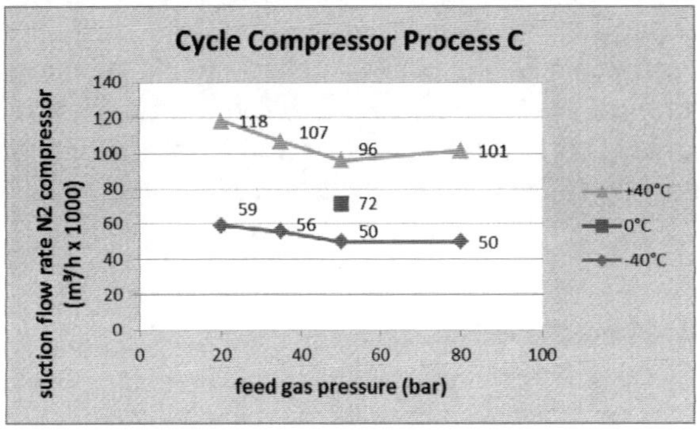

The investment cost of refrigerant cycle compressors depends not only on the shaft power, but also very much on the volumetric suction flow rate, as it governs the casing size.

Therefore, this parameter is relevant especially for smaller plants, where a low budget is crucial. The MR based Process A shows a high influence of the feed gas pressure on the suction flow rate, while this influence is small for the N_2 expander Process C. This different behavior can be explained by the more flexible concept of Process A (and B). The MR composition can be changed to reflect different operating conditions, while the simpler N_2 expander concept does not offer this degree of freedom for optimization.

In total Process A (and B) requires a relatively small compressor casing for high operating temperatures and high feed gas pressures, while Process C prefers the opposite: low operating temperature and low feed gas pressure give small compressor casings.

Conclusions

The table underneath summarizes the findings of this report.

	Process A	**Process B**	**Process C**
Efficiency	High, if plenty of heating surface is installed	High except for low feed gas pressure	Only fair, improves at low temperature level
Heat Exchanger Size	large number of cores makes larger trains challenging	moderate size compensates elevated price level of CWHEs	very small and competitive
Cycle Compressor Size	small casings for warm temperature and high pressure	small casings for warm temperature and high pressure	small casings for low temperature and low pressure

This comparison is focused on investment and operating cost drivers. No consideration was given to operability and safety. From an efficiency point of view Process B seems to be preferable for larger train sizes, while Process A has its merits for low feed gas pressure and smaller plants. Process C offers compact equipment under certain conditions and will be a strong contender, if safety and operability are of the essence. More details about StarLNG™ can be found at Linde's website[1].

References

1 http://www.lppusa.com/international/web/le/us/likeleuslbpp30.nsf/repositorybyalias/lpp_starlngbrochure/$file/2012%20StarLNG%20brochure_Issue%202.0.pdf

351061 Recent Development in Natural Gas Liquefaction Technology

Abstract:

The LNG industry is an ever-changing industry where we face new challenges and solve new problems every day. The shale gas boom, the search of off-shore natural gas resources, the use of LNG as transportation fuel, and the liquefaction process to be carried out in arctic area, are few examples of changes the industry is experiencing. These changes bring new challenges to modern LNG projects, because the design conditions, such as feed gas composition and variability, ocean motion and equipment movement, extreme ambient temperatures, are very different from and outside the experience of conventional LNG plants that were built many years ago. In this paper, we provide an overview of recent liquefaction technology development and discuss how these new challenges are addressed.

CONSISTENT METHODOLOGY FOR ESTIMATING GREENHOUSE GAS EMISSIONS FROM THE LNG VALUE CHAIN

Karin Ritter, API

Miriam Lev-On, The LEVON Group, LLC

14AIChE
Spring Meeting, New Orleans, LA

energy API

Natural Gas Resource Availability

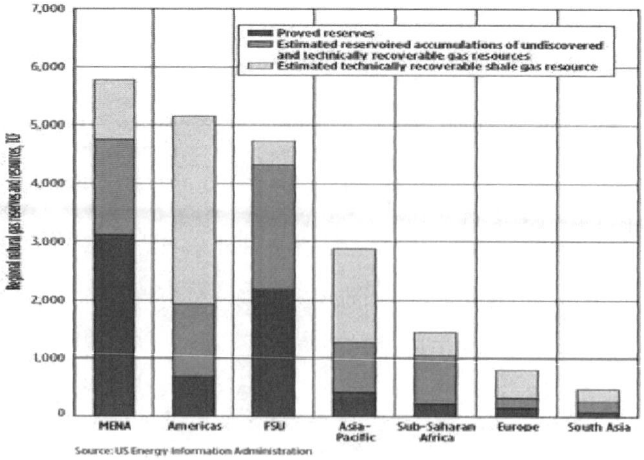

Source: US Energy Information Administration

14AIChE
Spring Meeting, New Orleans, LA

energy API

API LNG Guideline Document

- **Goals**
  - Develop consistent internationally accepted methodologies to estimate GHG emissions from operations in the Liquefied Natural Gas (LNG) value chain

- **Background**
  - Designed to align with other industry guidance, i.e. API Compendium, IPIECA Guidelines

14 AIChE Spring Meeting, New Orleans, LA

energy API

Document Development Process

- Established a U.S. based working group of companies technical experts
- Defined LNG segment boundaries and identified applicable emission sources
- Adopted available emissions estimation methodology for LNG operations
- Conducted several rounds of review and comment by a broader cadre of industry experts globally

14 AIChE Spring Meeting, New Orleans, LA

energy API

Document Contents

1. Overview
2. LNG Sector Background
3. GHG Emissions Inventory
4. Emission Estimation Methods
5. Methodology Implementation and Improvements

14 AIChE Spring Meeting, New Orleans, LA

energy API

LNG Overview and Background

1. **Overview**
   - o LNG Applications
   - o LNG Greenhouse Gas Emissions
2. **Sector Background**
   - o What is LNG
   - o LNG Operations Chain

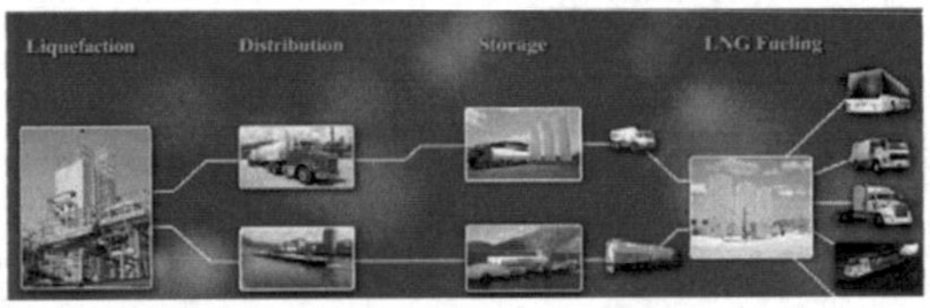

14 AIChE Spring Meeting, New Orleans, LA

energy API

Greenhouse Gas Emissions

3. **GHG Emission Inventories Boundaries**
   - Operational GHG Emissions
   - GHG Emission Sources
4. **Emission Estimation Methods**
   - Stationary Combustion
   - Vented Emissions
   - Fugitive Emissions
   - Transportation Emissions
   - Non-routine Emissions
5. **Methodology Implementation and Improvements**

14 AIChE Spring Meeting, New Orleans, LA

energy API

Description of LNG Operations

- LNG Operations boundaries
  - Starts at the entry of gas to the liquefaction plant
  - Terminates after regassification of liquids and transfer for transport and marketing
- Segments of the LNG Operations Chain
  - Liquefaction
  - Storage
  - Loading and Unloading
  - Shipping
  - Regassification

14 AIChE Spring Meeting, New Orleans, LA

energy API

GHG Inventory Boundaries

- This section links the reporting of GHG emissions to the IPIECA/API Industry GHG Reporting Guidelines
- GHG emission sources
  - Mapping of combustion sources
  - Mapping of Vented, Fugitives and Mobile emission Sources
- GHG compounds emitted
  - Emissions and quantification methods focus on CO_2, CH_4, and N_2O

14 AIChE Spring Meeting, New Orleans, LA

energy API

Key Contributors to GHG Emissions

- Emissions depend on facility design and operating practices
  - Liquefaction technology efficiency
  - Types of storage tanks
  - Level of compression
  - Vessel pre-cooling practices
  - Boil-off gas recovery
- Types of vaporizers used are key contributors to regassification emissions
  - Submerged Combustion Vaporizers (SCV)
  - Open Rack Vaporizers (ORV)
  - Shell & Tube Vaporizers (STV)
  - Ambient Air Vaporizers (AAV)

14 AIChE Spring Meeting, New Orleans, LA

energy API

GHG Emission Estimation Methods

- Focus on five categories of sources
  - Stationary combustion emissions;
  - Process vents and other vented emissions;
  - Fugitive emissions from equipment leaks;
  - Mobile source combustion emissions.
  - Non-routine emissions
- Methods that are included in the API Compendium are not repeated here
- Tables provide listing of applicable emission factors

Combustion Emission Estimation

- For CO_2 combustion estimation
  - EF and annual fuel use, w/default HHV; or
  - EF based on fuel use and average C-content; or
  - Measured annual fuel use and periodic measurements of the C-content of the fuel.
- For CH_4, N_2O combustion estimation
  - EF based on annual fuel use and fuel HHV;
  - Applicable equipment/technology based EF
- Emission Factors for Gas Flaring

Methane Emissions from Venting and Equipment Leaks

- Sources of Methane emissions
  - Compressor seals
  - CO_2 removal systems
  - Dehydration systems
  - Tank overpressure
  - Ship loading displacement vapors
  - Loading arm disconnection
- Venting of vapors
  - Tank overpressure venting
  - Leaks from pressure relief valves
  - Vapor recovery compressors
- Fugitive emissions from leaking components

Next Steps

- Pilot Draft document is available on API website

  http://www.api.org/environment-health-and-safety/climate-change/whats-new/consistent-methodology-for-estimating-ghgs-from-lng-operations
- Targeted investigations during pilot review period to collect updated information
  - E.g., study of GHG emissions associated with different Liquefaction technologies
- Undertake follow-up revisions and publish final document at the end of the 'pilot period'

Thank you for your attention

For additional information:

Karin Ritter, API, Washington DC, USA
ritterk@api.org

Dr. Miriam Lev-On, LEVON Group, California, USA
miriam@levongroup.net

14 AIChE Spring Meeting, New Orleans, LA

energy API

350382 Kinetic and Thermodynamic Aspects for CO2 Conversion to Methanol via Trireforming

Abstract:

Long term viability of the petroleum refinery sector must include a sustainable infrastructure whereby the triple bottom line – profit, people, and planet – is sought. One such avenue is development of alternate strategies of CO_2 conversion technologies, such as tri-reforming. Tri-reforming is a combination of methane steam reforming, CO_2 reforming, and methane oxidation that utilizes the steam and CO_2 released from flue stack gases and converts to syngas (CO and H_2). Unlike typical sequestration technologies that require CO2 to be separated from steam before being pumped underground, tri-reforming avoids this expense by using the mixed CO2 and steam. Conversion of carbon dioxide, if economically viable, extents our carbon resources and reduces our carbon footprint.

A thermodynamic analysis, along with a process simulation and optimization analysis, was conducted using Aspen simulation coupled to a General Algebraic Modeling System (GAMS), for the combined tri-reforming/methanol synthesis paradigm. Model parameters for the tri-reformer included temperature (400 – 1,200°C), pressure (1 – 5 atm), and CH4/Flue gas ratio (0.4 – 1.0). Results showed that high temperatures and low pressures favored high CO2 conversions with an optimum $H_2/CO = 2$. Furthermore, an economic analysis demonstrated substantial profits when the endothermic tri-reforming process was coupled to the highly exothermic methanol synthesis process.

In an additional study, a novel catalyst has been developed for the tri-reforming reaction that uses reverse micelles to create nano-sized Transition metal reactive sites onto the surface of TiO_2 anatase. The anatase, which has oxygen vacancies within its lattice structure, was found to be not just a surface area providing support but an active component within the overall operability of the catalyst. A kinetic evaluation and a time on stream analysis have shown to have better performance than a traditional steam methane reforming catalyst. The overall study demonstrates a viable, sustainable method that has the potential of economically reducing pinpoint CO_2 emissions. The complete catalyst development and kinetic evaluation, along with the Aspen simulation study, will be shown.

Study of Cobalt Copper nanoparticle as catalysts for ethanol synthesis from syngas

LOUISIANA STATE UNIVERSITY

Zi Wang James Spivey
Louisiana State University
03/31/2014

U.S. DEPARTMENT OF ENERGY
Center for Atomic-Level Catalyst Design
an Office of Basic Energy Sciences
Energy Frontier Research Center

Outline

- Background of Ethanol Synthesis and Cobalt Copper Catalysts
- Novel synthesize method: Microemulsion for Cobalt Copper composite nanospheres
- Catalyst Characterization Results (TEM TPR XPS analysis)
- CO Hydrogenation Activity Test

U.S. DEPARTMENT OF ENERGY | Center for Atomic-Level Catalyst Design | *an Office of Basic Energy Sciences Energy Frontier Research Center*

Outline

- Background of Ethanol Synthesis and Cobalt Copper Catalysts
- Novel synthesize method: Microemulsion for Cobalt Copper composite nanospheres
- Catalyst Characterization Results (TEM TPR XPS analysis)
- CO Hydrogenation Activity Test

3

U.S. DEPARTMENT OF ENERGY | Center for Atomic-Level Catalyst Design | *an Office of Basic Energy Sciences Energy Frontier Research Center*

Reaction Background

$$2CO + 4H_2 \rightarrow C_2H_5OH + H_2O$$

$$\Delta H_r^\circ = -61.2 \text{ kcal mol}^{-1}$$

$$\Delta G_r^\circ = -29.32 \text{ kcal mol}^{-1}$$

Reaction conditions: 230°C to 350°C and high pressure.

Side Products:
CH_4, CO_2, CH_3OH,
C_2+ hydrocarbons.

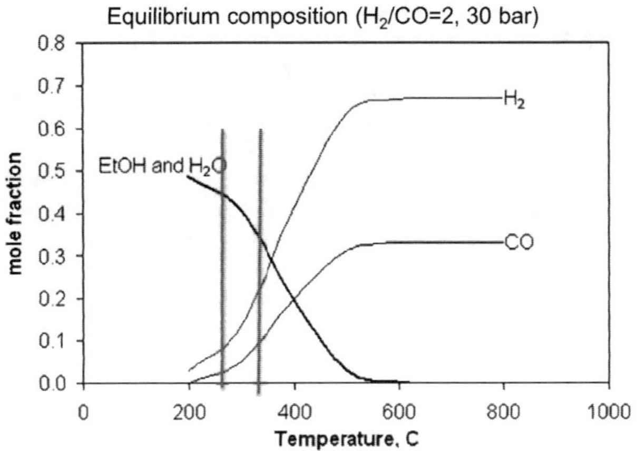

Spivey, J. J.; Egbebi, A.,. Chemical Society Reviews 2007, 36 (9), 1514-1528.

4

U.S. DEPARTMENT OF ENERGY | Center for Atomic-Level Catalyst Design | *an Office of Basic Energy Sciences* *Energy Frontier Research Center*

Reaction Mechanism on CuCo catalyst

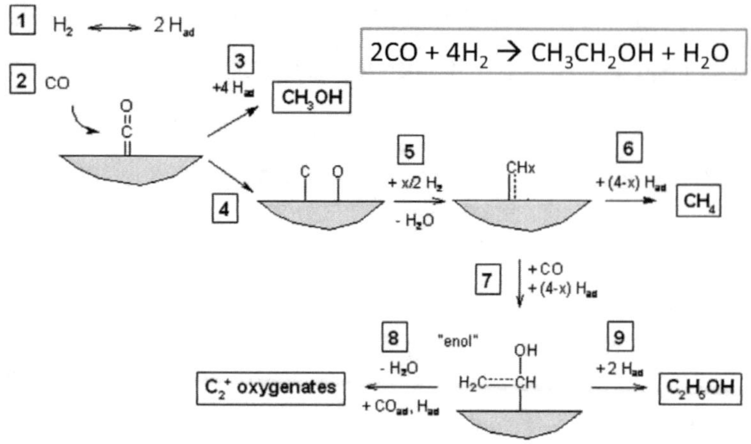

- 1) H_2 Adsorption
- 2) CO Adsorption
- 3) methanol formation
- 4) dissociative CO adsorption
- 5) CHx formation
- 6) Methane formation
- 7) CO insertion
- 8) C_2^+ Oxygenates formation
- 9) ethanol formation

Spivey, J. J.; Egbebi, A.,. Chemical Society Reviews 2007, 36 (9), 1514-1528

U.S. DEPARTMENT OF ENERGY | Center for Atomic-Level Catalyst Design | *an Office of Basic Energy Sciences* *Energy Frontier Research Center*

Reaction Mechanism on CuCo catalyst

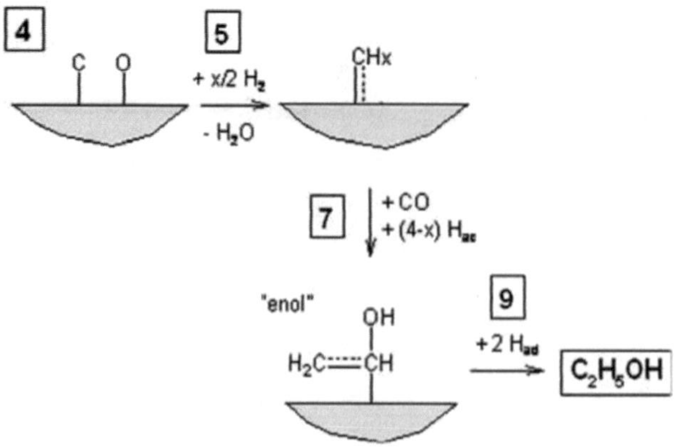

Ethanol synthesis requires two types of sites:

- One for dissociative CO adsorption Cobalt
- One for associative CO adsorption (insertion) Copper

- The effective site pairs: Co and Cu metal with intimate contact.

Spivey, J. J.; Egbebi, A.,. Chemical Society Reviews 2007, 36 (9), 1514-1528

Reaction Mechanism on CuCo catalyst

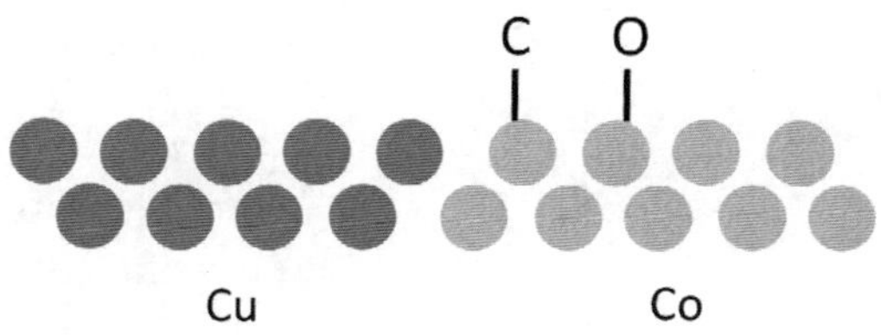

Ethanol synthesis requires two types of sites:

- One for dissociative CO adsorption Cobalt
- One for associative CO adsorption (insertion) Copper

- The effective site pairs: Co and Cu metal with intimate contact.

Spivey, J. J.; Egbebi, A.,. Chemical Society Reviews 2007, 36 (9), 1514-1528

U.S. DEPARTMENT OF ENERGY
Center for Atomic-Level Catalyst Design
an Office of Basic Energy Sciences
Energy Frontier Research Center

Reaction Mechanism on CuCo catalyst

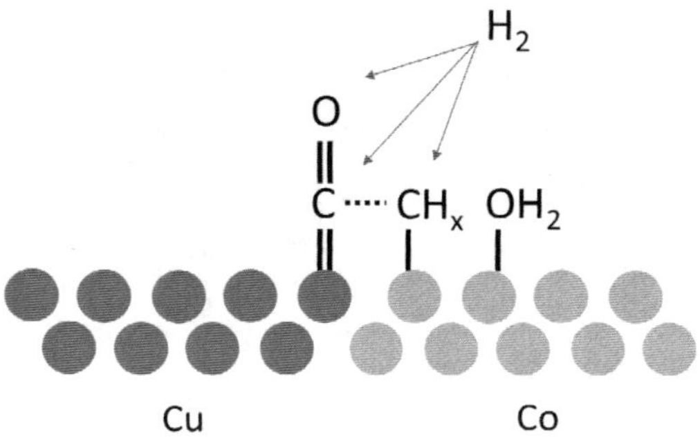

Ethanol synthesis requires two types of sites:

- One for dissociative CO adsorption Cobalt
- One for associative CO adsorption (insertion) Copper

- The effective site pairs: Co and Cu metal with intimate contact.

Spivey, J. J.; Egbebi, A.,. Chemical Society Reviews 2007, 36 (9), 1514-1528

Reaction Mechanism on CuCo catalyst

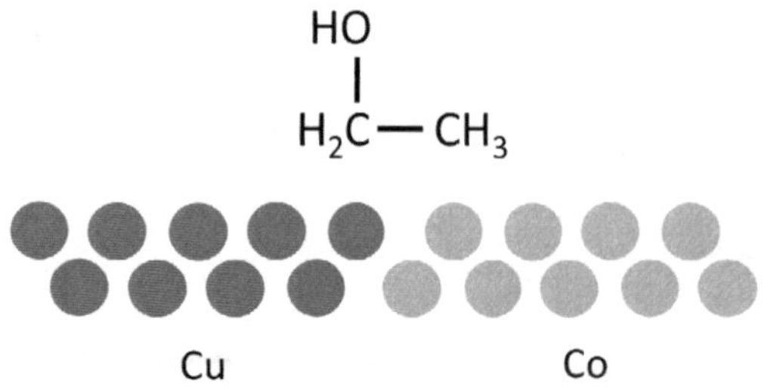

Ethanol synthesis requires two types of sites:

- One for dissociative CO adsorption Cobalt
- One for associative CO adsorption (insertion) Copper

- The effective site pairs: Co and Cu metal with intimate contact.

Spivey, J. J.; Egbebi, A.,. Chemical Society Reviews 2007, 36 (9), 1514-1528

Outline

- Background of Ethanol Synthesis and Cobalt Copper Catalysts
- Novel synthesize method: Microemulsion for Cobalt Copper composite nanospheres
- Catalyst Characterization Results (TEM TPR XRD XPS analysis)
- CO Hydrogenation Activity Test

Synthesis Method: Microemulsion

Advantages:

1. Nanoparticles of Cu and Co composite.
2. Ultrafine particle improved copper and cobalt metal interaction.
3. Large surface/volume ratio, altered surface structure creates more sites.

Synthesis Method: Microemulsion

Surfactant	$((C_{16}H_{33})N(CH_3)_3Br$
Base Solution	NH_4OH
Oil Phase	Cyclohexane
ICP Cu:Co atomic ratio	1 : 1.13

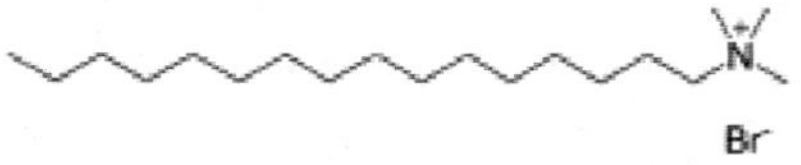

CTAB Cetrimonium bromide : $((C_{16}H_{33})N(CH_3)_3Br$

Synthesis Method: Microemulsion

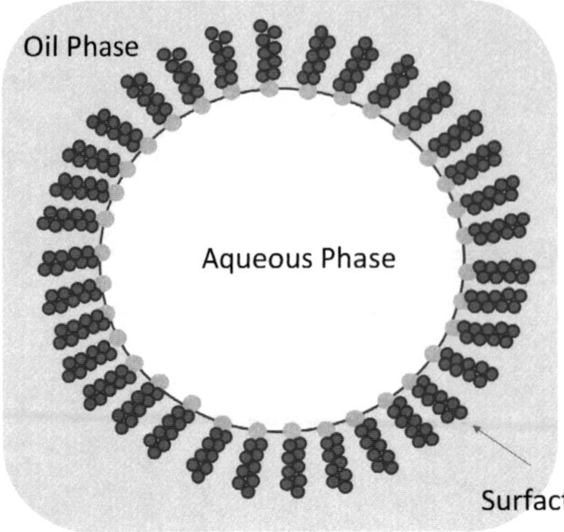

1. Microemulsion systems

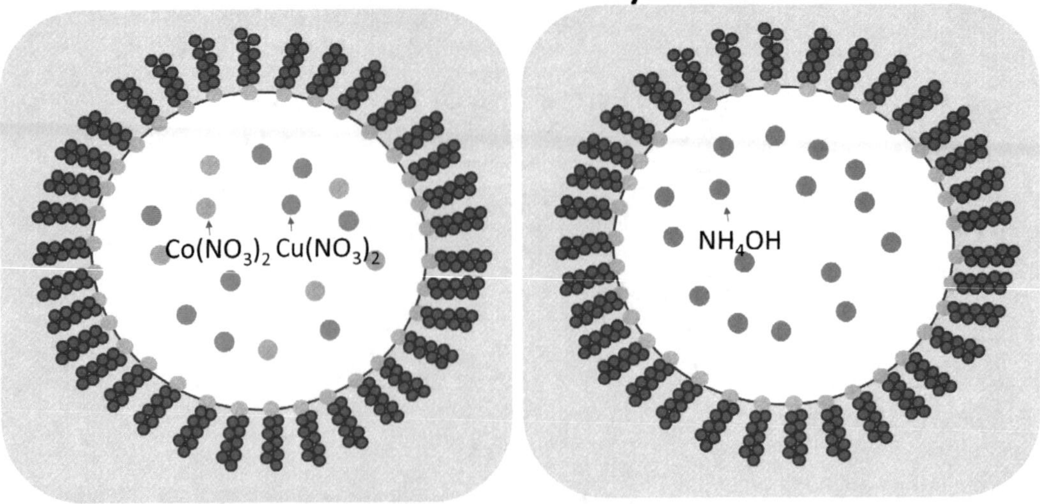

2. Coalescence of droplets

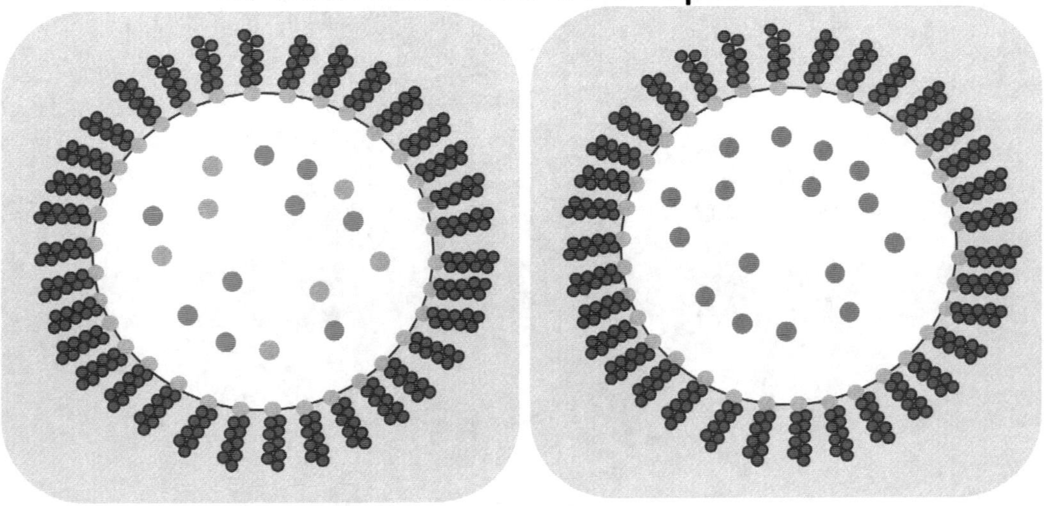

U.S. DEPARTMENT OF ENERGY
Center for Atomic-Level Catalyst Design
an Office of Basic Energy Sciences
Energy Frontier Research Center

3. Reaction

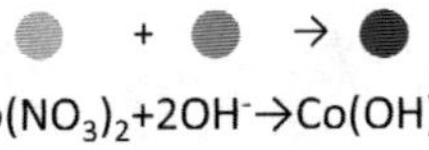

$Co(NO_3)_2 + 2OH^- \rightarrow Co(OH)_2\downarrow$

$Cu(NO_3)_2 + 2OH^- \rightarrow Cu(OH)_2\downarrow$

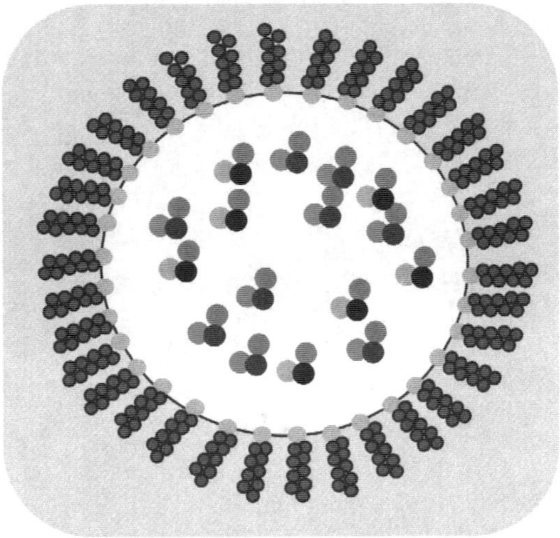

4. Precipitation

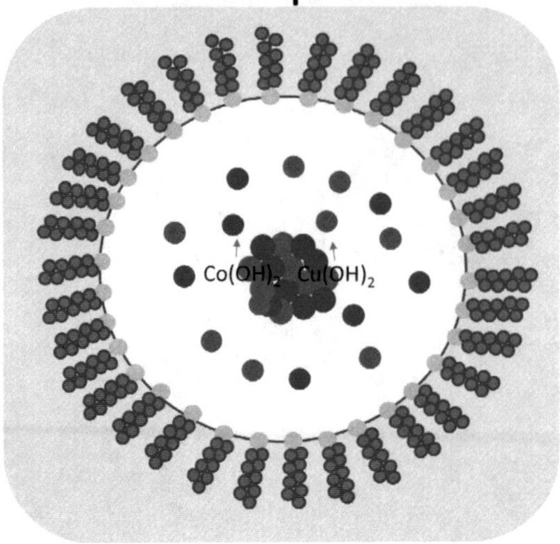

U.S. DEPARTMENT OF ENERGY
Center for Atomic-Level Catalyst Design
an Office of Basic Energy Sciences
Energy Frontier Research Center

5. Removing the ligands

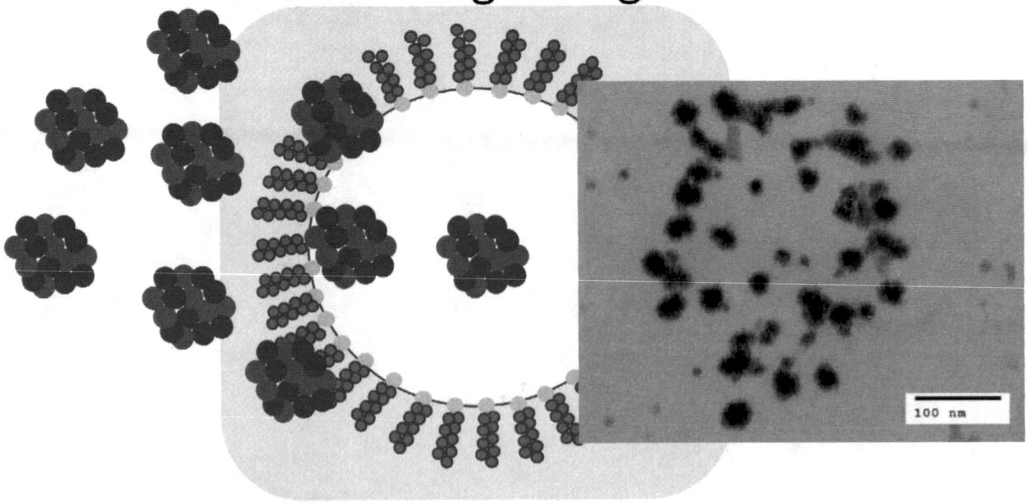

U.S. DEPARTMENT OF ENERGY | Center for Atomic-Level Catalyst Design | an Office of Basic Energy Sciences Energy Frontier Research Center

Outline

- Background of Ethanol Synthesis and Cobalt Copper Catalysts
- Novel synthesize method: Microemulsion for Cobalt Copper composite nanospheres
- Catalyst Characterization Results (TEM XPS TPR analysis)
- CO Hydrogenation Activity Test

19

U.S. DEPARTMENT OF ENERGY | Center for Atomic-Level Catalyst Design | an Office of Basic Energy Sciences Energy Frontier Research Center

Transmission electron microscopy- Microemulsion catalyst

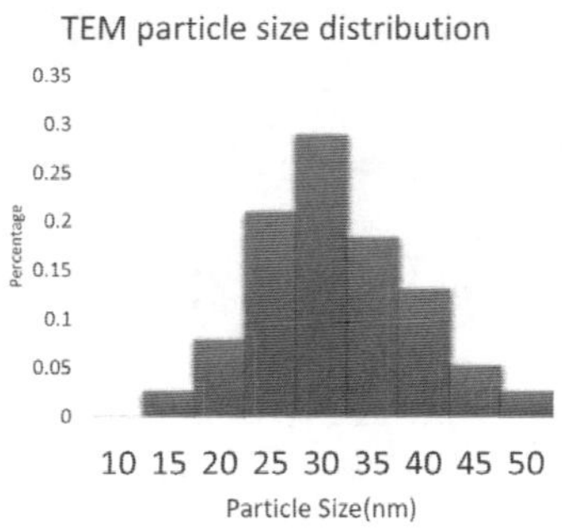

Over 80% of particle size between 20 to 40 nm

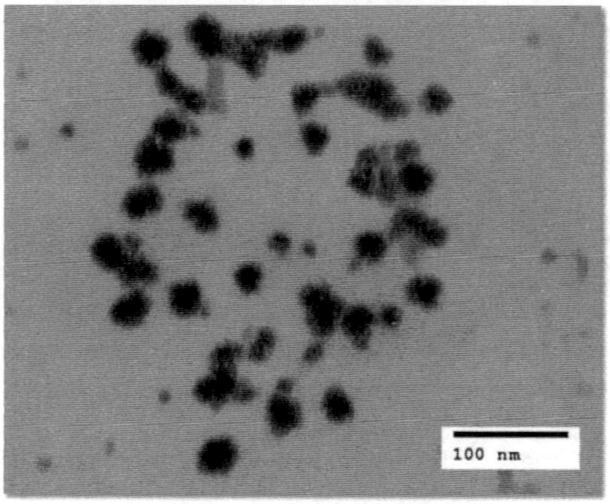

TEM on Microemulsion CuCo catalyst

20

U.S. DEPARTMENT OF ENERGY | Center for Atomic-Level Catalyst Design | *an Office of Basic Energy Sciences Energy Frontier Research Center*

Transmission electron microscopy- Microemulsion catalyst

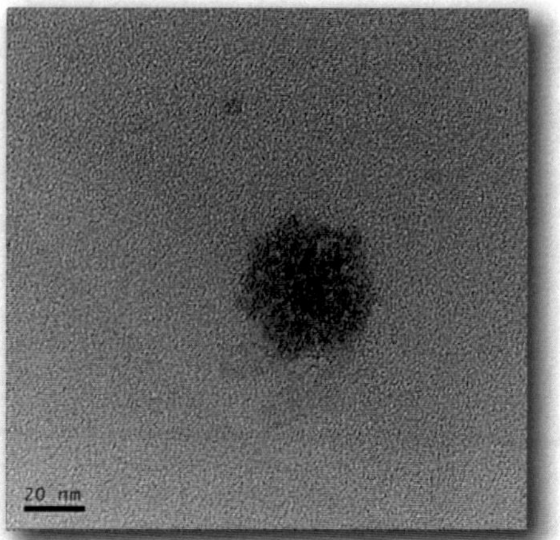

TEM on Microemulsion CuCo catalyst

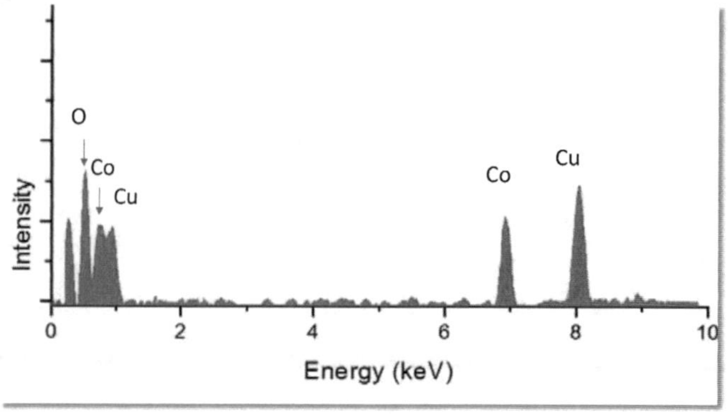

U.S. DEPARTMENT OF ENERGY | Center for Atomic-Level Catalyst Design | *an Office of Basic Energy Sciences Energy Frontier Research Center*

X-ray photoelectron spectroscopy

Catalyst	Binding Energies		Cu:Co atomic ratio by XPS	Cu:Co atomic ratio by ICP
	Cu $2p_{3/2}$	Co $2p_{3/2}$		
CuCo Microemulsion	933.3	780.2	1 : 0.99	1:1.13

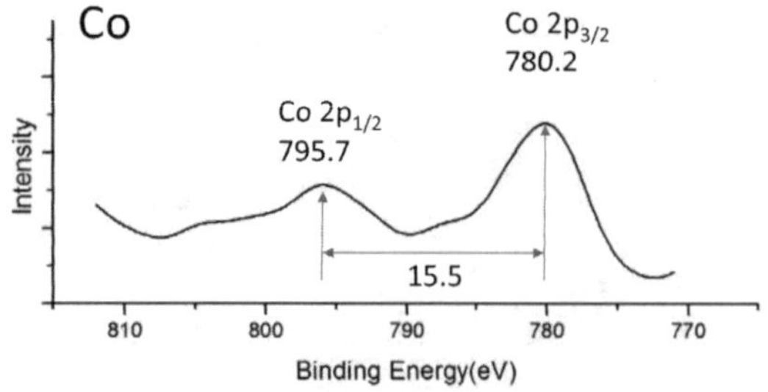

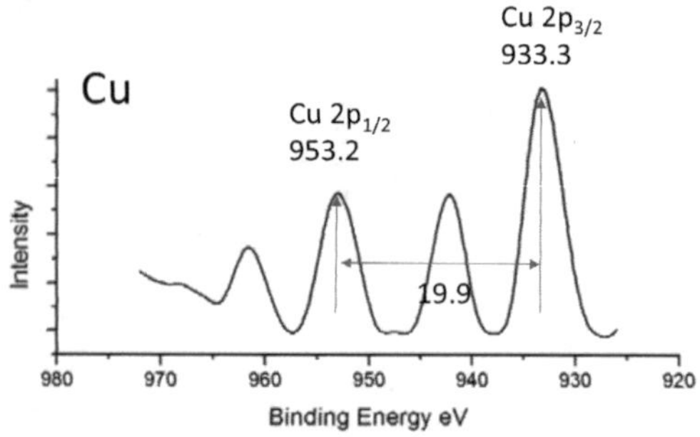

X-ray photoelectron spectroscopy

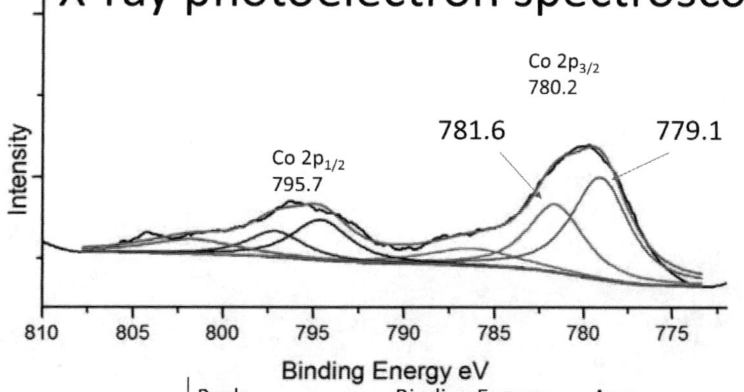

	Peak	Binding Energy eV	Area
Co $2p_{3/2}$	Co^{3+}	779.1	1775.6
	Co^{2+}	781.6	1068.0

Temperature Programed Reduction

1:1 Cobalt Copper catalyst by coprecipitation

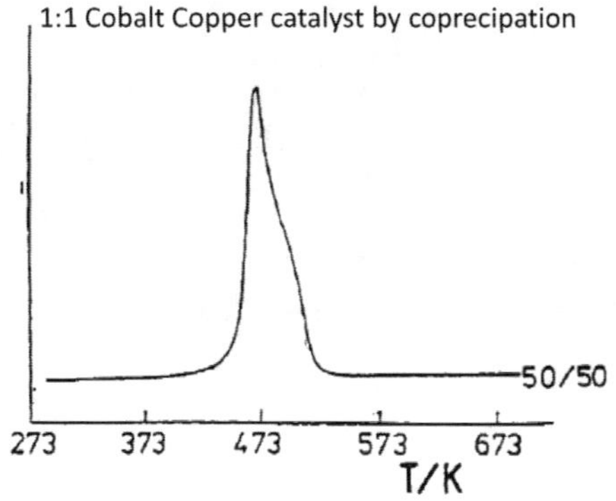

1:1 Cobalt core copper shell

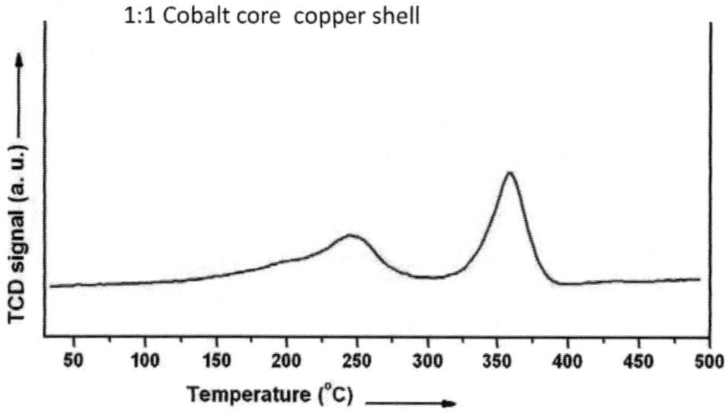

Fierro, G.; Lo Jacono, M.; Inversi, M.; Dragone, R.; Porta, P., *Topics in Catalysis* **2000,** *10* (1-2), 39-48.

Subramanian, N. D.; Balaji, G.; Kumar, C. S. S. R.; Spivey, J. J., *Catalysis Today* **2009,** *147* (2), 100-106.

Temperature Programed Reduction

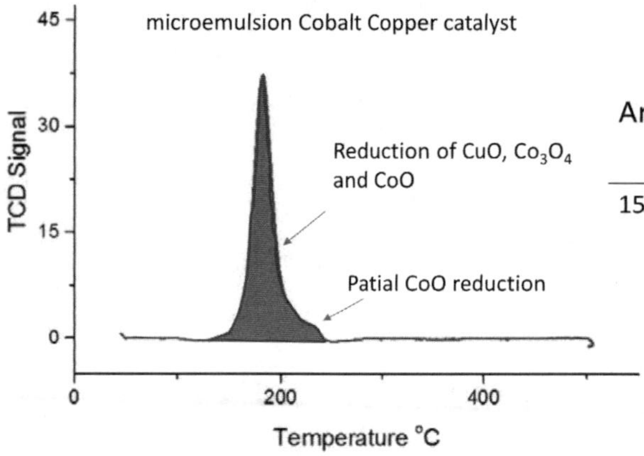

Area	H_2 consumed (μmol/g)	Expected (μmol/g)
150°C~240°C	1480±160	1380

- Cobalt and Copper are fully reduced in the catalyst.
- Copper promotes cobalt reduction in microemulsion catalyst.

U.S. DEPARTMENT OF ENERGY
Center for Atomic-Level Catalyst Design
an Office of Basic Energy Sciences
Energy Frontier Research Center

Temperature Programed Reduction

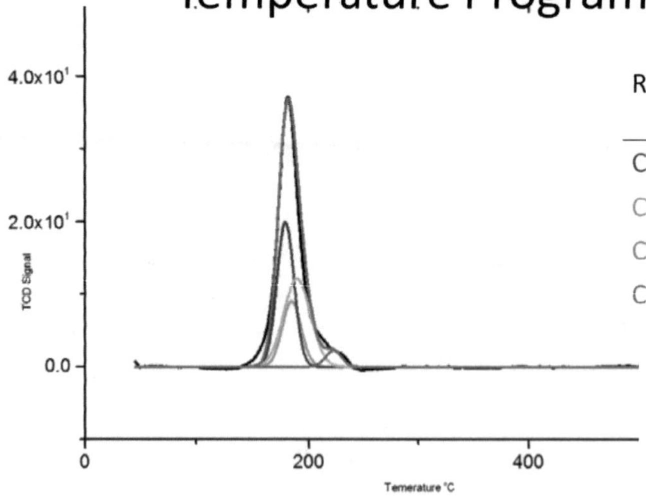

Reactions	H_2 consumed (μmol/g)	Expected (μmol/g)
$CuO \rightarrow Cu^0$	552±74	550
$Co_3O_4 \rightarrow CoO$	212±30	207
$CoO \rightarrow Co^0$	580±82	622
$CoO \rightarrow Co^0$	60±8	

U.S. DEPARTMENT OF ENERGY | Center for Atomic-Level Catalyst Design | *an Office of Basic Energy Sciences Energy Frontier Research Center*

Outline

- Background of Ethanol Synthesis and Cobalt Copper Catalysts
- Novel synthesize method: Microemulsion for Cobalt Copper composite nanospheres
- Catalyst Characterization Results (TEM TPR XPS analysis)
- **CO Hydrogenation Activity Test**

27

U.S. DEPARTMENT OF ENERGY | Center for Atomic-Level Catalyst Design | *an Office of Basic Energy Sciences Energy Frontier Research Center*

CO hydrogenation Activity Test

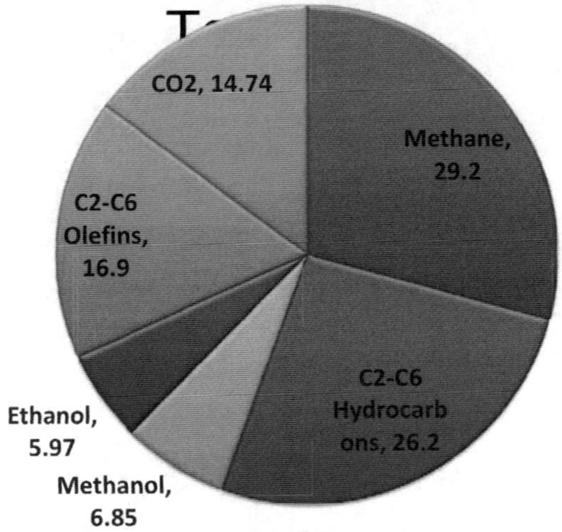

Reduced at 400°C in H_2/He
Reaction condition: 40bar, 250°C, GHSV=36000 $scc/g_{cat}/h$, $H_2:CO=2:1$
CO Conversion: 0.12%

24

U.S. DEPARTMENT OF ENERGY | Center for Atomic-Level Catalyst Design | an Office of Basic Energy Sciences Energy Frontier Research Center

CO hydrogenation Activity Test

Catalyst	Selectivity MeOH	EtOH	CH4	CO2	C2+HC	Pressure bar	Temperature C	Conversion %
Microemulsion	6.85	5.97	29.2	14.74	26.2	40	250	0.12
Co Core Cu Shell [1]	2.5	2.7	47.1	19.1	24.9	20	270	0.16
Cu(111) on Co(100)[2]	5.24	1.06	24	41.9	18.9	60	240	3.5

1. Subramanian, N. D.; Balaji, G.; Kumar, C. S. S. R.; Spivey, J. J., *Catalysis Today* **2009,** *147* (2), 100-106.
2. Lü, D.; Zhu, Y.; Sun, Y., *Chinese Journal of Catalysis* **2013,** *34* (11), 1998-2003.

U.S. DEPARTMENT OF ENERGY | Center for Atomic-Level Catalyst Design | an Office of Basic Energy Sciences Energy Frontier Research Center

Conclusions

1. Copper Cobalt nanospheres are synthesized using microemulsion, particle size between 20 to 40 nm

2. Close proximity between Cu and Co are formed in Cu-Co nanospheres. Co oxide and Cu oxide reduce at low temperature.

3. Microemulsion catalysts has higher ethanol selectivity in CO hydrogenation.

30

Acknowledgments

- This study is based upon work supported as part of the Center for Atomic Level Catalyst Design, an Energy Frontier Research Center funded by the U.S. Department of Energy, Office of Science, Office of Basic Energy Sciences under Award Number DE-SC0001058.

14 AIChE
Spring Meeting, New Orleans, LA

Syngas Utilization through High Pressure Solvent Assisted Fischer-Tropsch Synthesis

Jan Hendrik Blank, Samah Warag, Shaik Afzal, Layal Bani Nasser, Rehan Hussain Nimir O. Elbashir

Texas A&M University at Qatar

CHEMICAL ENGINEERING PROGRAM | TEXAS A&M UNIVERSITY at QATAR

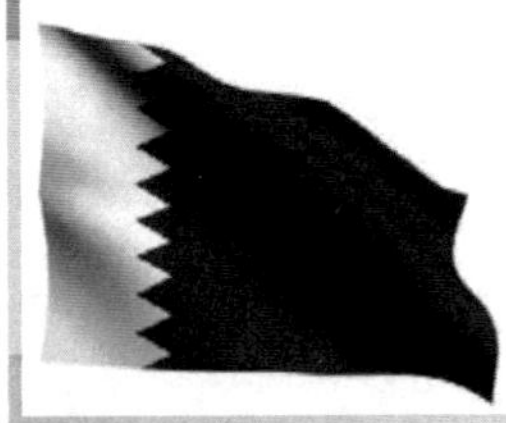

State of Qatar

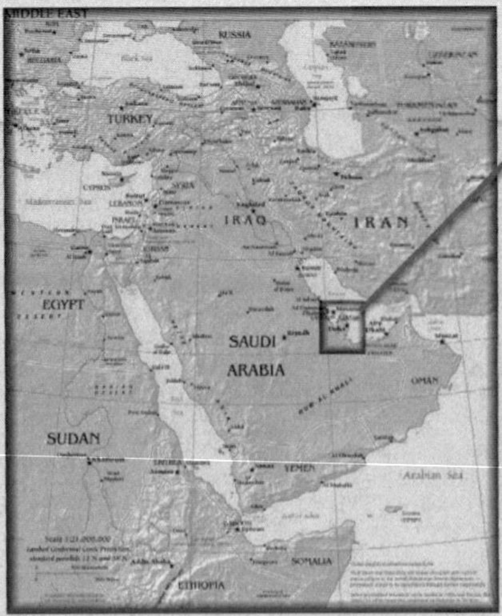

Area: 4,416 sq mi
Water: negligible
Population: 1,853,563
Density: 319.1/sq mile

- Qatar has the world's largest per capita production and proven reserves of both oil and natural gas.
- Qatar had the world's highest GDP per capita, while the economy grew by **18.8%**, the fastest in the world (2011 est.)

CHEMICAL ENGINEERING PROGRAM 2 | TEXAS A&M UNIVERSITY at QATAR

Qatar's Role in the Energy Market

World's Primary Energy Supply

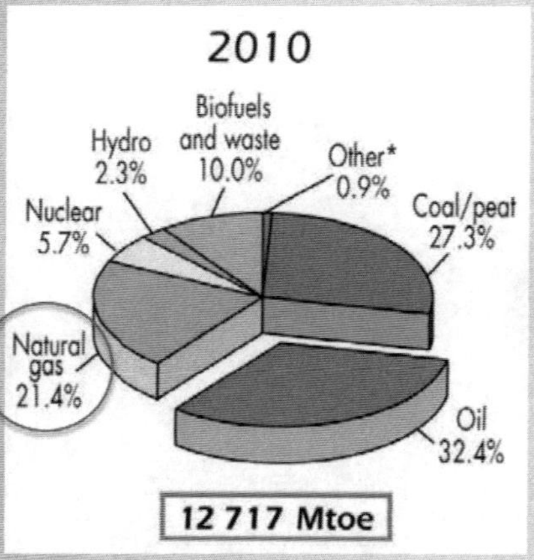

Natural gas

Producers	bcm	% of world total
Russian Federation	677	20.0
United States	651	19.2
Canada	160	4.7
Qatar	151	4.5
Islamic Rep. of Iran	149	4.4
Norway	106	3.1
People's Rep. of China	103	3.0
Saudi Arabia	92	2.7
Indonesia	92	2.7
Netherlands	81	2.4
Rest of the world	1 126	33.3
World	**3 388**	**100.0**

2011 data

Net exporters	bcm
Russian Federation	196
Qatar	119
Norway	99
Canada	63
Algeria	49
Indonesia	46
Netherlands	33
Turkmenistan	29
Nigeria	26
Malaysia	22
Others	152
Total	**834**

2011 data

Source: International Energy Agency
www.iea.org

CHEMICAL ENGINEERING PROGRAM 3 TEXAS A&M UNIVERSITY at QATAR

Qatar contribution to the "Energy Market"

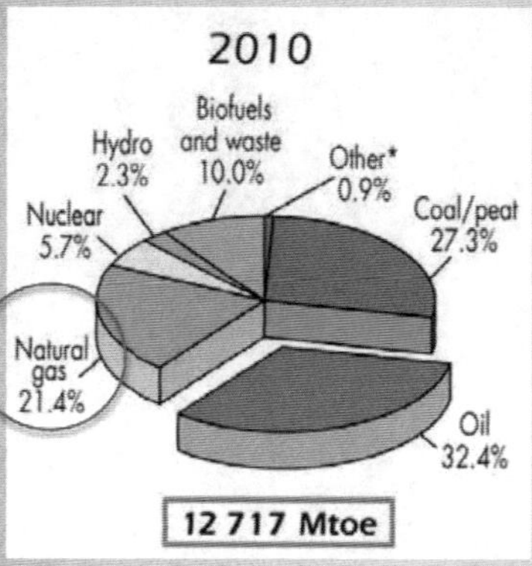

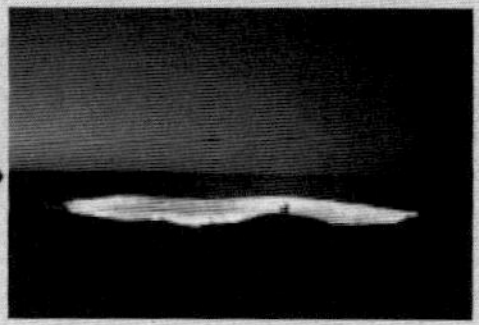

Natural Gas

Transportation

Physical

1/600 volume

Chemical

Qatar's aspiration to become the "World Gas Capital" led to the building the largest GTL and LNG plants in the world.

CHEMICAL ENGINEERING PROGRAM 4 TEXAS A&M UNIVERSITY at QATAR

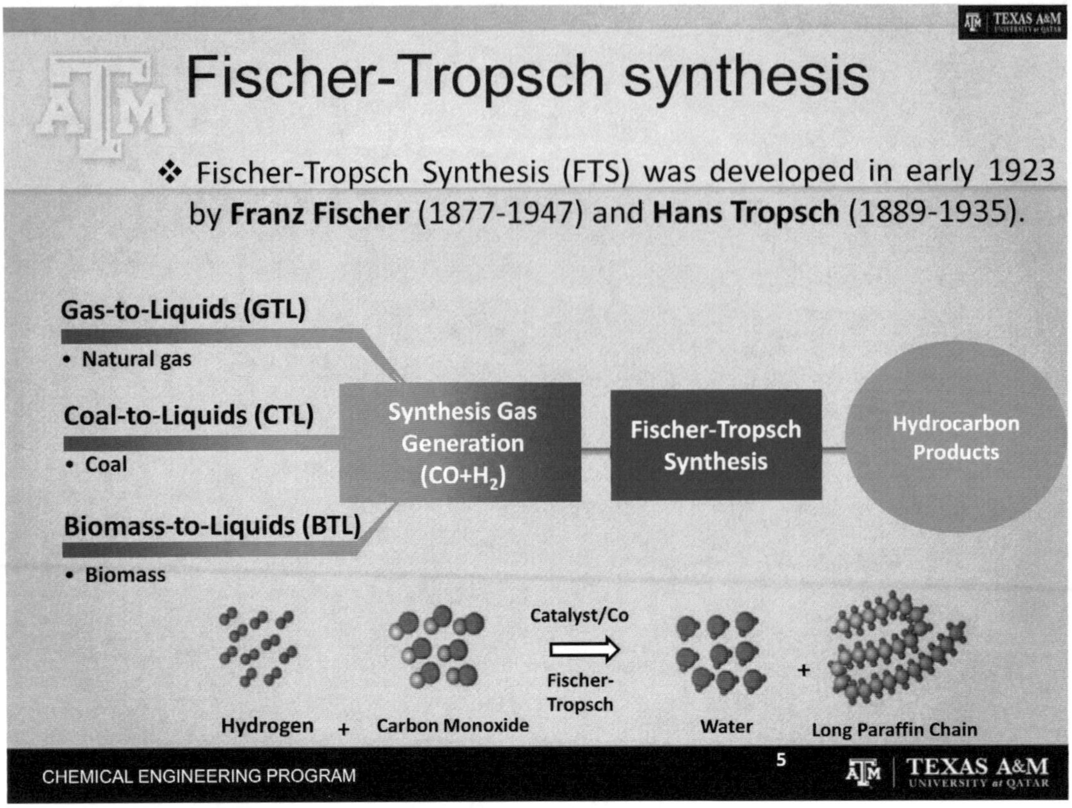
TEXAS A&M UNIVERSITY at QATAR
Fischer-Tropsch synthesis
❖ Fischer-Tropsch Synthesis (FTS) was developed in early 1923 by Franz Fischer (1877-1947) and Hans Tropsch (1889-1935).
Gas-to-Liquids (GTL)
• Natural gas
Coal-to-Liquids (CTL)
• Coal
Biomass-to-Liquids (BTL)
• Biomass
Synthesis Gas Generation ($CO+H_2$)
Fischer-Tropsch Synthesis
Hydrocarbon Products
Catalyst/Co
Fischer-Tropsch
Hydrogen + Carbon Monoxide
Water + Long Paraffin Chain
CHEMICAL ENGINEERING PROGRAM
5
TEXAS A&M UNIVERSITY at QATAR

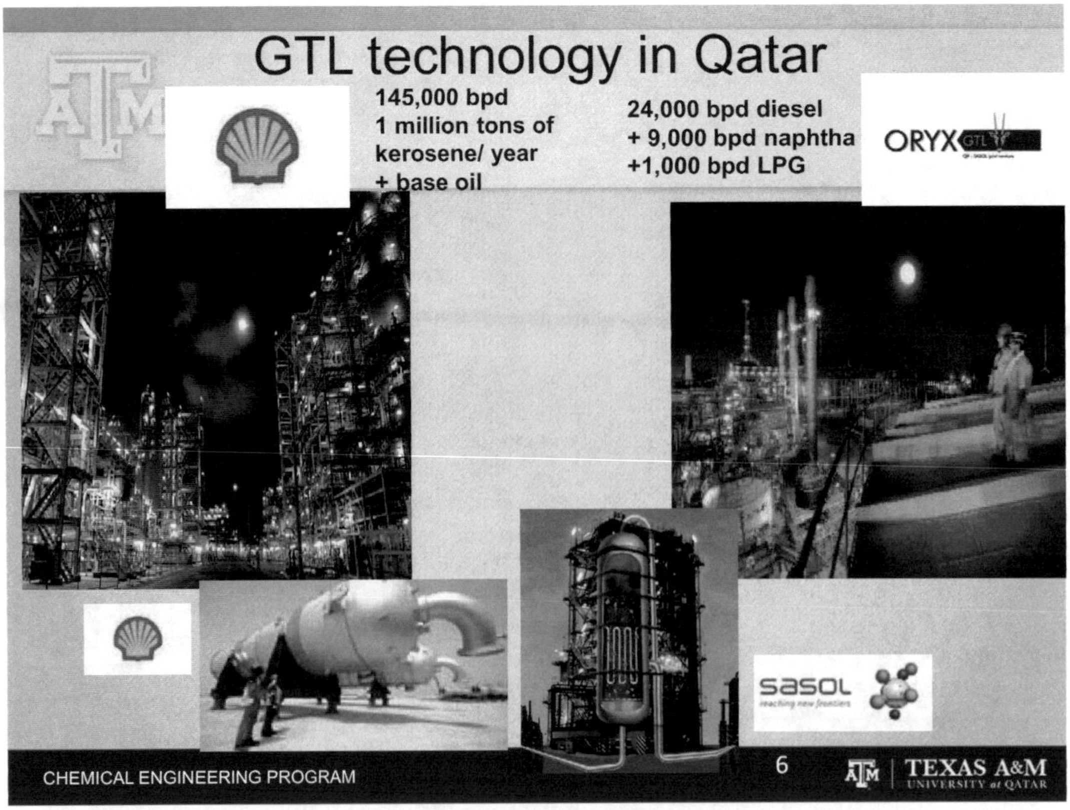
GTL technology in Qatar
145,000 bpd
1 million tons of kerosene/ year
+ base oil
24,000 bpd diesel
+ 9,000 bpd naphtha
+1,000 bpd LPG
ORYX GTL
sasol
reaching new frontiers
CHEMICAL ENGINEERING PROGRAM
6
TEXAS A&M UNIVERSITY at QATAR

Common commercial Fischer-Tropsch reactors

Fixed-Bed Reactor
(Gas-Phase reaction media)

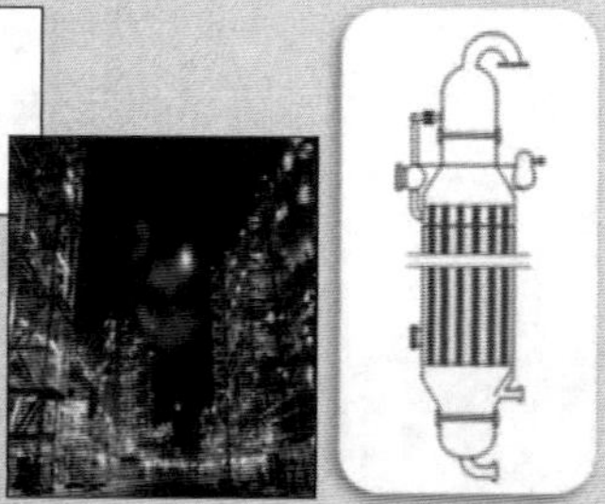

- Local overheating of the catalyst
- Mass transfer limitation (pore blockage)
- Faster reaction rate

Slurry- Phase Reactor
(Liquid-Phase reaction media)

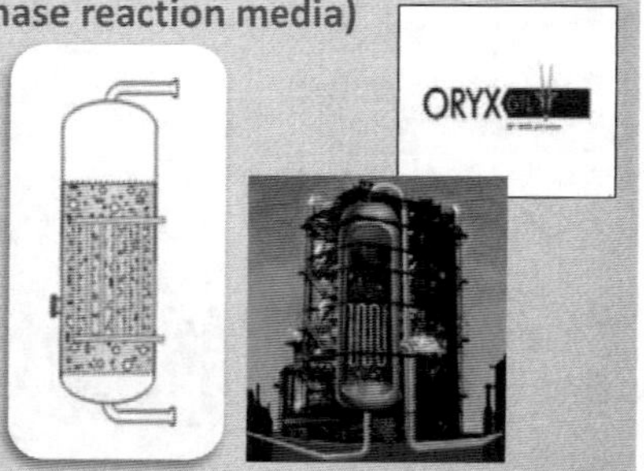

- Better heat dissipation
- Rate of reaction is lower
- Difficult catalyst separation

Gas-Phase FTS in fixed bed reactor

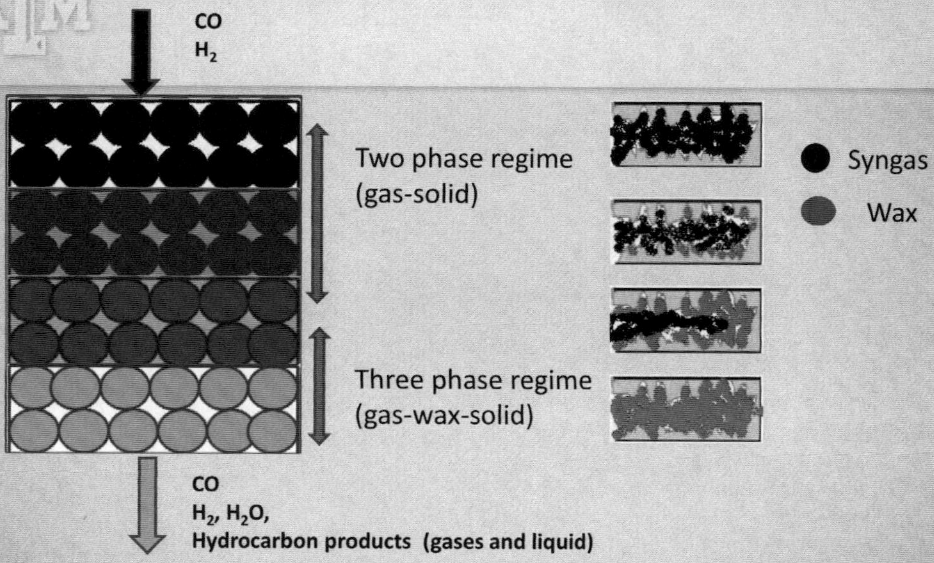

Ref: L.Caldwell, D.S.van Vuuren. *Chemical Engineering Science* 41(1) (**1986**) 89-96

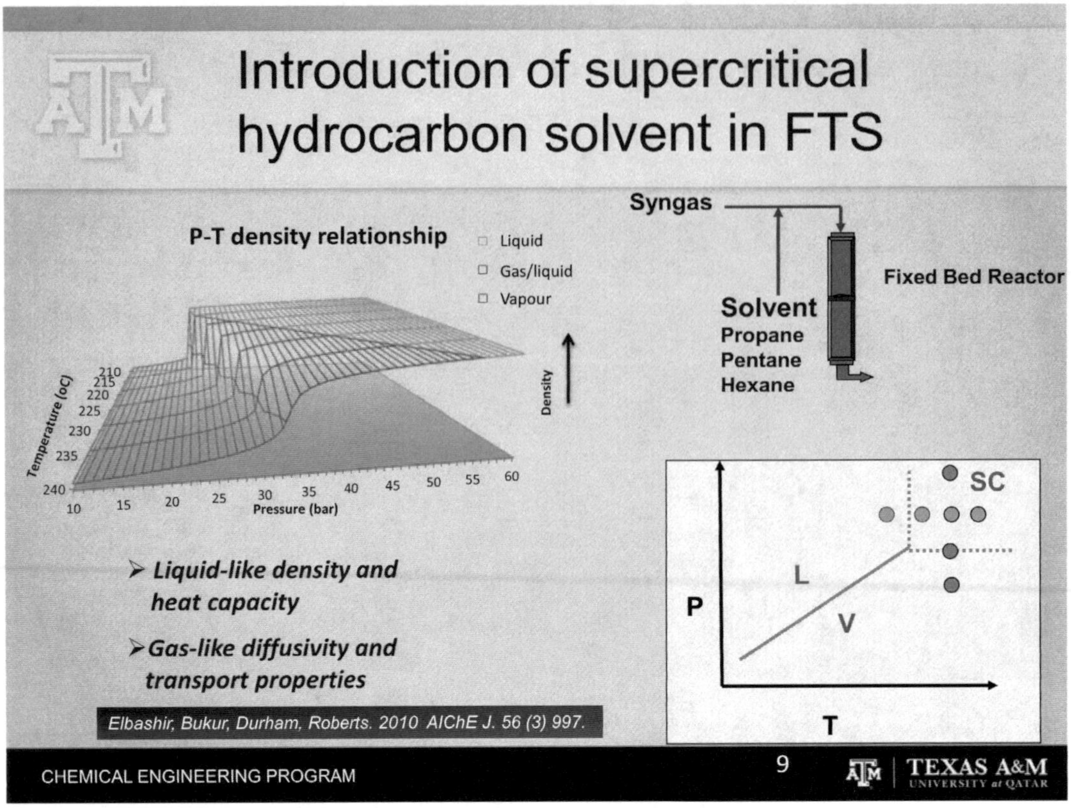
Introduction of supercritical hydrocarbon solvent in FTS
P-T density relationship
Liquid
Gas/liquid
Vapour
Temperature (oC)
210
215
220
225
230
235
240
10
15
20
25
30
35
40
45
50
55
60
Pressure (bar)
Density
Syngas
Fixed Bed Reactor
Solvent
Propane
Pentane
Hexane
SC
L
V
P
T
➢ Liquid-like density and heat capacity
➢ Gas-like diffusivity and transport properties
Elbashir, Bukur, Durham, Roberts. 2010 AIChE J. 56 (3) 997.
CHEMICAL ENGINEERING PROGRAM
9
TEXAS A&M
UNIVERSITY at QATAR

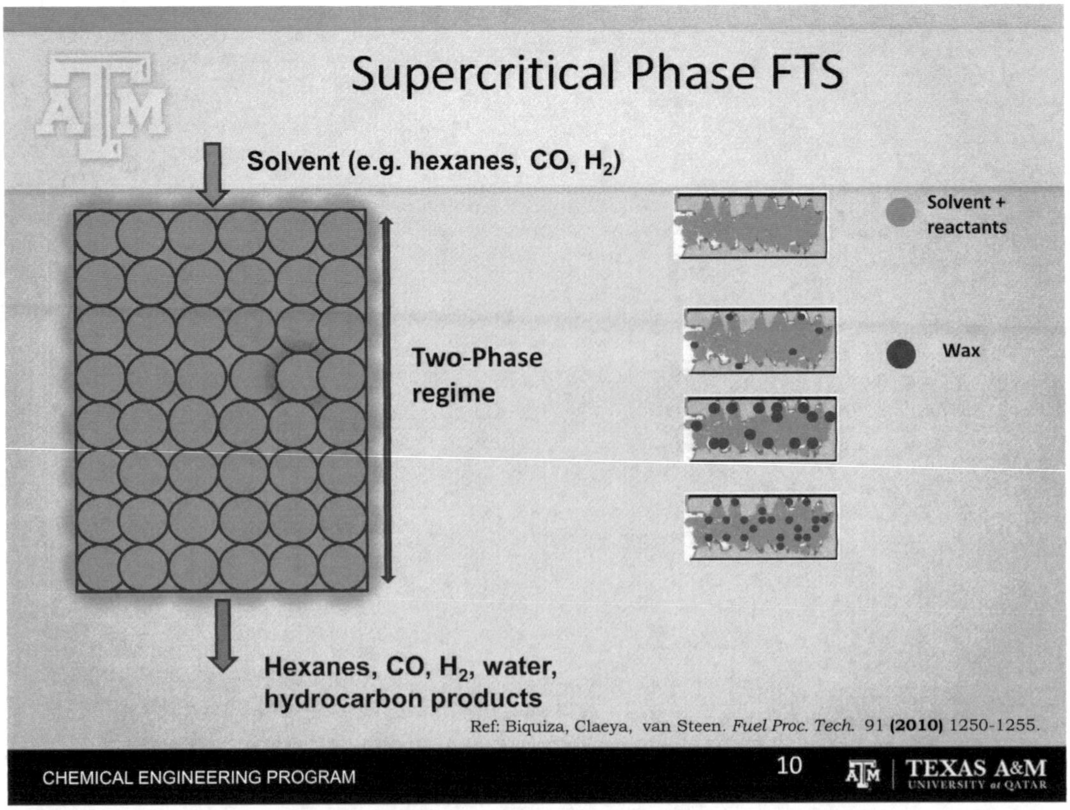
Supercritical Phase FTS
Solvent (e.g. hexanes, CO, H2)
Solvent + reactants
Two-Phase regime
Wax
Hexanes, CO, H2, water, hydrocarbon products
Ref: Biquiza, Claeya, van Steen. Fuel Proc. Tech. 91 (2010) 1250-1255.
CHEMICAL ENGINEERING PROGRAM
10
TEXAS A&M
UNIVERSITY at QATAR

Enhancement in SCF-FTS: wax extraction

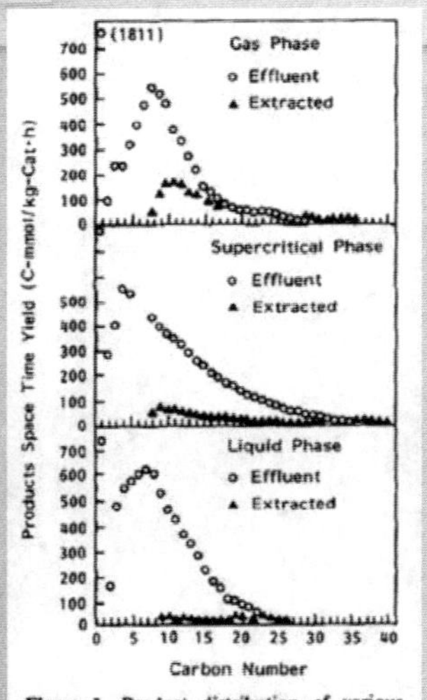

Figure 1 Product distribution of various phase Fischer-Tropsch synthesis

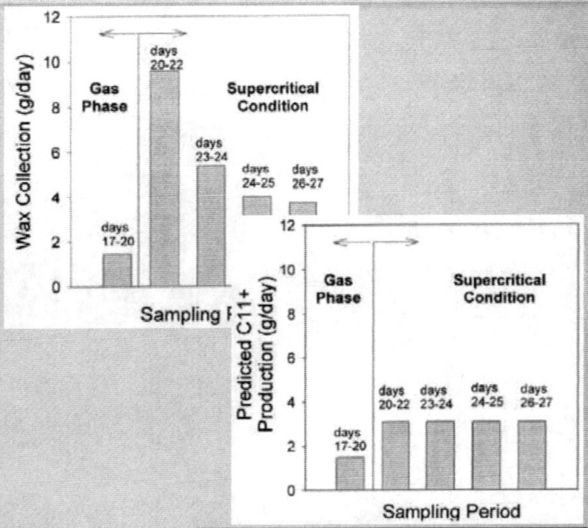

Yokota, K.; Fujimoto, K., *Ind. Eng. Chem. Res.* 30 (1991)

G. Jacobs et al. / Fuel 82 (2003) 1251–1260

Enhancement in SCF-FTS: selective control of products

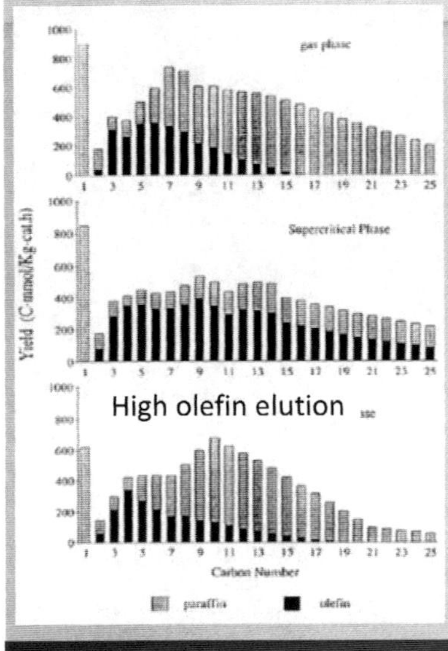

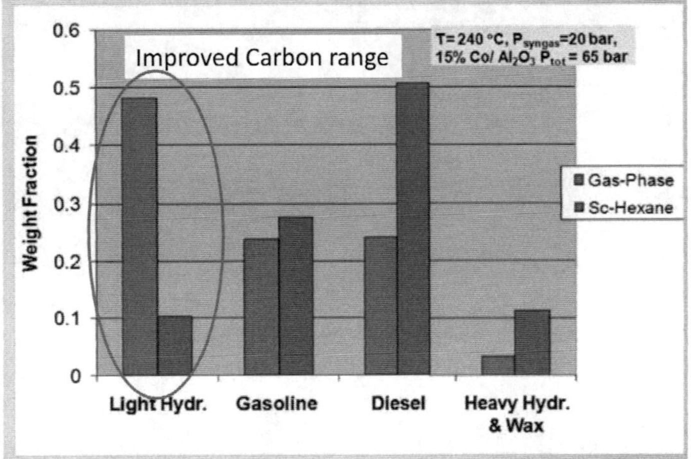

Fan, L.; Yokota, K.; Fujimoto, K., *Top. Catal.* 2 (1995)

Elbashir, N.O.; Roberts, B.,Prep ACS, Div, Pet. Chem. 49. 157

What it takes to design a novel reactor technology?

TEXAS A&M UNIVERSITY at QATAR

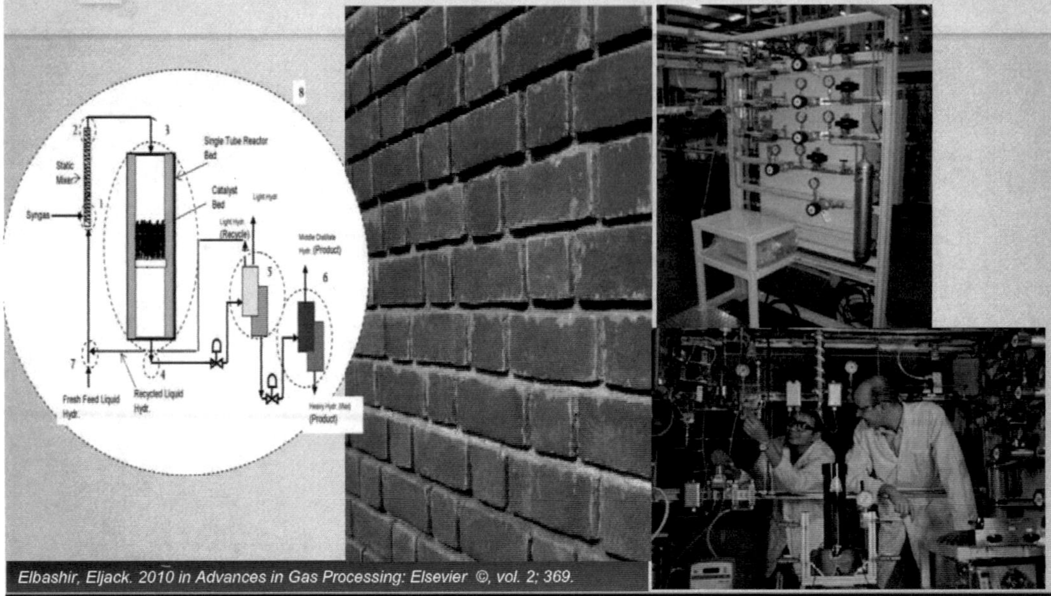

Elbashir, Eljack. 2010 in Advances in Gas Processing: Elsevier ©, vol. 2; 369.

CHEMICAL ENGINEERING PROGRAM 13 TEXAS A&M UNIVERSITY at QATAR

The design phases of a novel reactor

Process Design: Energy Integration & Optim TEXAS A&M UNIVERSITY

Phase Behavior & Thermodynamics TEXAS A&M UNIVERSITY at QATAR

Process Control TEXAS A&M UNIVERSITY

TEXAS A&M UNIVERSITY Products Processi DTU TEXAS A&M UNIVERSITY at QATAR

Kinetics & Product Distribution TEXAS A&M UNIVERSITY at QATAR

Reactor Design & Experimental Camp TEXAS A&M UNIVERSITY at QATAR

UNIVERSITY OF CAMBRIDGE *In situ* Reactor Behavior TEXAS A&M UNIVERSITY at QATAR

sasol ExxonMobil Advisory

CHEMICAL ENGINEERING PROGRAM Elbashir, Eljack. 2010 in Advances in Gas Processing: Elsevier ©, vol. 2; 369. 14 TEXAS A&M UNIVERSITY at QATAR

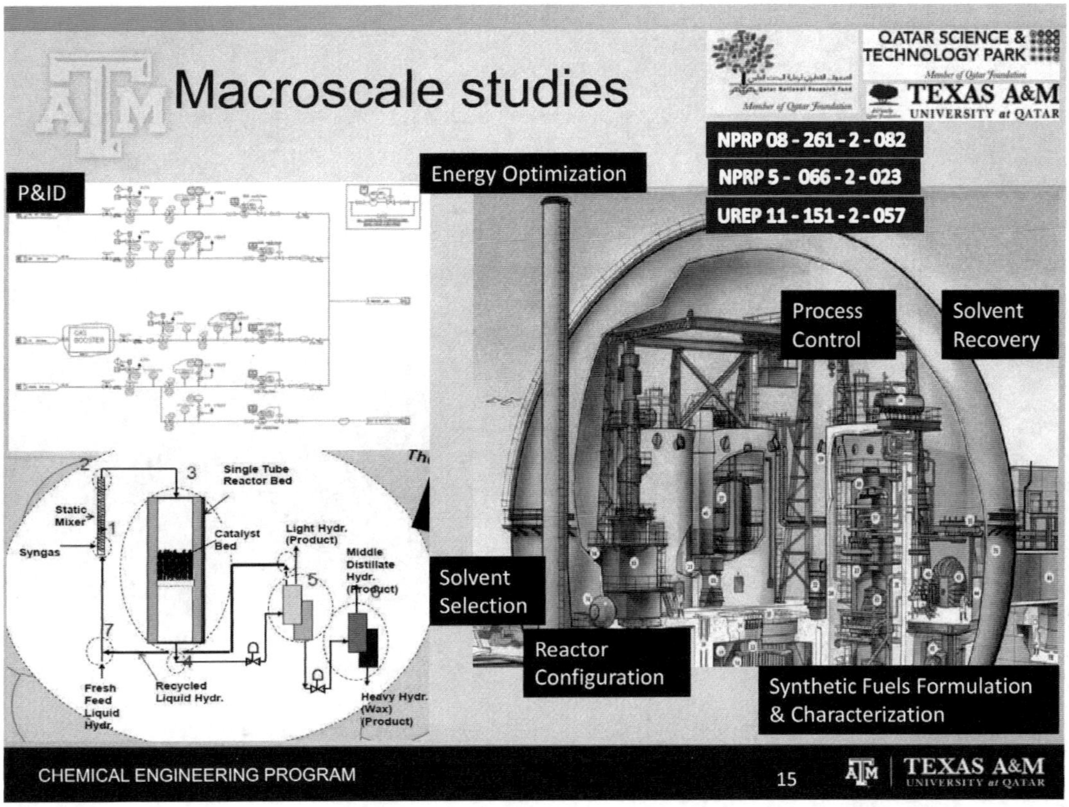
Macroscale studies
QATAR SCIENCE & TECHNOLOGY PARK
TEXAS A&M UNIVERSITY at QATAR
NPRP 08 - 261 - 2 - 082
NPRP 5 - 066 - 2 - 023
UREP 11 - 151 - 2 - 057
P&ID
Energy Optimization
Process Control
Solvent Recovery
Solvent Selection
Reactor Configuration
Synthetic Fuels Formulation & Characterization
Single Tube Reactor Bed
Static Mixer
Syngas
Catalyst Bed
Light Hydr. (Product)
Middle Distillate Hydr. (Product)
Fresh Feed Liquid Hydr.
Recycled Liquid Hydr.
Heavy Hydr. (Wax) (Product)
CHEMICAL ENGINEERING PROGRAM
15

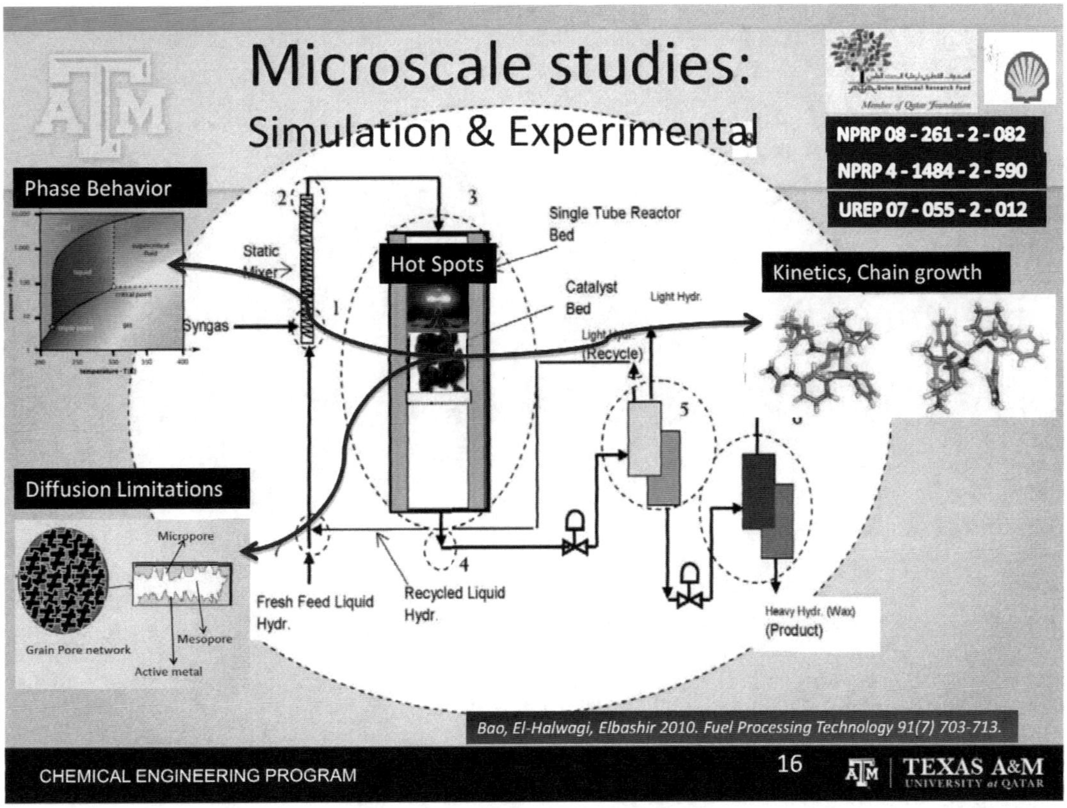
Microscale studies:
Simulation & Experimental
NPRP 08 - 261 - 2 - 082
NPRP 4 - 1484 - 2 - 590
UREP 07 - 055 - 2 - 012
Phase Behavior
Hot Spots
Kinetics, Chain growth
Diffusion Limitations
Single Tube Reactor Bed
Static Mixer
Syngas
Catalyst Bed
Light Hydr.
Light Hydr. (Recycle)
Fresh Feed Liquid Hydr.
Recycled Liquid Hydr.
Heavy Hydr. (Wax) (Product)
Micropore
Mesopore
Grain Pore network
Active metal
Bao, El-Halwagi, Elbashir 2010. Fuel Processing Technology 91(7) 703-713.
CHEMICAL ENGINEERING PROGRAM
16

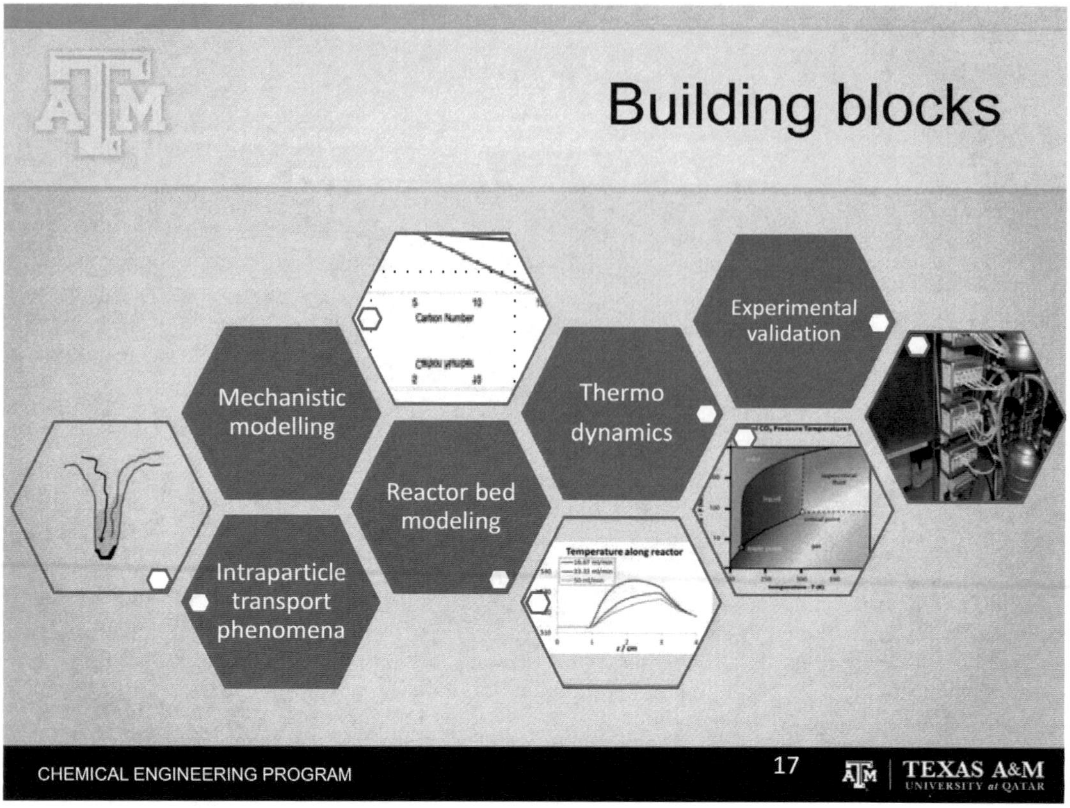
Building blocks
Mechanistic modelling
Intraparticle transport phenomena
Reactor bed modeling
Thermo dynamics
Experimental validation
Temperature along reactor
CHEMICAL ENGINEERING PROGRAM
17
TEXAS A&M UNIVERSITY at QATAR

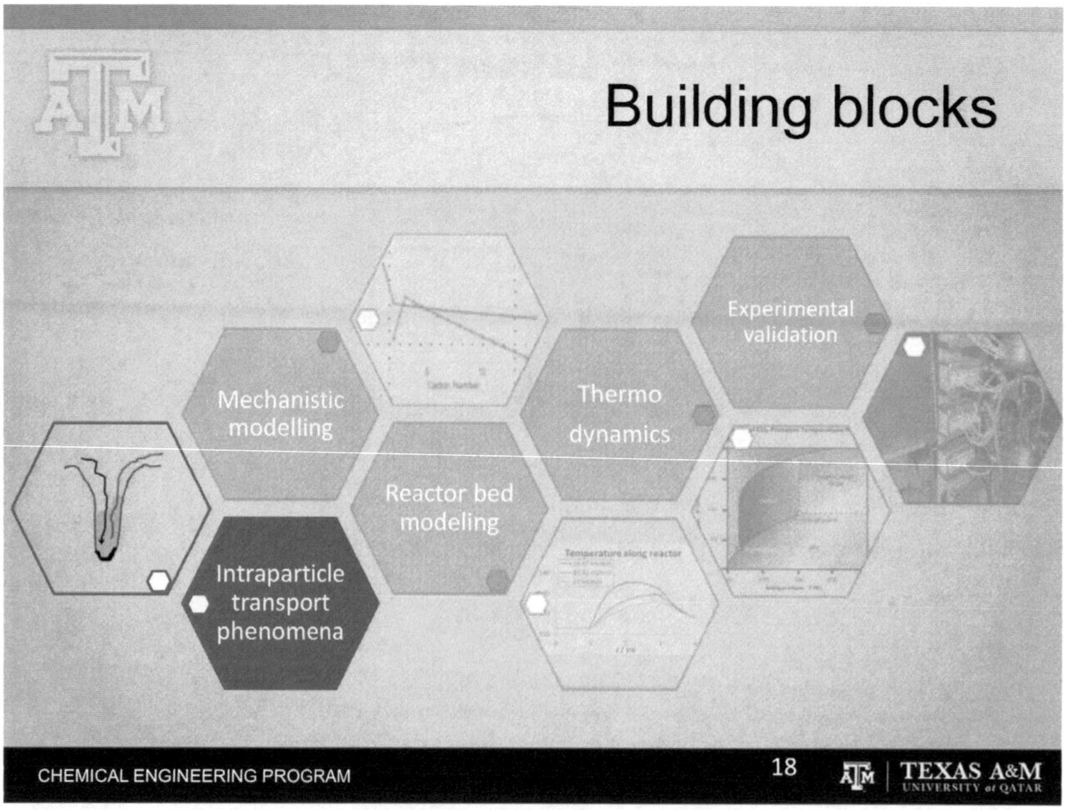
Building blocks
Mechanistic modelling
Intraparticle transport phenomena
Reactor bed modeling
Thermo dynamics
Experimental validation
CHEMICAL ENGINEERING PROGRAM
18
TEXAS A&M UNIVERSITY at QATAR

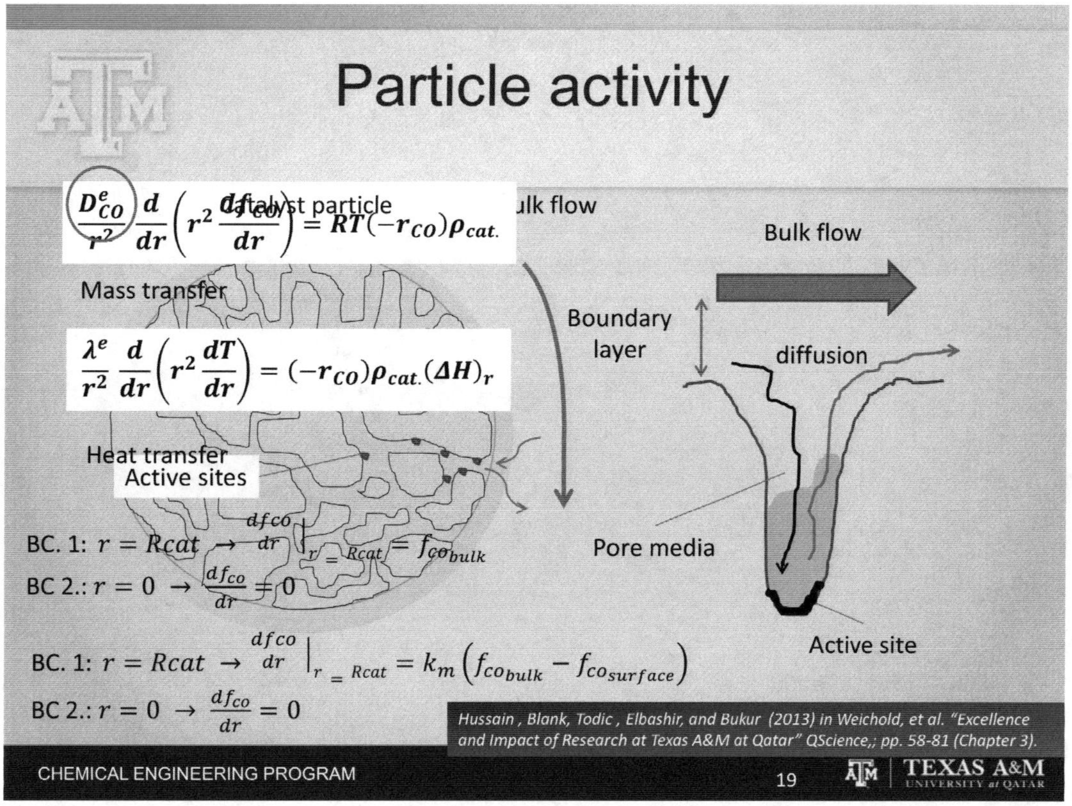
Particle activity
$\frac{D^e_{CO}}{r^2}\frac{d}{dr}\left(r^2\frac{df_{CO}}{dr}\right) = RT(-r_{CO})\rho_{cat.}$
Catalyst particle
Bulk flow
Mass transfer
$\frac{\lambda^e}{r^2}\frac{d}{dr}\left(r^2\frac{dT}{dr}\right) = (-r_{CO})\rho_{cat.}(\Delta H)_r$
Heat transfer
Active sites
Boundary layer
diffusion
Pore media
Active site
BC. 1: $r = Rcat \rightarrow \frac{dfco}{dr}\Big|_{r = Rcat} = f_{co_{bulk}}$
BC 2.: $r = 0 \rightarrow \frac{df_{co}}{dr} = 0$
BC. 1: $r = Rcat \rightarrow \frac{dfco}{dr}\Big|_{r = Rcat} = k_m\left(f_{co_{bulk}} - f_{co_{surface}}\right)$
BC 2.: $r = 0 \rightarrow \frac{df_{co}}{dr} = 0$
Hussain, Blank, Todic, Elbashir, and Bukur (2013) in Weichold, et al. "Excellence and Impact of Research at Texas A&M at Qatar" QScience,; pp. 58-81 (Chapter 3).
CHEMICAL ENGINEERING PROGRAM
19
TEXAS A&M UNIVERSITY at QATAR

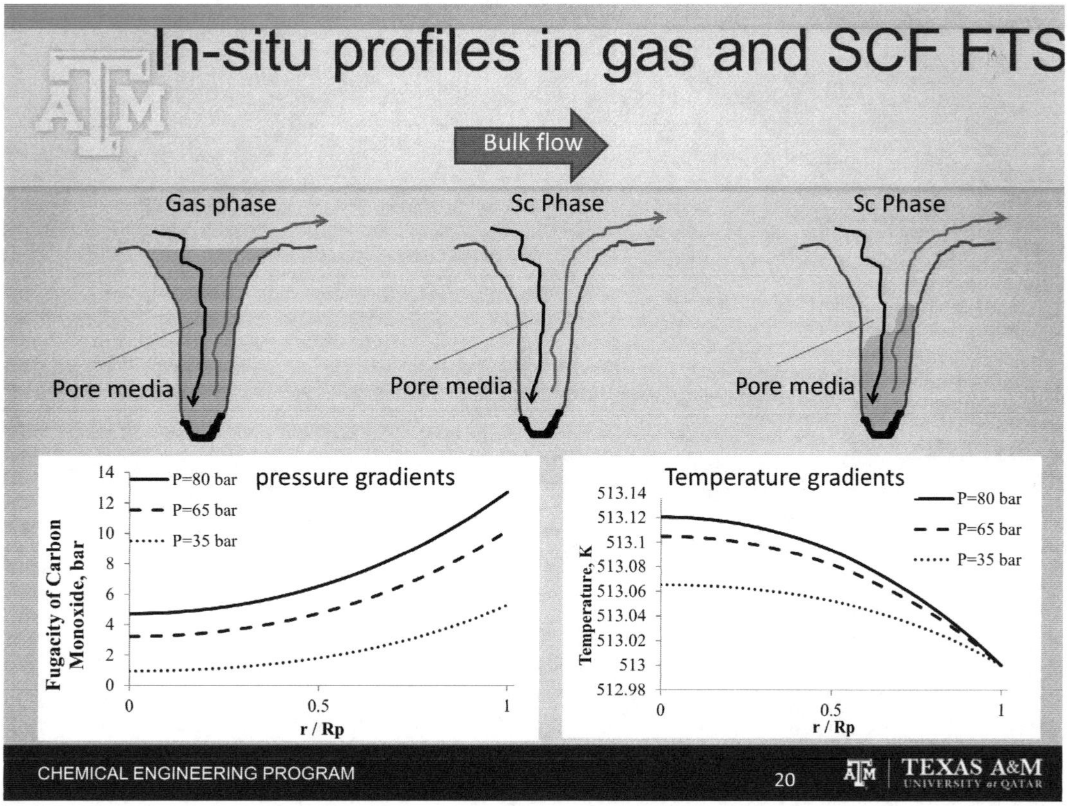
In-situ profiles in gas and SCF FTS
Bulk flow
Gas phase
Sc Phase
Sc Phase
Pore media
Pore media
Pore media
pressure gradients
P=80 bar
P=65 bar
P=35 bar
Fugacity of Carbon Monoxide, bar
r / Rp
Temperature gradients
P=80 bar
P=65 bar
P=35 bar
Temperature, K
r / Rp
CHEMICAL ENGINEERING PROGRAM
20
TEXAS A&M UNIVERSITY at QATAR

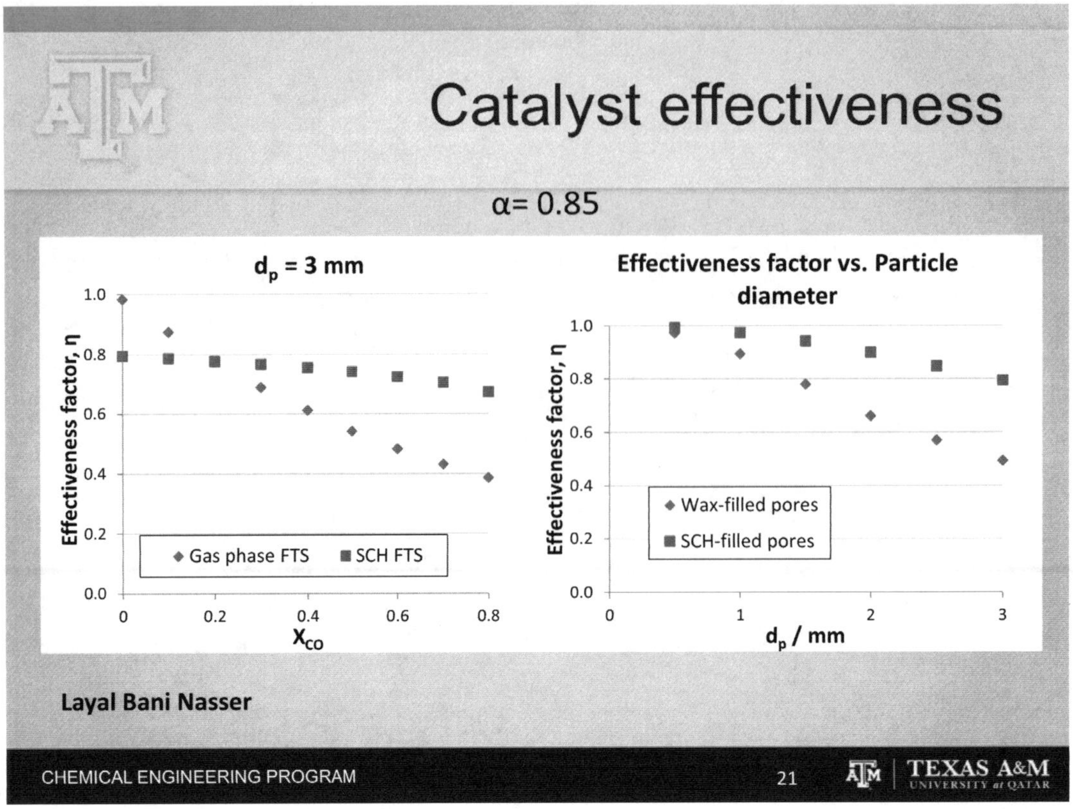
Catalyst effectiveness
α= 0.85
d$_p$ = 3 mm
Effectiveness factor, η
1.0
0.8
0.6
0.4
0.2
0.0
0
0.2
0.4
0.6
0.8
X$_{CO}$
Gas phase FTS
SCH FTS
Effectiveness factor vs. Particle diameter
Effectiveness factor, η
1.0
0.8
0.6
0.4
0.2
0.0
0
1
2
3
d$_p$ / mm
Wax-filled pores
SCH-filled pores
Layal Bani Nasser
CHEMICAL ENGINEERING PROGRAM
21
TEXAS A&M
UNIVERSITY at QATAR

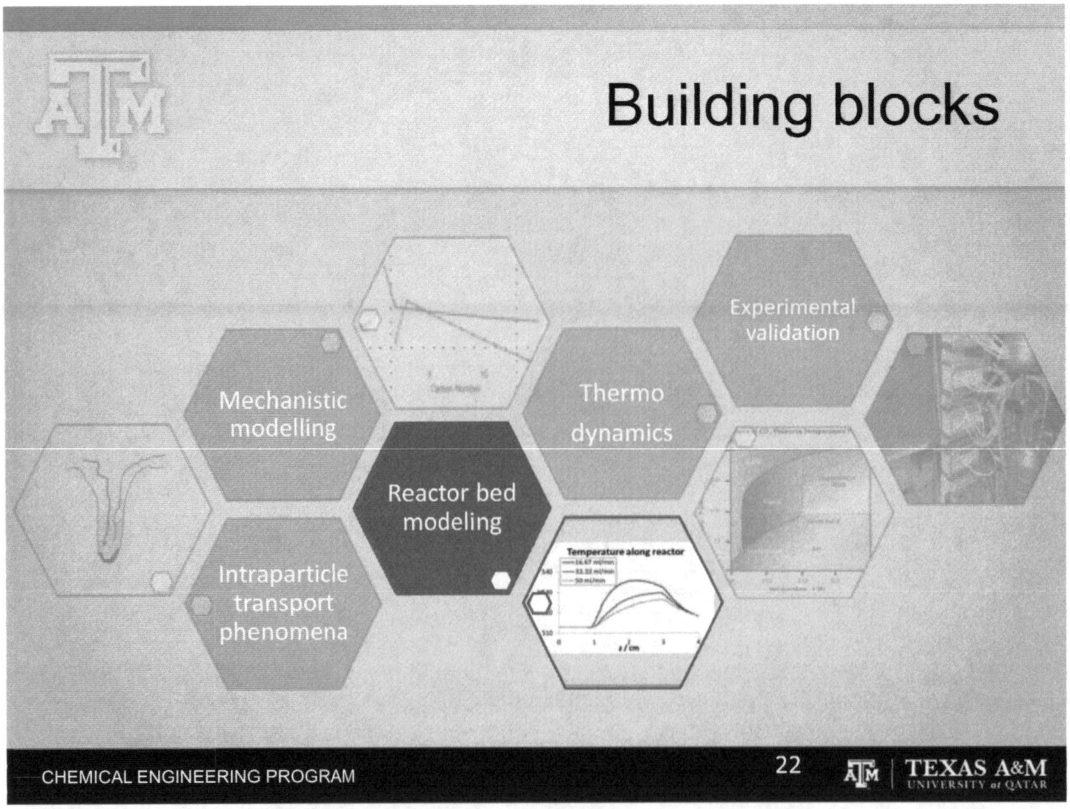
Building blocks
Experimental validation
Mechanistic modelling
Thermo dynamics
Reactor bed modeling
Intraparticle transport phenomena
Temperature along reactor
CHEMICAL ENGINEERING PROGRAM
22
TEXAS A&M
UNIVERSITY at QATAR

1-D Reactor bed modeling

- $-\frac{d(f_{CO}u_z)}{dz} = r_A\rho_{bed}\eta$ Mass balance
- $\frac{dT}{dz} = \frac{(-\Delta H)r_A\eta\rho_{bed} - 4\frac{U}{d_t}(T-T_r)}{u_s\rho_g c_p}$ Heat generation
- $-\frac{dp_t}{dz} = f\frac{\rho_g u_s^2}{d_p}$ Momentum balance

Correct for "local" conditions

1-D Reactor bed modeling

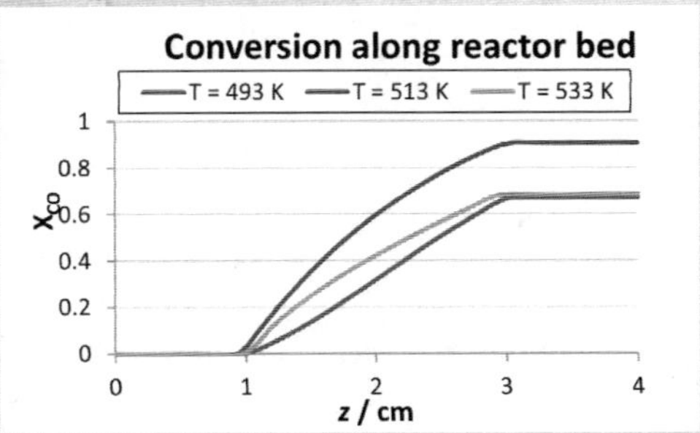

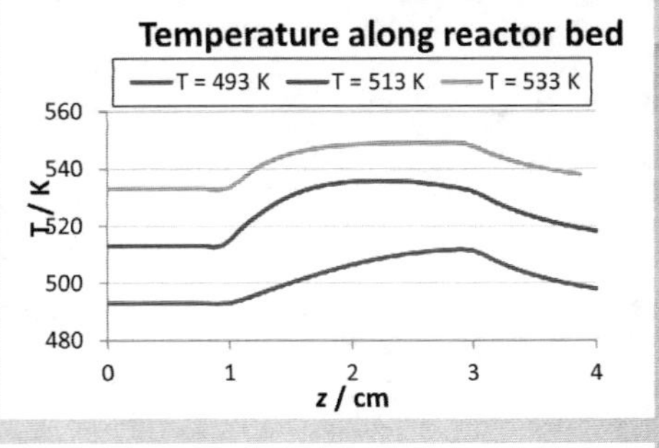

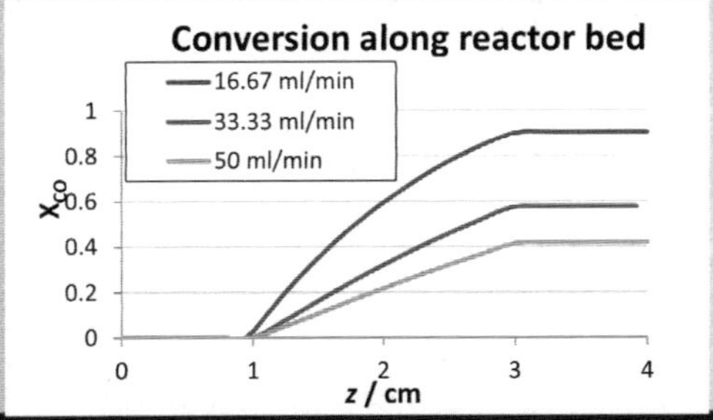

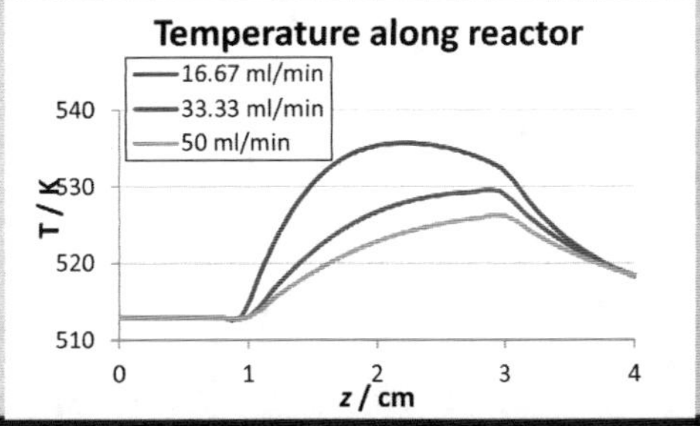

Building blocks

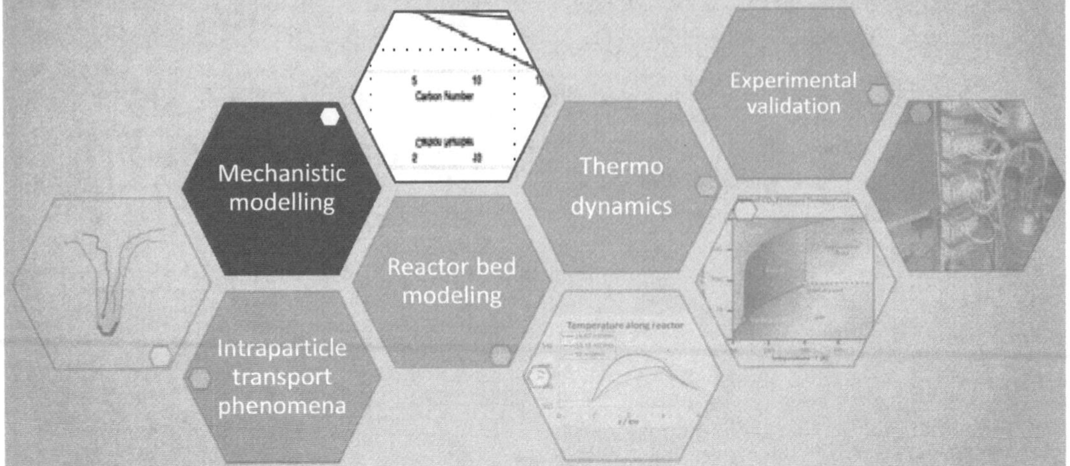

Macro-kinetic models for Co catalysts

	Kinetic expression	Mechanistic Implications	References
Cobalt Catalysts			
C1	$-r_{CO} = aP_{CO}^{0.2}P_{H_2}^{0.7}$	CO inhibits reaction, CO strongly adsorbed, high θ_{CO}	[illegible] *et al.*, 1997; Zennaro *et al.*, 2000
C2	$-r_{CO} = \frac{aP_{CO}P_{H_2}}{(1+bP_{CO})^2}$	high θ_{CO}; enolic mechanism, hydrogenation of HCOH (ads)	Yates and Satterfield, 1991; Zennaro *et al.*, 2000
C3	$-r_{CO} = \frac{aP_{CO}^{m}P_{H_2}^{n}}{1+bP_{CO}}$, $m = 0.5–0.6, n = 0.6–0.9$	moderate θ_{CO}; Eley-Rideal; stepwise hydrogenation of C_s	Iglesia *et al.*, 1993b; [illegible] *et al.*, 2001
C4	$-r_{CO} = \frac{aP_{CO}^{0.5}P_{H_2}^{0.5}}{(1+bP_{CO}^{0.5})^2}$	high θ_{CO}; $C_s + H_s$ and $[illegible] + H_s$ are rds's	Sarup, 1989; Keyser *et al.*, 2000; [illegible], 2000; [illegible] and Bartholomew 2005

Carbide mechanism for FTS

No.	Elementary reaction	kinetic constant
1^{RDS}	$CO + H\text{-}S \rightarrow H\text{-}S\text{-}CO$ $CO + CH_3\text{-}S \rightarrow CH_3\text{-}S\text{-}CO$ $CO + C_nH_{2n+1}\text{-}S \rightarrow C_nH_{2n+1}\text{-}S\text{-}CO$	k_1
2	$H\text{-}S\text{-}CO + H_2 \leftrightarrow H\text{-}S\text{-}C + H_2O$ $CH_3\text{-}S\text{-}CO + H_2 \leftrightarrow CH_3\text{-}S\text{-}C + H_2O$ $C_nH_{2n+1}\text{-}S\text{-}CO + H_2 \leftrightarrow C_nH_{2n+1}\text{-}S\text{-}C + H_2O$	K_2
3	$H\text{-}S\text{-}C + H_2 \leftrightarrow H\text{-}S\text{-}CH_2$ $CH_3\text{-}S\text{-}C + H_2 \leftrightarrow CH_3\text{-}S\text{-}CH_2$ $C_nH_{2n+1}\text{-}S\text{-}C + H_2 \leftrightarrow C_nH_{2n+1}\text{-}S\text{-}CH_2$	K_3
4	$C_nH_{2n+1}\text{-}S\text{-}CH_2 \leftrightarrow C_nH_{2n+1}CH_2\text{-}S$	K_4
5^{RDS}	$CH_3\text{-}S + H_2 \rightarrow CH_4 + H\text{-}S$ $C_nH_{2n+1}\text{-}S + H_2 \rightarrow C_nH_{2n+2} + H\text{-}S$	k_{5M} k_5
6^{RDS}	$C_2H_5\text{-}S \rightarrow C_2H_4 + H\text{-}S$ $C_nH_{2n+1}\text{-}S \rightarrow C_nH_{2n} + H\text{-}S$	k_{6E} k_{6n}
7	$H_2 + 2S \leftrightarrow 2H\text{-}S$	K_7

I&EC research

Kinetic Model of Fischer–Tropsch Synthesis in a Slurry Reactor on Co–Re/Al_2O_3 Catalyst

dx.doi.org/10.1021/ie3028312 | *Ind. Eng. Chem. Res.* 2013, 52, 669–679

Validation

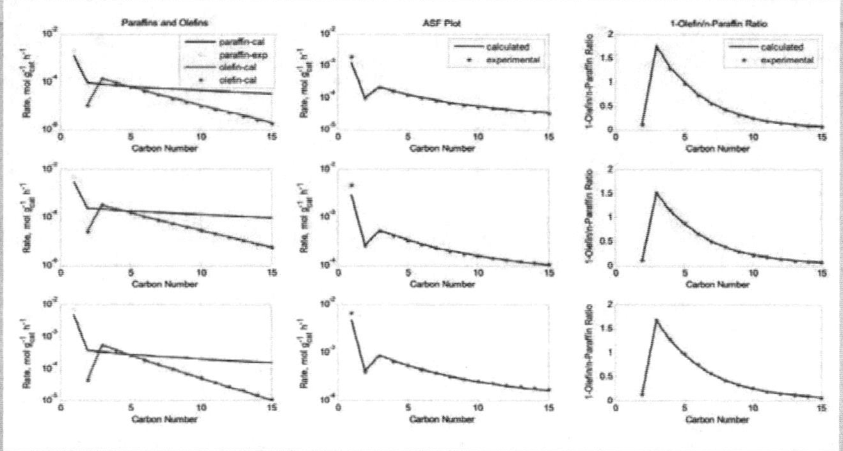

- Model data from Todic et al. used as experimental data input to the GA code

Shaik Afzal

dx.doi.org/10.1021/ie3028312 | *Ind. Eng. Chem. Res.* 2013, 52, 669–679

Building blocks

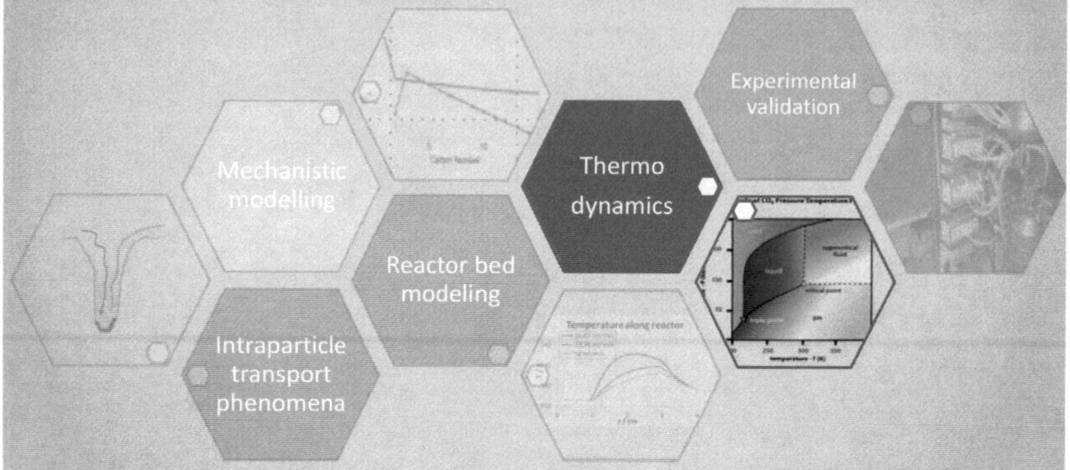

CHEMICAL ENGINEERING PROGRAM 29 TEXAS A&M UNIVERSITY at QATAR

Experimental comparisons

	Syngas	Nitrogen	Hexane	total
Gas Phase	20	-	-	20
Supercritical	20	-	60	80
Nitrogen	20	60	-	80

IDEAL

Determination of "real" conditions using Peng Robinson EoS leads to a deviation

NON-IDEAL

	Syngas		Nitrogen	Hexane	total
	CO	H_2			
Gas Phase	6.6	13.3	-	-	20
Nitrogen	7.02	14.4	63.3	-	80
Supercritical	12.5	33.2	-	23.3*	80

*The non-ideal hexane compresses significantly, making space for the relatively ideal syngas

CHEMICAL ENGINEERING PROGRAM 31 TEXAS A&M UNIVERSITY at QATAR

Comparing conditions in gas-phase and SCF-FTS

	Syngas	Nitrogen	Hexane	total
Gas Phase	20	-	-	20
Supercritical	20	-	60	80
Nitrogen	20	60	-	80

IDEAL

Determination of "real" conditions using Peng Robinson EoS leads to a deviation

NON-IDEAL

	Syngas		Nitrogen	Hexane	total
	CO	H_2			
Gas Phase	6.6	13.3	-	-	20
Nitrogen	7.02	14.4	63.3	-	80
Supercritical	12.5	33.2	-	23.3*	80

Building blocks

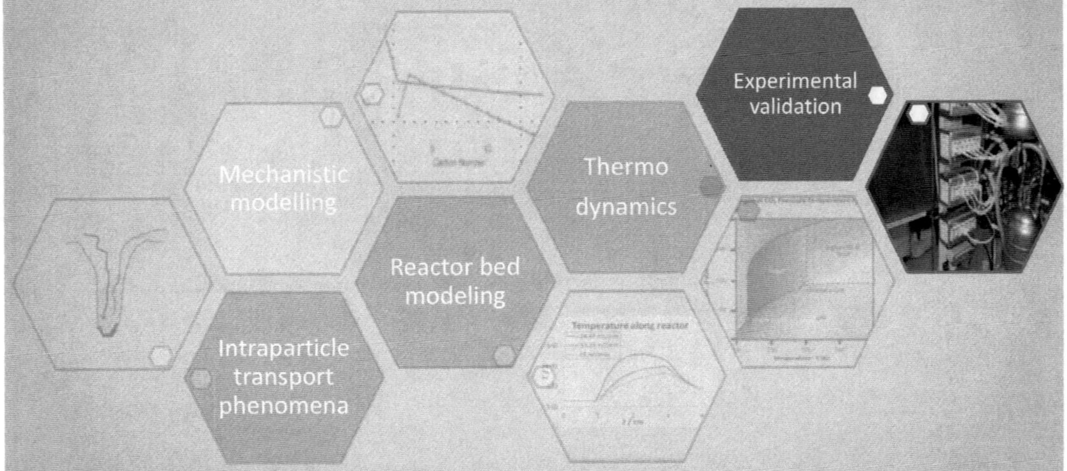

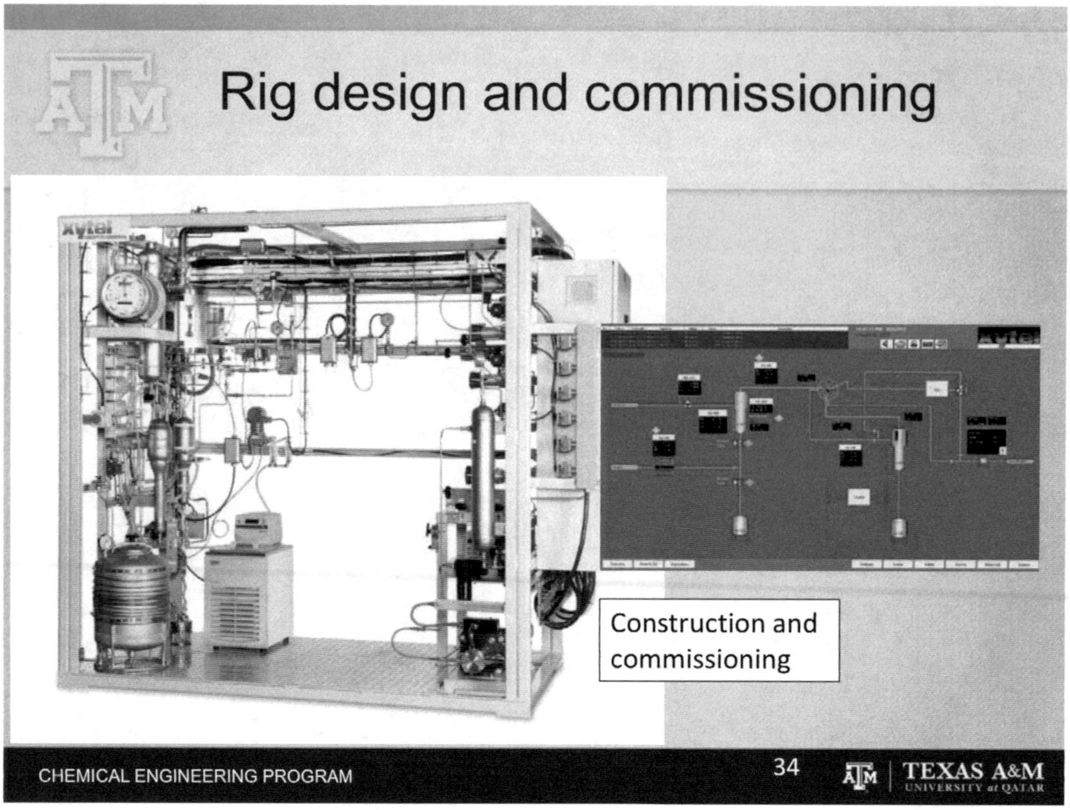
Rig design and commissioning
Construction and commissioning
CHEMICAL ENGINEERING PROGRAM
34
TEXAS A&M
UNIVERSITY at QATAR

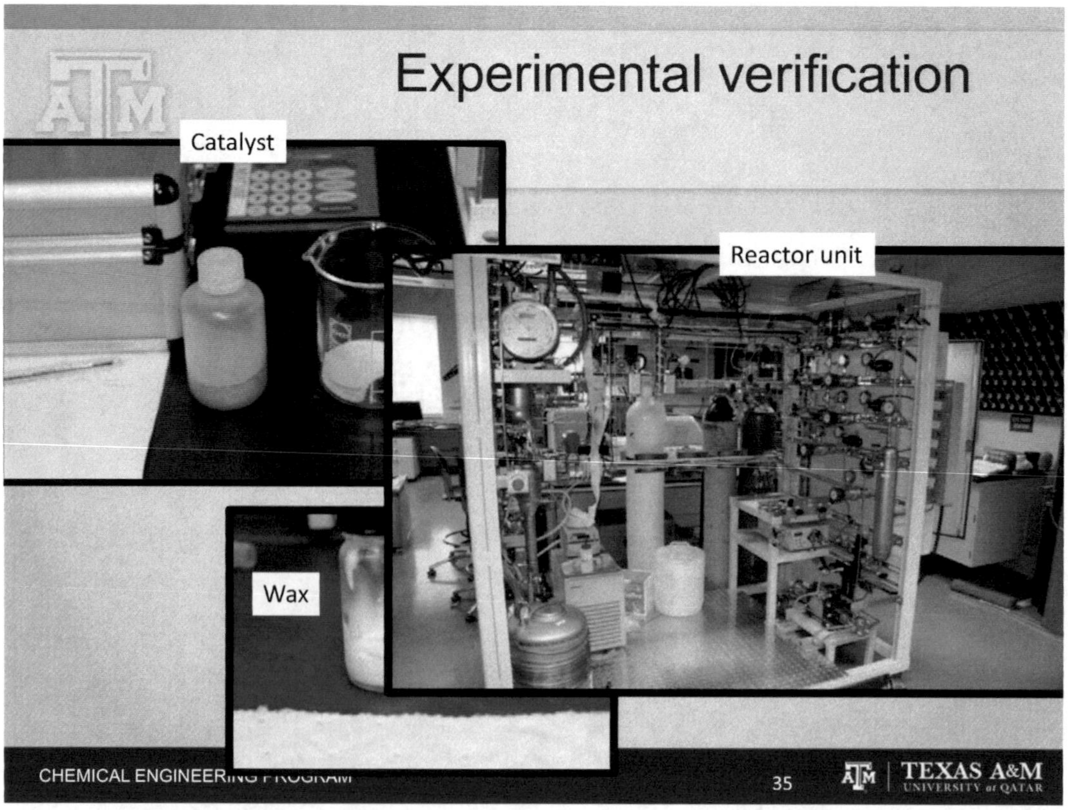
Experimental verification
Catalyst
Reactor unit
Wax
CHEMICAL ENGINEER
35
TEXAS A&M
UNIVERSITY at QATAR

Catalyst activity data

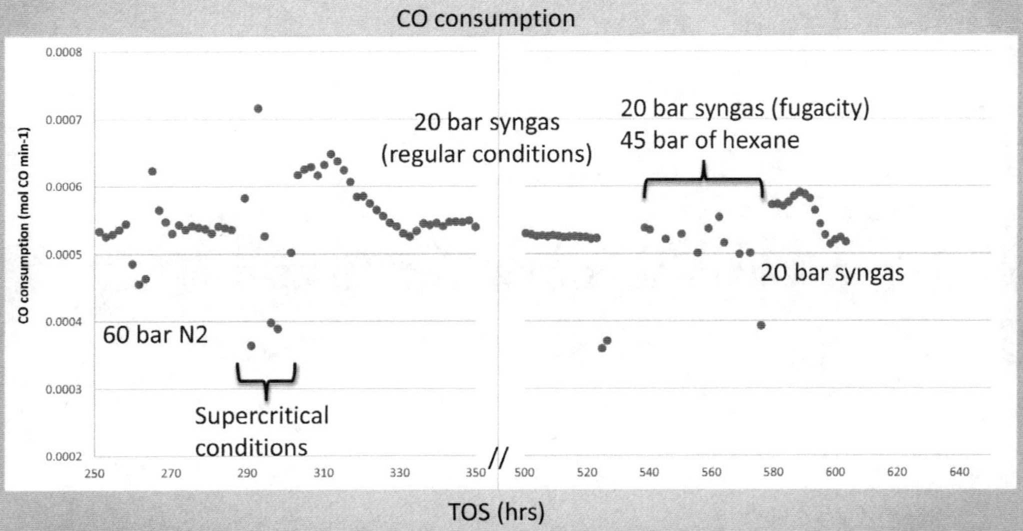

Methane selectivity

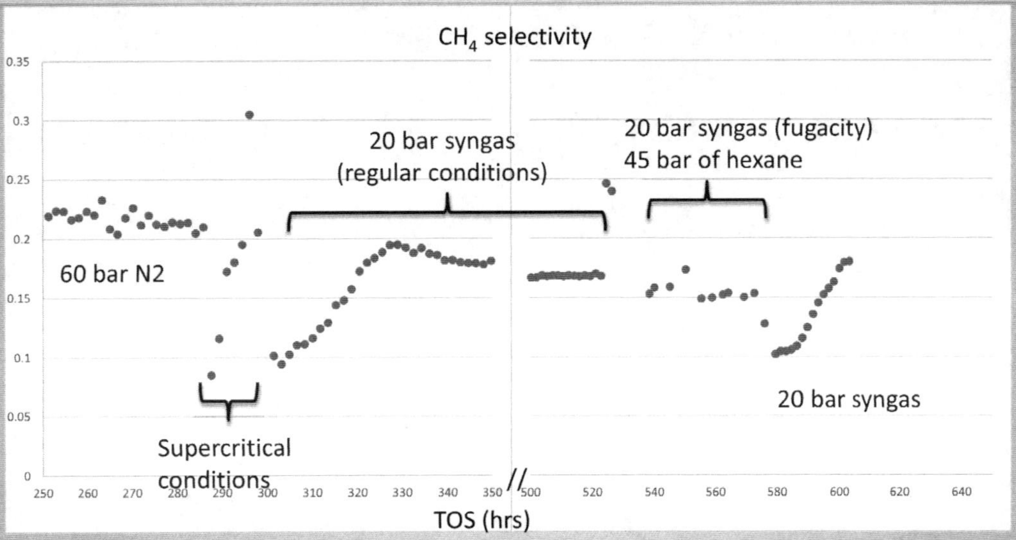

Conclusions

- Our aim is to develop a comprehensive model to describe many aspects of the Fischer-Tropsch reactor bed
- Thermodynamics
- Intraparticle transfer phenomena
- 1-D reactor bed modelling
- Kinetic mechanistic modelling
- Detailed experimental verification

CHEMICAL ENGINEERING PROGRAM 38 TEXAS A&M UNIVERSITY at QATAR

Work in progress

- Determination of kinetic parameters for the Fischer-Trospch reaction
- Determination of thermal and mass diffusion coefficients
- Determination of the composition of the catalyst pore
- Define fundamental differences between the use of SCF's and regular gas phase reaction
- Detailed thermodynamic description of fluid behavior inside the catalyst pore

CHEMICAL ENGINEERING PROGRAM 39 TEXAS A&M UNIVERSITY at QATAR

Peng-Robinson EOS extended to confined fluids

$$Q(T,V,N)=\frac{\left(q\lambda^{-3}V_f\right)^N}{N!}\exp\left(\int_{\infty}^{T}\frac{E_{conf}}{kT^2}dT\right)$$

❖ **Free volume V_f : repulsive part of the EOS**

❖ **Configurational energy E_{conf} : attractive part of the EOS**

Assumptions:

- Molecules of the fluid are hard spheres.
- Pores are cylinders with continuous inner surface.
- Molecules of the fluid interact with others and with the pore wall via square-well potential.

Molecule – Wall interactions

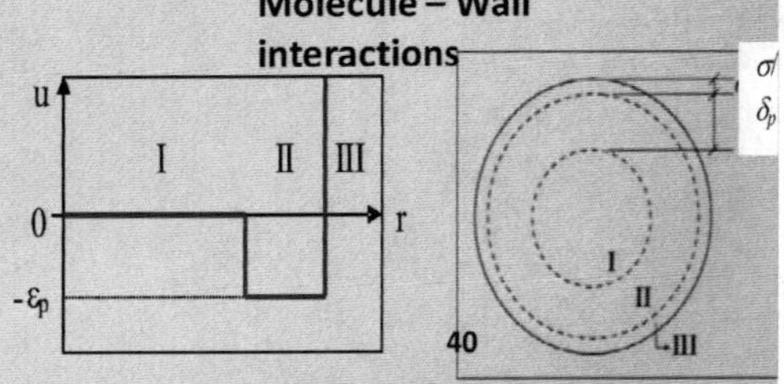

40

**L.Travalloni, M. Castier, and F.W. Tavares,. Fluid Phase Equilibria, 362(0): p. 335-341 (2014).

TEXAS A&M UNIVERSITY at QATAR

MRI studies

Velocity Profile and Phase Distribution

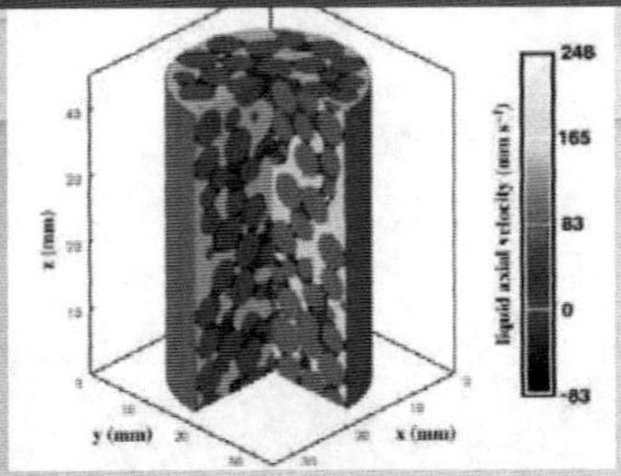

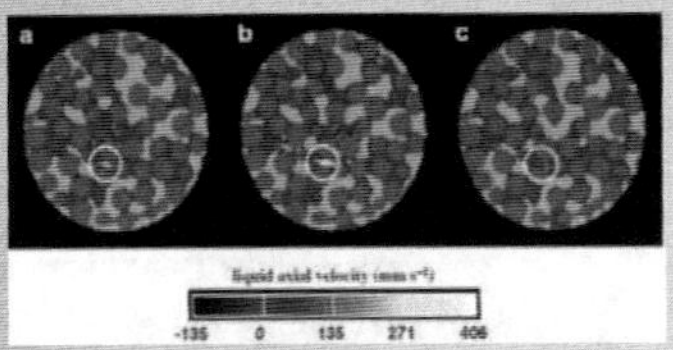

NMR Relaxometry

Gladden, et al. 2009. J Phys. Chem. C 113, 6610.

Adsorption/Desorption

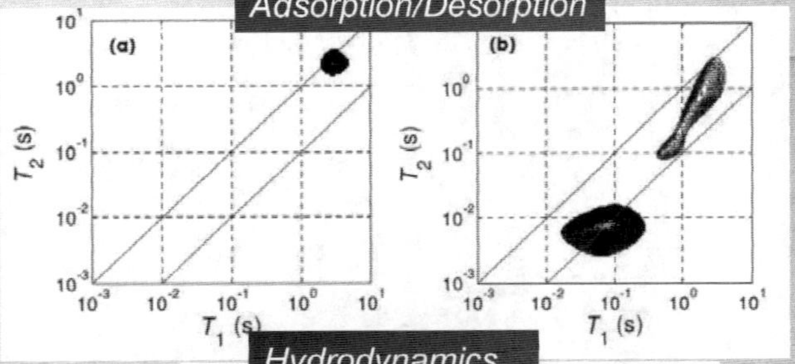

Hydrodynamics

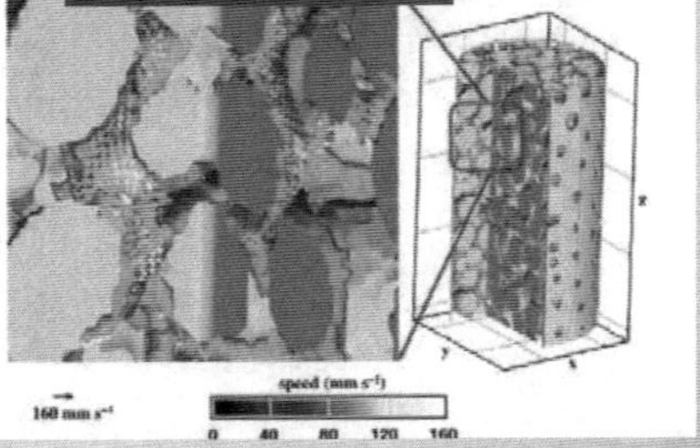

Liquid & gas velocity profile 3D visualization

Gladden et al. 2009. JMR 196, 142.

CHEMICAL ENGINEERING PROGRAM 41 TEXAS A&M UNIVERSITY at QATAR

Acknowledgements
Collaborators
TEXAS A&M UNIVERSITY
Prof. Christopher Roberts
Prof. Mahmoud El-Halwagi
Prof. Juergen Hahn (currently RPI)
Prof. Fadwa Eljack
UNIVERSITY OF CAMBRIDGE
Prof. Lynn Gladden
TEXAS A&M
Prof. Dragomir Bukur
Prof. Marcelo Castier
Industry Advisory Board
Willem Scholten
ExxonMobil
Taking on the world's toughest energy challenges.
Dr. Jim Rigby
sasol
reaching new frontiers
Dr. Ernest De Toit (former)
QATARGAS
Rashid Al-Rashdi
CHEMICAL ENGINEERING PROGRAM
42
TEXAS A&M UNIVERSITY at QATAR

Research Outcome:
Development of Research Capacity in Qatar
L to R: Moiz Bohra, Maria Orillano
Undergraduates
Elfatih Elmalik
Research Associate
(Rensselaer Polytechnic)
L to R: Laial Bani Nassr, Ibrahim Al-Nuaimi, Samah Warrag
Graduate Researchers Qatar
Dr. Nimir Elbashir
Associate Professor, TAMUQ
Team leader
Additional Researchers:
Dr. Bao Buping
Mohamed Noureldin
TEXAS A&M
Dr. Jan H. Blank
Postdoctoral Researcher
(University of St. Andrews)
Shaik Afzal
Graduate Researcher ,India
Rehan Hussain
Research Associate
(University of Cambridge)
CHEMICAL ENGINEERING PROGRAM
43
TEXAS A&M UNIVERSITY at QATAR

Acknowledgements

This publication was made possible by following Grants from the Qatar National Research Fund (a member of Qatar Foundation). The statements made herein are solely the responsibility of the authors.

NPRP 08 - 261 - 2 - 082

NPRP 4 - 1484 - 2 - 590

UREP 07 - 055 - 2 - 012

Funding : Qatar National Research Fund, Member of Qatar Foundation; National Priorities Research Program NPRP; UREP

351247 Improve Energy Efficiency & Capacity by Integrating Synloop & Process Compressors in Ammonia Plants

Abstract:

Abstract A revamp study of a stretched Ammonia plant is presented (already revamped twice from 1100 stpd to 1500 stpd to 1750 stpd) for an additional 11% capacity along with 8% reduction in Energy consumption. The current operation at 1750 stpd maxed out the operation of Primary Reformer as well all the major compressors- to the limits of their design rating- including Process Air Compressor, Make Up Gas compressor & Refrigeration Compressor. A new integrated scheme between Synloop & Compressors allowed offloading both the Primary Reformer & all major compressors without any increase in compression power or any hardware upgrades. Only a few proven upgrades in the Primary Reformer were required. As a result, the production could be further increased by additional 11% along with energy improvement of ~8% for a stretched Ammonia plant- with an attractive payback. The key operating parameters of the Base & Revamp case along with the major hardware changes will be presented along with overall economics.

PROSYS™ Maximize Your Plant.

Operator Errors and What Can be Done to Minimize

Global Congress on Process Safety

March 31, 2014

10TH GLOBAL CONGRESS ON PROCESS SAFETY

PROSYS™ Maximize Your Plant.™

10TH GLOBAL CONGRESS ON PROCESS SAFETY

Presenter

Tom Nolan

- **Graduated from Ohio University with B.S. in Chemical Engineering**
- **24 years experience in chemical process industry in a variety of roles**
- **MAIC and DFSS Six Sigma Black Belt**

PROSYS Maximize Your Plant.™

10th GLOBAL CONGRESS ON PROCESS SAFETY

The Cost of Errors

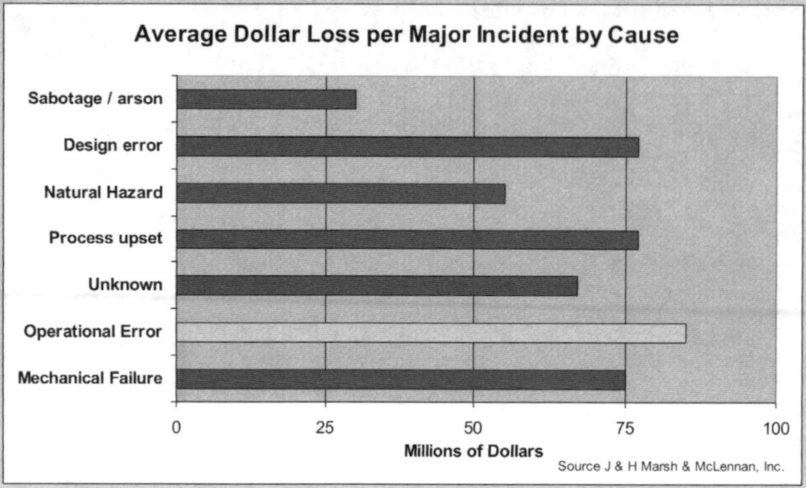

PROSYS Maximize Your Plant.™

10th GLOBAL CONGRESS ON PROCESS SAFETY

The Cost of Operator Errors

- **ASM estimates total loss due to operator error is $8B per year**
- **Chemical Safety Topical Committee - average of one chemical incident per day - cost of over $2 million per incident to comply with requirements**
- **Errors cause 42% of unscheduled shutdowns**
- **70% of process incidents occur during start-up or shutdown**

PROSYS Maximize Your Plant.℠

10th Global Congress on Process Safety

Reasons for Errors

- Lack of Skill
- Lack of Knowledge
- Carelessness
- System Design
  - Operator set up to make errors by inappropriate design – built in errors
  - Errors that are predictable are preventable by better design

PROSYS Maximize Your Plant.℠

10th Global Congress on Process Safety

Focus of this Discussion

- System Design
  - **Alarm Management**
  - **Operator Graphics**
  - Operator set up to make errors by inappropriate design – built in errors
  - Errors that are predictable are preventable by better design

PROSYS Maximize Your Plant.™

10th Global Congress on Process Safety

Alarm Floods

In a number of industrial incidents, alarm floods were identified as a significant contributing cause to the incident...

As found by EEMUA in 1999 and CSB

Alarm Flood defined by ISA 18.2 as -
"10 or more annunciated alarms in any 10 minute period per operator"

PROSYS Maximize Your Plant.™

10th Global Congress on Process Safety

Why do Alarm Floods Occur?

One reason is not providing dynamic alarm management

- **Alarms need to indicate abnormal situations that require operator action**
- **Processes do not operate in one state**
- **What is normal vs abnormal changes with state**
- **Alarms are typically configured for run therefore many alarms are triggered upon a Change of State: Run to Shutdown**
- **Many of the alarms are not applicable or actionable for the new state impleading the operators ability to act quickly on what is important**

PROSYS Maximize Your Plant.™

10th GLOBAL CONGRESS ON PROCESS SAFETY

What is Impacted by Alarm Floods?

- **Product quality**
- **Operability or profitability of the process**
- **Loss of equipment**
- **Loss of containment – environmental releases**
- **Injury and loss of life in plant or community**

PROSYS Maximize Your Plant.™

10th GLOBAL CONGRESS ON PROCESS SAFETY

What Makes Alarm Floods so Dangerous?

Can be a problem for three reasons:

- **A deluge of alarms can cause critical alarms to be missed**
- **Floods can be a significant distraction when dealing with process upsets**
- **Can be an indicator of larger systemic safety issues**

Impacting Alarm Management Design

- **Alarm rationalization is not a process to eliminate alarms - its about quality**
- **Good rationalization will add alarms when appropriate**
- **Typical (Static) rationalization is only for run mode**
- **Dynamic rationalization considers all plant modes**

PROSYS Maximize Your Plant.™ 10th Global Congress on Process Safety

Dynamic Alarm Management

- **Dynamic rationalization does everything a static rationalization does plus asks "when" for each alarm**
- **Requires dynamic software to make changes based upon operating mode of the plant**
- **Eliminates redundant alarms and lowers operator loading during transitions**
- **Only alarms what is abnormal and actionable for the given state**

PROSYS Maximize Your Plant.™

10th GLOBAL CONGRESS on PROCESS SAFETY

Actual Performance Metrics vs ISA 18.2

Type	Avg Alarm Rate/hr	Low/Hi of Avg Rate/hr	Peak Alarm Rate/hr	Low/Hi of Peak Rate/hr	% Time in Flood	Standing Alarms >24 hrs
Before Rationalization	30	Low=6.2 Hi=61	638	Low=152 Hi=2402	17.2%	9
ISA 18.2 Target Metrics	6	---	<60	---	<1%	<5
After Dynamic Alarm Management	2	Low=0.09 Hi=2.5	25	Low=13 Hi=42	0.25%	4

PROSYS Maximize Your Plant.™

10th GLOBAL CONGRESS on PROCESS SAFETY

Results After Dynamic Alarm Management

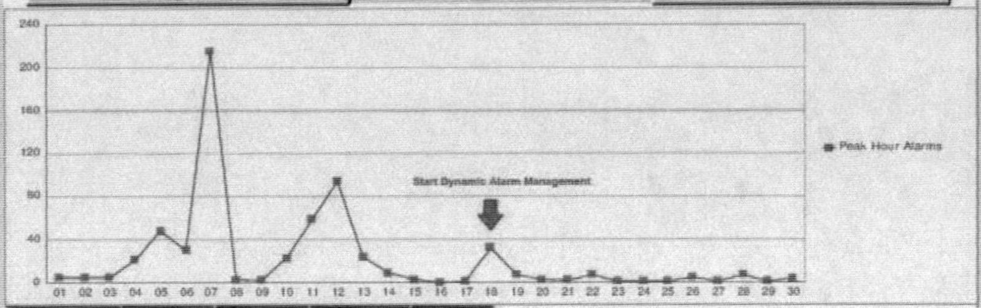

Day	Min Hour Alarms	Avg Hour Alarms	Peak Hour Alarms	Total Alarms
01-Jun-2013	0	0.3	5	6
02-Jun-2013	0	0.4	5	9
03-Jun-2013	0	0.5	5	13
04-Jun-2013	0	1.1	21	27
05-Jun-2013	0	13.3	48	319
06-Jun-2013	0	3.6	30	86
07-Jun-2013	0	11.9	215	285
08-Jun-2013	0	0.4	3	10
09-Jun-2013	0	0.3	2	7
10-Jun-2013	0	1.5	23	35
11-Jun-2013	0	4.7	59	112
12-Jun-2013	0	11.9	94	285
13-Jun-2013	0	3.8	24	90
14-Jun-2013	0	1.3	9	32
15-Jun-2013	0	0.3	3	8

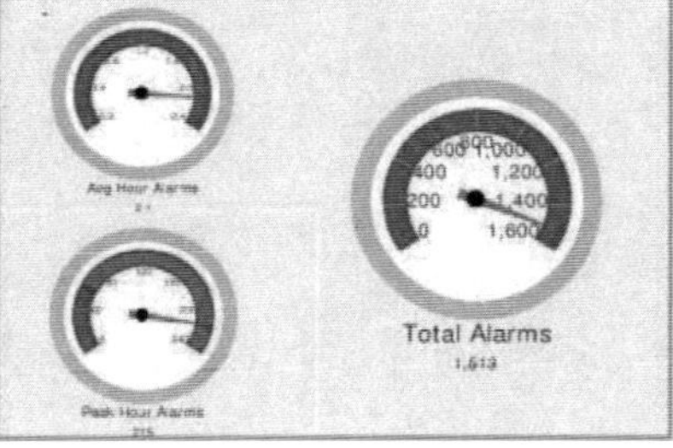

PROSYS™ Maximize Your Plant.™

10™ GLOBAL CONGRESS ON PROCESS SAFETY

Operator Graphics – The Risk of Changing Focus

- **Distraction of changing focus from process graphic to faceplate window**
- **Added workload from managing multiple open windows**
- **Increased probability of errors when changing values for un-intended tag with multiple faceplates open**
- **All of above become multiplied and more complex when process is transitioning from one state to another**

PROSYS™ Maximize Your Plant.™

10™ GLOBAL CONGRESS ON PROCESS SAFETY

Faceplate Operation

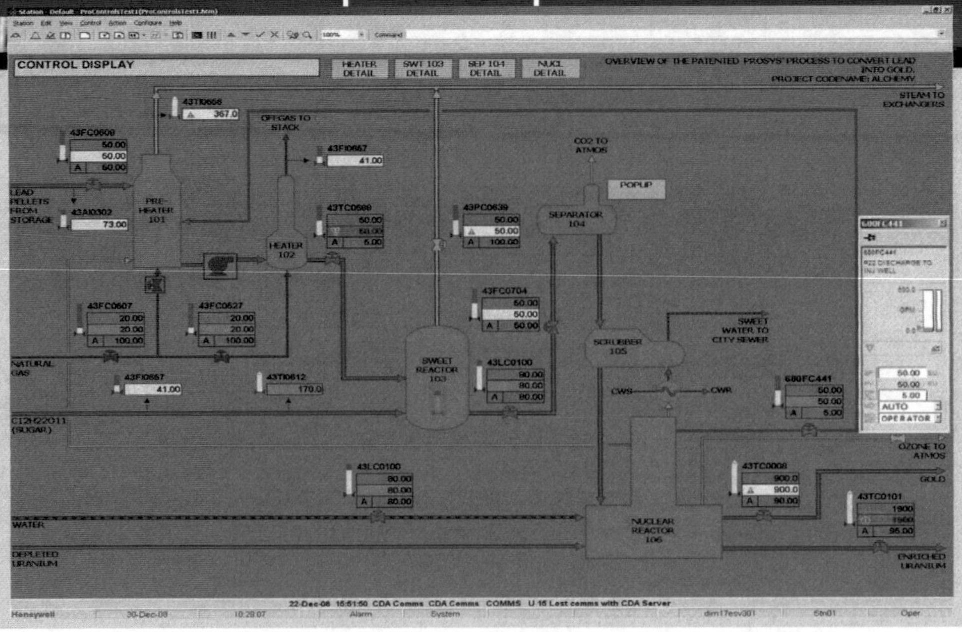

PROSYS Maximize Your Plant.™

10TH GLOBAL CONGRESS ON PROCESS SAFETY

Potential for Errors

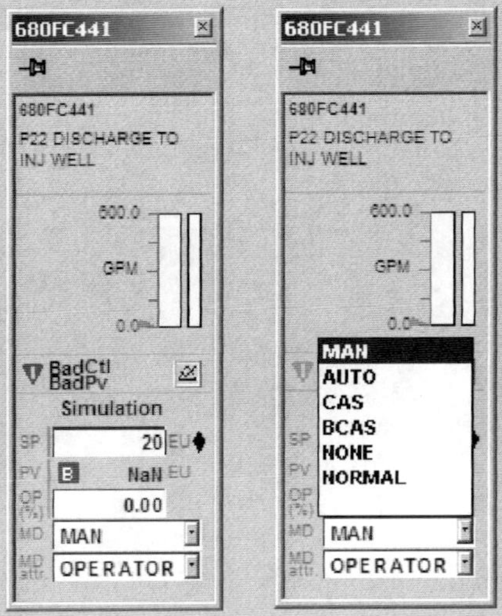

Faceplate Design -can introduce additional errors

- Mode drop down list covers SP, PV and OP values
- Operator may select wrong Mode if SP, PV and OP values are hidden from view
- Mode list offers more options than necessary

PROSYS Maximize Your Plant.™

10TH GLOBAL CONGRESS ON PROCESS SAFETY

Direct Entry Operation

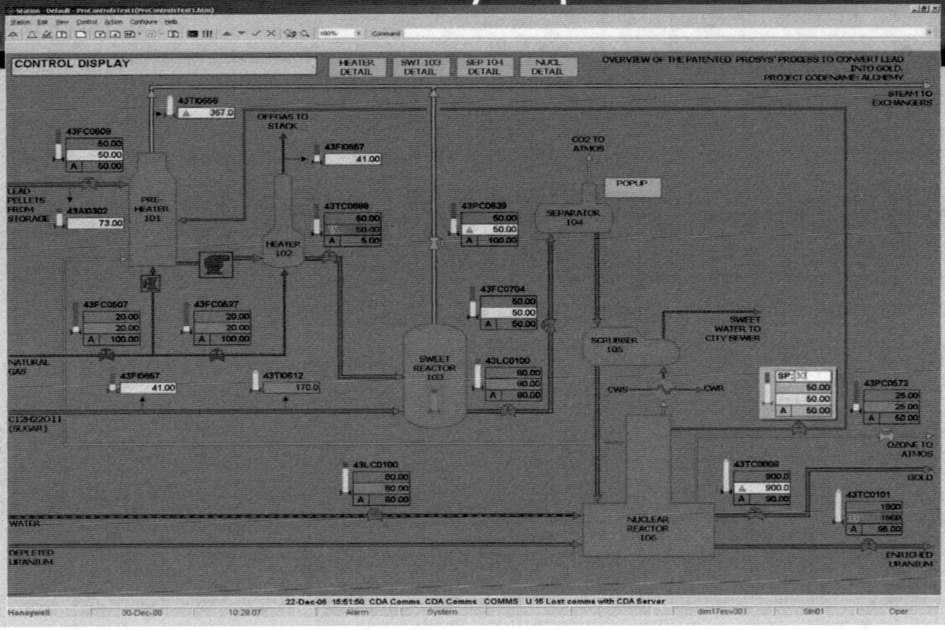

PROSYS Maximize Your Plant.™

10th GLOBAL CONGRESS ON PROCESS SAFETY

Direct Entry Fields

Values can be entered by:

- Select point
- Typing value via keypad
- Pressing enter

Key Factors:

- By maintaining focus of points after value change, it improves safety for input errors, multiple clicking functions and is more time efficient
- Easily recognizable operator enterable fields

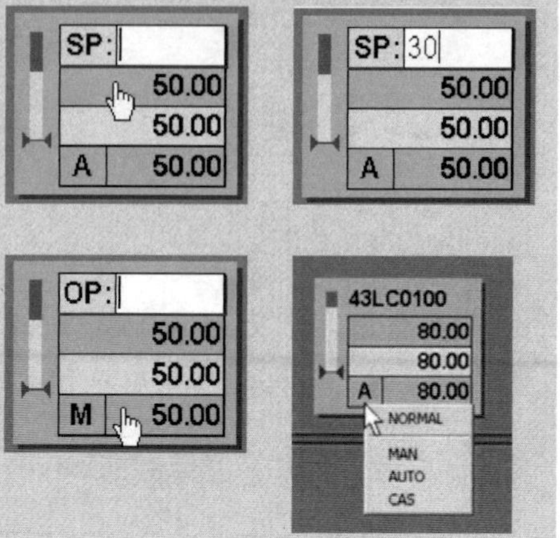

PROSYS Maximize Your Plant.™

10th GLOBAL CONGRESS ON PROCESS SAFETY

Evaluation of Methods

Comparison of Faceplate to Direct Entry

- **Keystroke Level Analysis**
- **Operator Loading Analysis**
- **Risk Analysis**

PROSYS™ Maximize Your Plant.™

Keystroke Level Model

Symbol	Time (s)	Description
K	0.28	Keystroke
P	1.1	Point to object
BB	0.2	Click on object
H	0.4	Home hands on keyboard or mouse
M	1.2+	Mental act or routine thinking

PROSYS™ Maximize Your Plant.™

Estimate Execution Time Analysis

- **Evaluates execution time by an operator comparing faceplates to direct entry for SP/OP and Mode Changes**

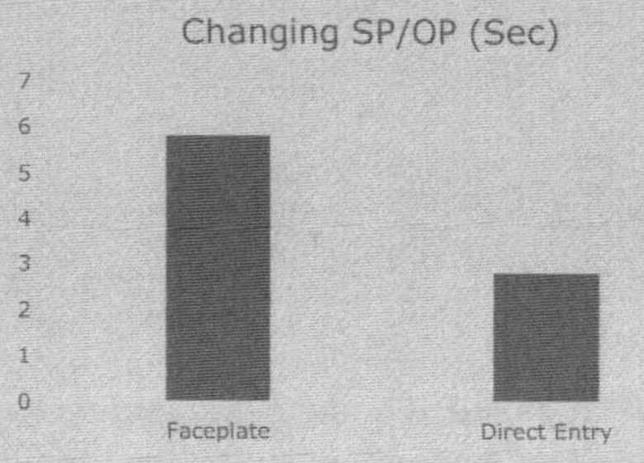

- **50 % Reduction**

PROSYS Maximize Your Plant.℠

10th GLOBAL CONGRESS ON PROCESS SAFETY

Estimated Operator Loading Analysis

- **Evaluates Mental and Physical Load on Operator**
  - Tasks that require very little thought are removed – such as (BB), (H)

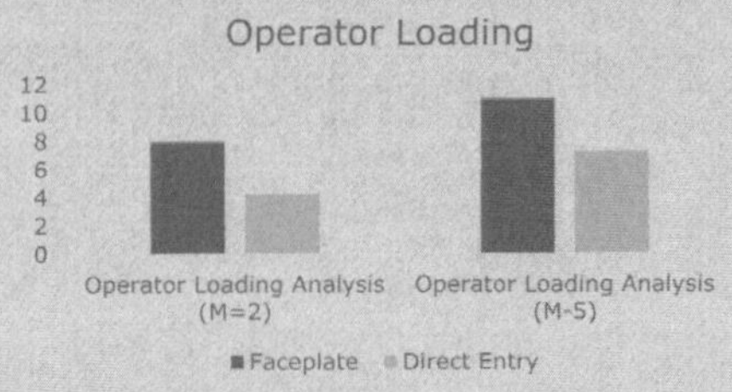

  - Mental and Physical tasks are weighted based on the amount of fatigue they cause – assigned a weight of M=2.0 to 5.0

PROSYS Maximize Your Plant.℠

10th GLOBAL CONGRESS ON PROCESS SAFETY

Risk Analysis

- **Identifies actions where operator entry errors can occur**

PROSYS Maximize Your Plant.™

10th GLOBAL CONGRESS ON PROCESS SAFETY

Risk Analysis

Faceplate operation KLM for changing SP/OP	Potential Error
Move to shape (P)	Select wrong point
Click on shape (BB)	
Move to faceplate (P)	Point in faceplate is previous point
Click on SP/OP field (BB)	Change wrong parameter
Move hands to keyboard (H)	
Type in value and press enter (4K)	Mistype value and press enter
Move hands to mouse (H)	
Move to close faceplate (P)	
Click to close faceplate (BB)	
Potential error count	**4**

PROSYS Maximize Your Plant.™

10th GLOBAL CONGRESS ON PROCESS SAFETY

Risk Analysis

Direct Entry operation KLM for changing SP/OP	Potential Error
Move to SP/OP (P)	Select wrong parameter or point
Click on SP/OP field (BB)	
Move hands to keyboard (H)	
Type in value and press enter (4K)	Mistype value and press enter
Potential error count	**2**

PROSYS Maximize Your Plant.™

10th GLOBAL CONGRESS ON PROCESS SAFETY

Direct Entry Vs Faceplate

Summary of Benefits

- **Execution Time SP/OP Changes - 51.5% less**
- **Execution Time for Mode Changes – 50% less**
- **Operator Loading – 33 % to 47% less**
- **Risk of Errors – 50% less**

PROSYS Maximize Your Plant.™

10th GLOBAL CONGRESS ON PROCESS SAFETY

Conclusion

- **Many industrial errors are a result of operators using control systems with flawed designs**
  - Poor design and performance by alarm management distracts operators and/or occludes critical alarms
  - Operator graphics with multiple open faceplates can cause a change intended for one controller to be entered into the faceplate of another
- **The two mechanisms listed above are predictable and therefore preventable through better design available today**

PROSYS™ Maximize Your Plant.
Questions?
Comments?
10TH GLOBAL CONGRESS ON PROCESS SAFETY

Problem Statement

visium KMS
Process Safety Management Solution

- Any organisation can improve by leveraging knowledge and lessons learned.
- Knowledge can be externally or hidden locally, in a global enterprise, where lessons learned are difficult to share beyond their initial impact.
- There is so much data, information and knowledge (opportunities) coming at your company that can make you stronger or make you look like your not paying attention.
- Not paying attention or managing the available knowledge, can sink your ship… nobody wants that!

Expert Insight Delivered 2

Osys
Part of the Rolls-Royce Group

What is Knowledge Management?

visium KMS
Process Safety Management Solution

- Knowledge management is the concept in which an enterprise consciously and comprehensively gathers, organizes, shares, and analyses its knowledge to further its aims.
- Turning data into information is nice.
- Turning Information into Knowledge is good!
- Turning Knowledge into Wisdom is great!
- But something is missing…
- The CAPA (doing) loop…

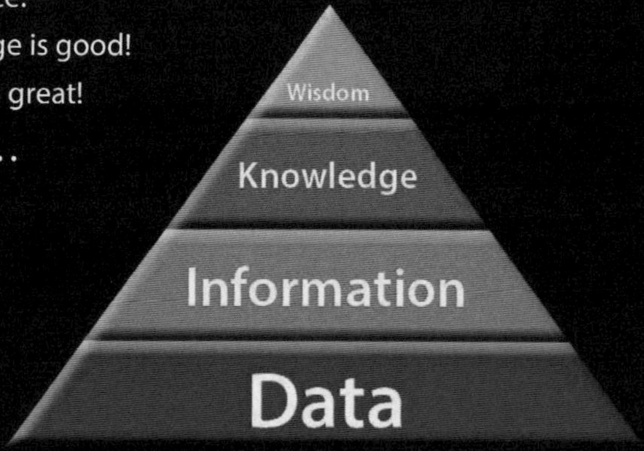

Expert Insight Delivered 3 Osys Part of the Rolls-Royce Group

What we need is this…The HUAA MS

visium KMS
Process Safety Management Solution

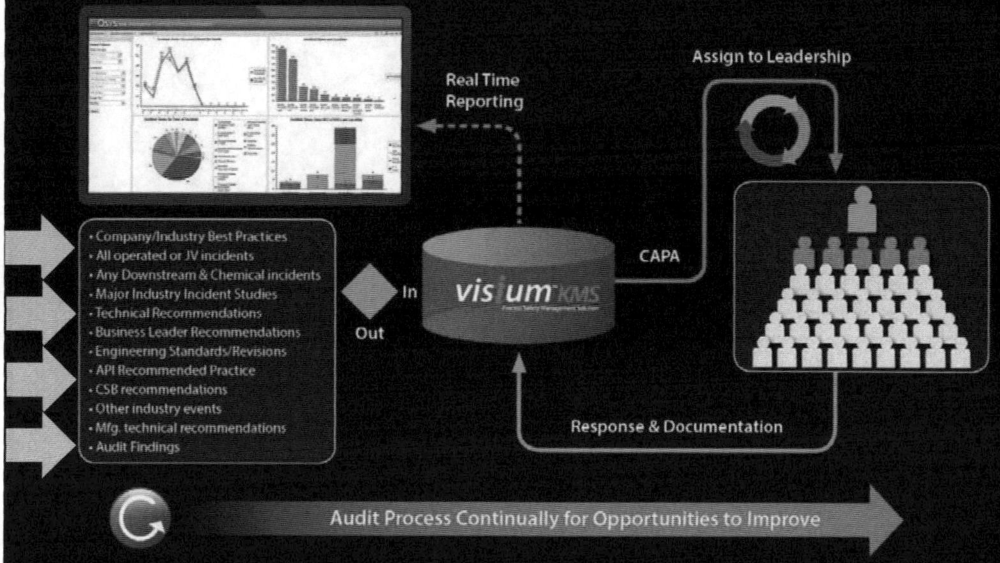

Expert Insight Delivered 4 Osys Part of the Rolls-Royce Group

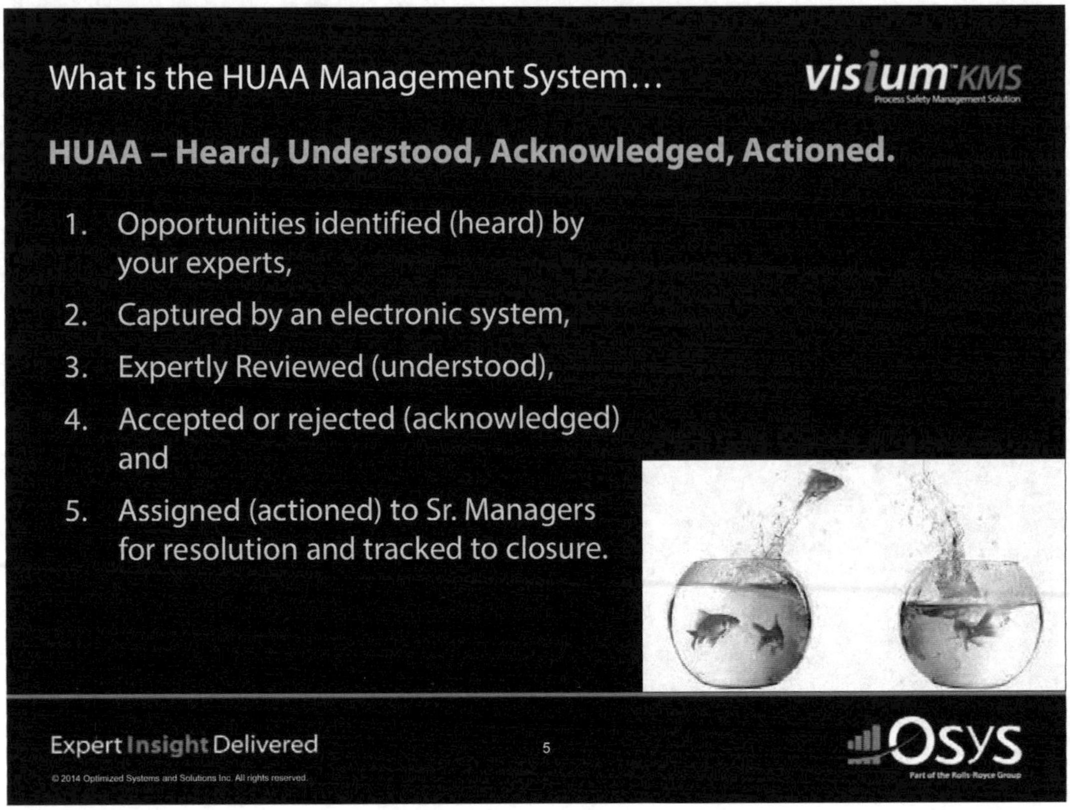
What is the HUAA Management System…
visium KMS
Process Safety Management Solution
HUAA – Heard, Understood, Acknowledged, Actioned.
1. Opportunities identified (heard) by your experts,
2. Captured by an electronic system,
3. Expertly Reviewed (understood),
4. Accepted or rejected (acknowledged) and
5. Assigned (actioned) to Sr. Managers for resolution and tracked to closure.
Expert Insight Delivered
5
© 2014 Optimized Systems and Solutions Inc. All rights reserved.
Osys
Part of the Rolls-Royce Group

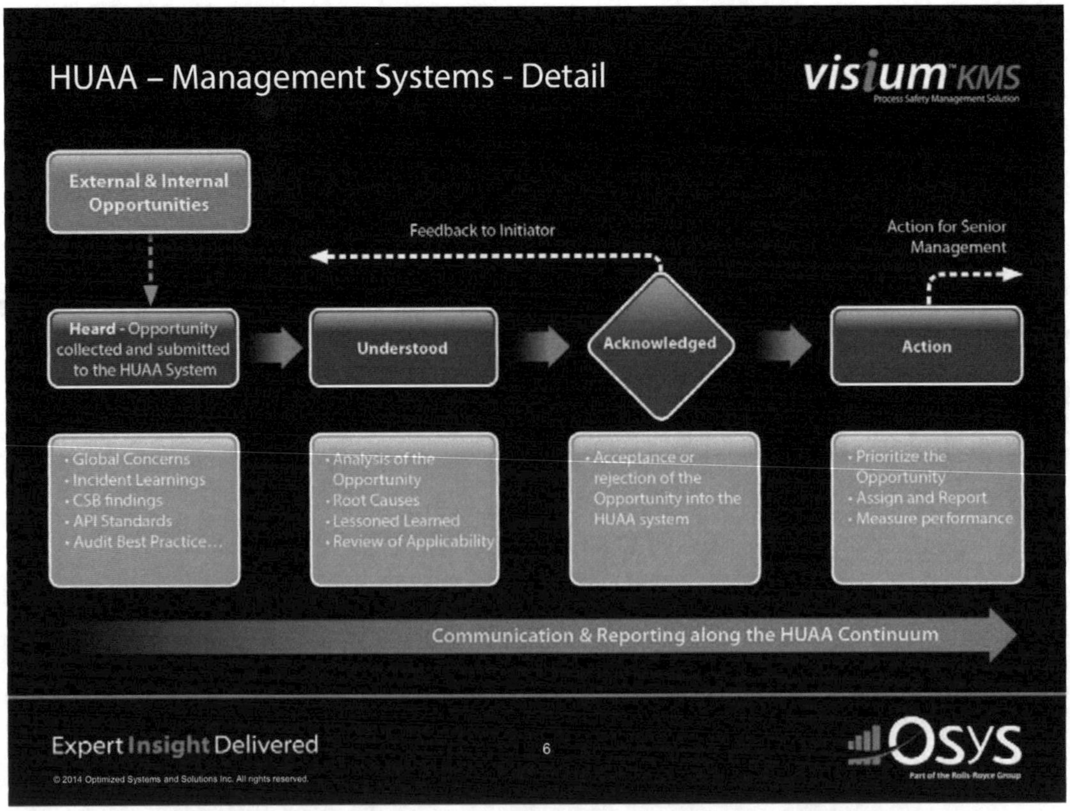
HUAA – Management Systems - Detail
visium KMS
Process Safety Management Solution
External & Internal Opportunities
Feedback to Initiator
Action for Senior Management
Heard - Opportunity collected and submitted to the HUAA System
Understood
Acknowledged
Action
• Global Concerns
• Incident Learnings
• CSB findings
• API Standards
• Audit Best Practice…
• Analysis of the Opportunity
• Root Causes
• Lessoned Learned
• Review of Applicability
• Acceptance or rejection of the Opportunity into the HUAA system
• Prioritize the Opportunity
• Assign and Report
• Measure performance
Communication & Reporting along the HUAA Continuum
Expert Insight Delivered
6
© 2014 Optimized Systems and Solutions Inc. All rights reserved.
Osys
Part of the Rolls-Royce Group

HUAA Results

visium KMS
Process Safety Management Solution

- Efficient Collection - Important Opportunities are identified and collected from many sources (internal and external), and entered into the HUAA management system.
- Expert Vetting - These opportunities are vetted, parsed and resolved for inclusion or exclusion; documenting & communicating both.
- Accountable Assignment - Included opportunities are systematically assigned to Sr. Leaders and managed across the enterprise at SVP and C levels.
- Action - Improvement will be assured through the accountable actions of these leaders.
- Results - The organization will gain competitive advantage by active listening, learning from their experience, as well as the experience of others, and by taking ACTION!

Expert Insight Delivered 7 OSYS Part of the Rolls-Royce Group

Sharing is not enough to keep your ship afloat!

visium KMS
Process Safety Management Solution

Expert Insight Delivered 8 OSYS Part of the Rolls-Royce Group

352631 Autocad - Smart Grid – Agriculture - Architecture and the United States Government

Abstract:

The computational methods, process control, process design, process safety, public policy of the smart chip for agriculture industry, and architecture industry in the United States utilizing the computational methods of AutoCAD in the structural development stages of the smart grid for device/component/equipment and geographical development and public policy for grant funding.

352747 Production, Characterization and Catalytic Studies of Biobased Carbon Materials

Abstract:

Carbon-based nano materials were produced from wood chips, lignin and bio-char. The influence of reaction conditions on the production of nano-carbons was investigated. Carbon-based nanomaterials were characterized using scanning electron microscopy (SEM), transmission electron microscopy (TEM), high resolution transmission electron microscopy (HRTEM), and X-ray diffraction (XRD). The catalytic performance of these nanomaterials as syngas cleaning adsorbents and chemical conversion catalysts for bio-fuel processes has been evaluated.

347835 Uncertainty in Sour Gas Viscosities Estimation, What Is the Effect on Your Reservoir Inflow and Tubing Performance

Abstract:

In light of the need to find an applicable, consistent and promising correlation to accurately calculating sour gases viscosity under several reservoir conditions, this study aims to compare the existing correlations and other industry standard methods, check their accuracy and range of applicability for sour/acidic gases. Moreover, investigating the impact of uncertainties in estimating sour gas viscosities on reservoir flow performance relationship (IPR), reserves and resources estimations, tubing performance, pipeline and facilities design, the design of compressors and more importantly the need to find such correlation to be used in further advancements of reservoir simulation.

354052 Chromonic Nanocarriers for Chemotherapeutics: Size Distribution and Control Release Studies

Abstract:

Conventional chemotherapy fails to release the drug in the controlled manner; rather they release the drug instantly due to which re-administration is required as therapeutic index falls. Moreover, they target both the cancer cells and non-cancer cells which make the treatment of the cancer cells highly ineffective due to excessive toxicities. Recently, the nanoparticle mediated drug delivery for chemotherapeutics has gained much attention. The present study reports the use of a novel class of materials – *Chromonics* in cancer drug delivery. Folic acid as a chromonic material has been explored to develop nanocarrier for sustained release of chemotherapeutic drugs *in-vitro*. With the help of a biocompatible polymer as a tool, folate nano-domains were formed from self-assembled liquid crystalline folate solutions. These nano-domains were crosslinked to form a stable nanoparticle system. The effect of different factors like crosslinking agent, relative polymer and folate concentration, on the size distribution was investigated in detail. Based on the observations, we have reported that not only the synthesis but also the size of the developed nanoparticles can be controlled by varying the folate: HPMC weight ratios. The present study then explored the release of drugs from these nanocarriers. We have investigated the release of folic acid as a model drug from these nanoparticles as well as release of methotrexate and cytarabine as an anticancer drug for the period of 60 days. It was observed that release of drug was nanoparticle's size-dependent, release medium dependent, pH of the release medium dependent, crosslinking agent and its concentration dependent. The significant part of the current study is the synthesis of chromonic nanoparticles, how the size can be controlled, how controlled release can be obtained as well as how the release profiles can be modified depending on the dosage forms. Therefore, we can say that folate nanoparticles can be a potent nanocarrier for the sustained release of cancer drugs.

IPEX
Committed to Excellence

Double Containment Piping Systems: A Fail-Safe, Leak-Free Solution

Darin Johnson, P.Eng.
Applications Engineer
IPEX USA LLC

AIChE – 2014 Spring Meeting
New Orleans, LA

Monday, March 31, 2014

IPEX
Committed to Excellence

Why Double Containment Piping Systems?

- December 1988 – law change within the ***Resource, Conversation, and Recovery Act*** in the United States regarding underground piping.
- U.S. Environmental Protection Agency (EPA) requirement: Code of Federal Regulations (CFR), Title 40, Part 280.
  - Petroleum Products
  - Hazardous Substances (over 3200, CFR Title 40 Part 302.4)
  - Hazardous Wastes

Why Double Containment Piping Systems?

- Buried tank - all buried piping connected to it must be double contained.
- Above ground tank – associated buried piping consists of 10% or more of the tank volume, it also must be double contained.

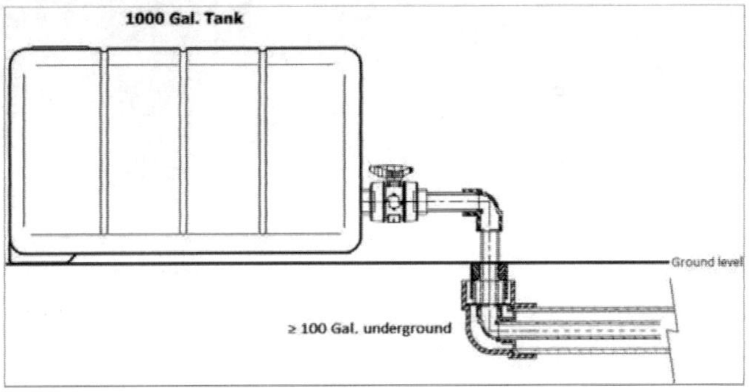

Why Double Containment Piping Systems?

- If the piping is less than 10% of the volume of the aboveground tank, and you choose not to double contain the buried piping, ***responsibility for any clean-up or remediation resulting from a release will still fall on the Client.***

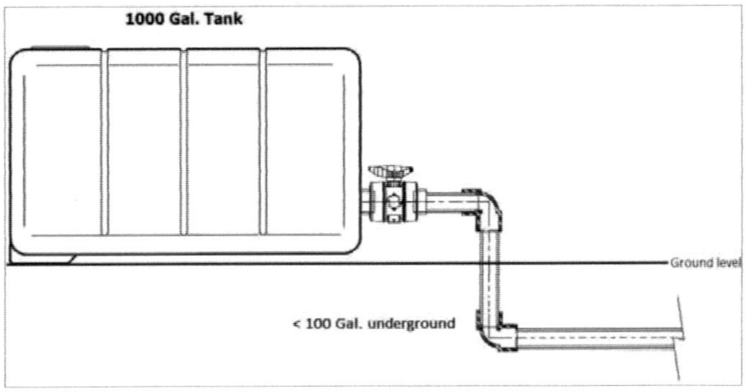

Why Double Containment Piping Systems?

- **Above Ground Piping Systems**
  - Different jurisdictions have to ability to require chemicals be double contained.

Nomenclature of Double Containment Piping

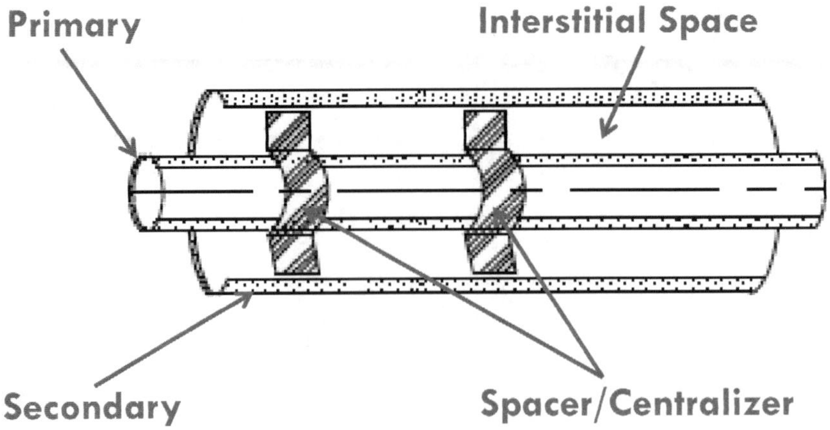

IPEX
Committed to Excellence

Double Containment Piping Systems: Material Combinations

SYSTEM Carrier	Containment	Carrier SIZE inches
PVC 80	PVC 80	1/2 - 12
PVC 80	PVC 40	1/2 - 12
PVC 40	PVC 40	1/2 - 12
PVC DWV	PVC DWV	1-1/2 - 12
PVC 80	CS SCH 10	1/2 - 12
CPVC 80	CPVC 80	1/2 - 8
CPVC 80	PVC 80	1/2 - 12
CPVC 80	PVC 40	1/2 - 12
CPVC 80	CS SCH 10	1/2 - 12
NAT PP 80	PVC 80	1/2 - 2
NAT PP 80	PVC 40	1/2 - 2
RED PVDF 80	CPVC 80	1/2 - 2
RED PVDF 80	PVC 80	1/2 - 2
NAT PVDF 80	PVC 80	1/2 - 2
ENFUSION PP	PVC DWV	1/2 - 6
S40 A53B CS BW	PVC 80	1/2 - 8
S40 A106B CS BW	PVC 80	1/2 - 8

SYSTEM Carrier	Containment	Carrier SIZE inches
S40 A53 BLK SW	FRP	1/2 - 2
S40 A53 BLK SW	PVC 80	1/2 - 2
S80 A53B CS SW	PVC 80	1/2 - 2
S10 T304L SS BW	PVC 80	1/2 - 8
S10 T304L SS BW	PVC 40	1/2 - 8
S10 T0304L SS BW	CPVC 80	1/2 - 8
S10 T304L SS BW	FRP	1/2 - 12
S10 T304L SS BW	S10 T304L	1/2 - 12
S40 T316L SS BW	PVC 80	1/2 - 8
S40 T316L SS BW	CPVC 80	1/2 - 8
S40 T316L SS BW	FRP	1/2 - 12
S40 T304L SS SW	PVC 80	1/2 - 2
S40 T304L SS SW	PVC 40	1/2 - 2
S40 T304L SS SW	CPVC 80	1/2 - 2
S40 T304L SS SW	FRP	1/2 - 2
S10 T316L SS SW	PVC 80	1/2 - 2
S10 T316L SS SW	CPVC 80	1/2 - 2

IPEX
Committed to Excellence

Double Containment Piping Systems: Design Considerations

- **Material Selection**
  - Primary Pipe
    - Chemical Compatibility
    - Pressure Capabilities
    - Thermal Properties
  - Secondary Pipe
    - Chemical Compatibility
    - Loading (Static & Live) for buried applications
    - Hanging requirements for above ground
    - Leak detection requirements
    - UV requirements

IPEX
Committed to Excellence

Double Containment Piping Systems: Design Considerations

- **Thermal Expansion & Contraction**
  - Two basic thermal relationships may be present:
    - When only one of the two systems experiences thermal expansion/contraction.
    - When both systems experience different magnitudes of thermal expansion/contraction.
  - The Floating System
  - The Restrained System

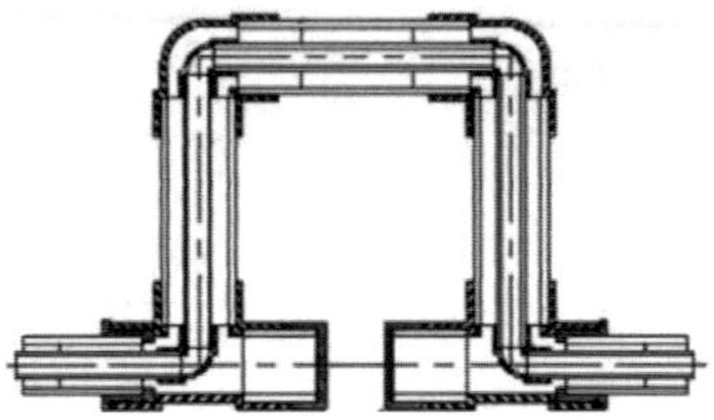

IPEX
Committed to Excellence

Double Containment Piping Systems: Leak Detection Requirements

- EPA requirement: (CFR), Title 40, Part 280
  - Pressurized systems require electronic leak detection that must be checked once per year.
  - Drainage systems (gravity flow) require visual leak detection with monthly manual inspection for a media leak.

IPEX
Committed to Excellence

Double Containment Piping Systems: Leak Detection Requirements

- The following forms of leak detection are acceptable as per regulations

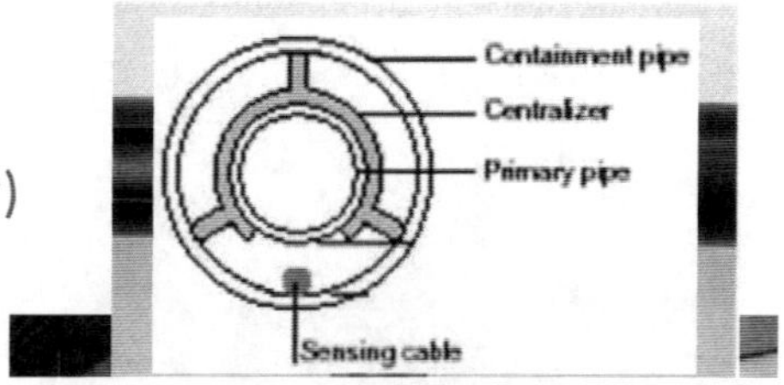

  - Visual (drainage applications only)
  - Electronic low point probes
  - Continuous cable

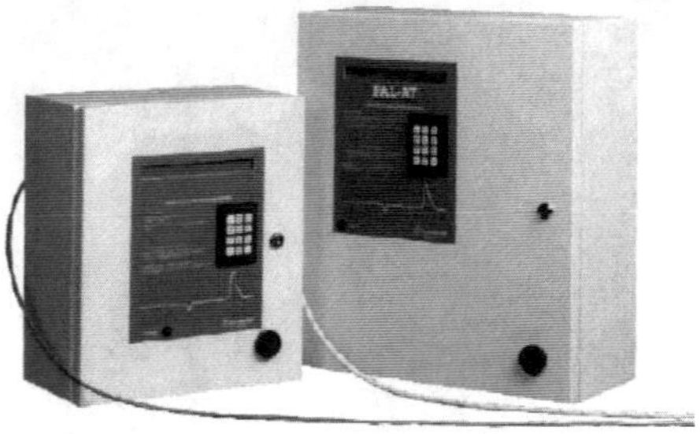

IPEX
Committed to Excellence

Double Containment Piping Systems: Applications

- Underground Applications
- WTP & WWTP
- Oil & Gas
- Hospitals
- Museums
- Food and Beverage
- Pharmaceuticals
- Laboratories
- Data Centers

IPEX
Committed to Excellence

Double Containment Piping Systems: A Fail-Safe, Leak-Free Solution

Questions?

Thank you!

DTP Process

"Now Ready to License"

JGC Corporation
Mitsuo Morita

2014

JGC CORPORATION | MITSUBISHI CHEMICAL

What is DTP Process ?

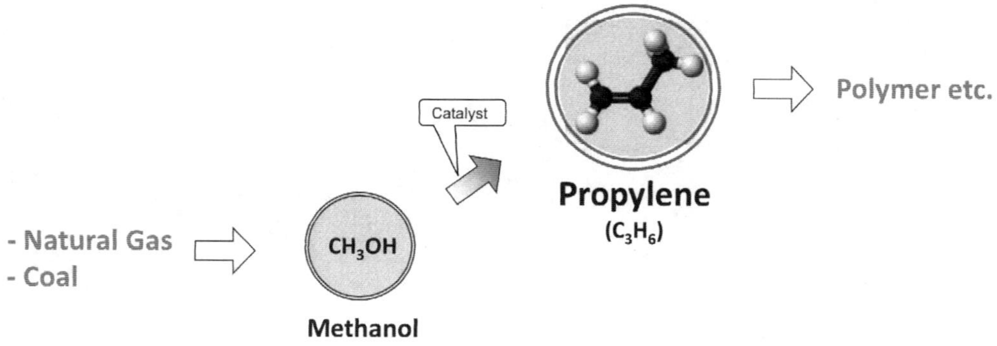

-On-Purpose Propylene Production from Methanol

JGC CORPORATION | - 2 - | MITSUBISHI CHEMICAL

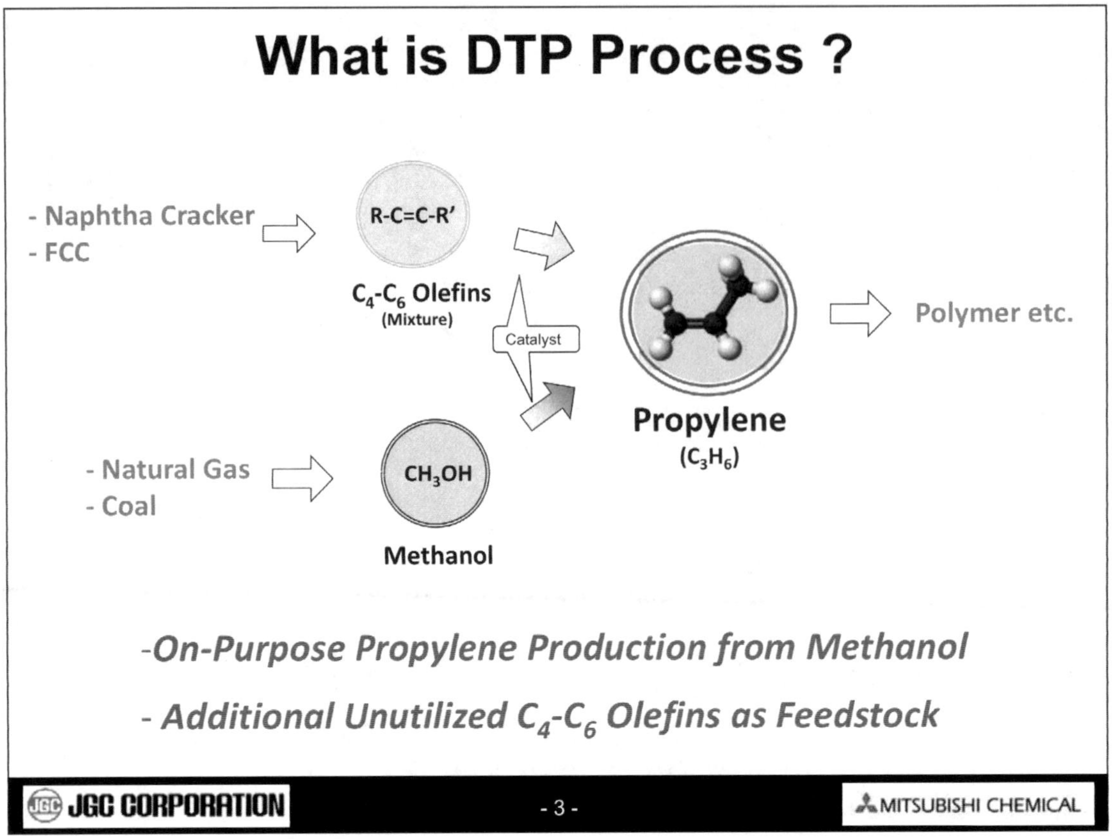
What is DTP Process ?
- Naphtha Cracker
- FCC
R-C=C-R'
C4-C6 Olefins
(Mixture)
Catalyst
Propylene
(C3H6)
Polymer etc.
- Natural Gas
- Coal
CH3OH
Methanol
-On-Purpose Propylene Production from Methanol
- Additional Unutilized C4-C6 Olefins as Feedstock
JGC CORPORATION
- 3 -
MITSUBISHI CHEMICAL

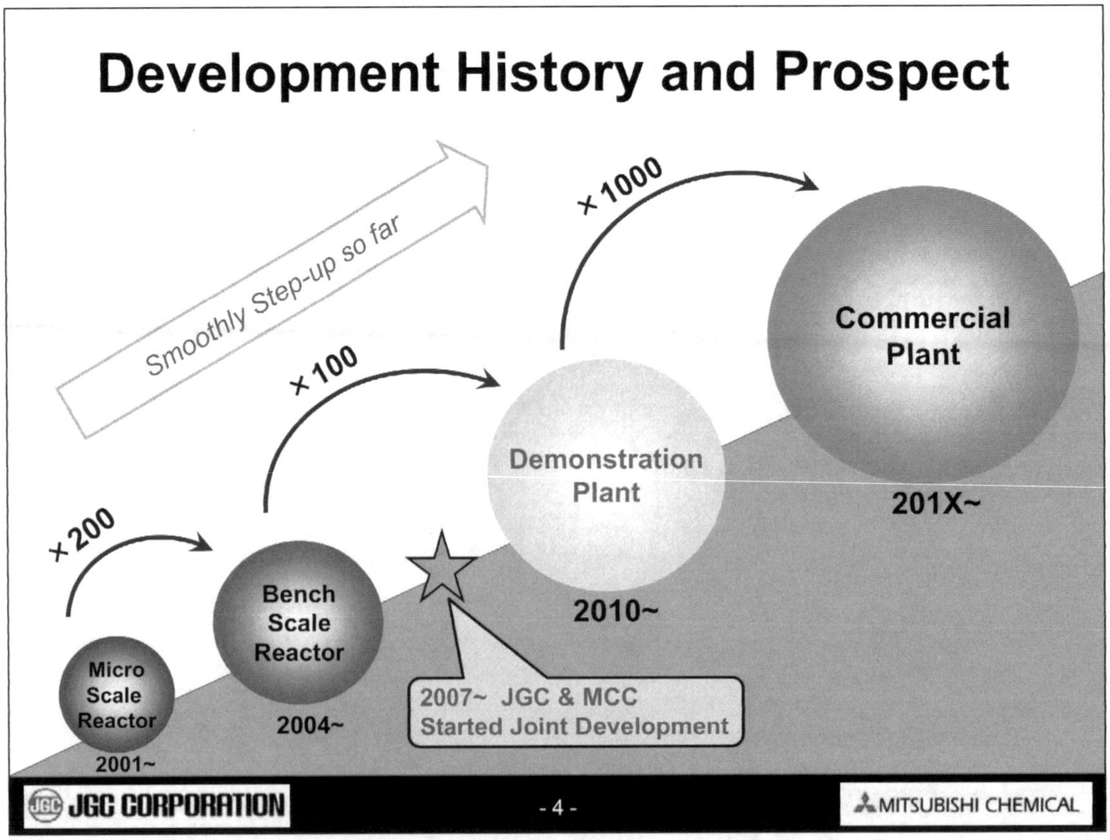
Development History and Prospect
Smoothly Step-up so far
×1000
×100
×200
Commercial Plant
201X~
Demonstration Plant
2010~
Bench Scale Reactor
2004~
Micro Scale Reactor
2001~
2007~ JGC & MCC
Started Joint Development
JGC CORPORATION
- 4 -
MITSUBISHI CHEMICAL

Demonstration Plant Overview

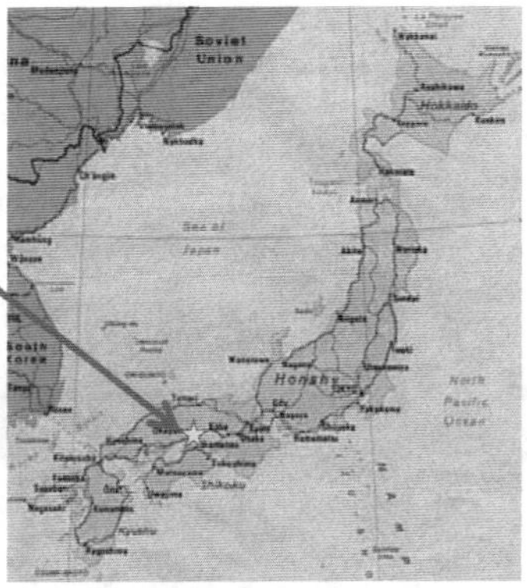

- **Located in Mitsubishi Chemical Mizushima Plant**
- **Successfully Start-up in August 2010**
- **Continuous Operation for Two Years**

Process Flow (Standalone Case)

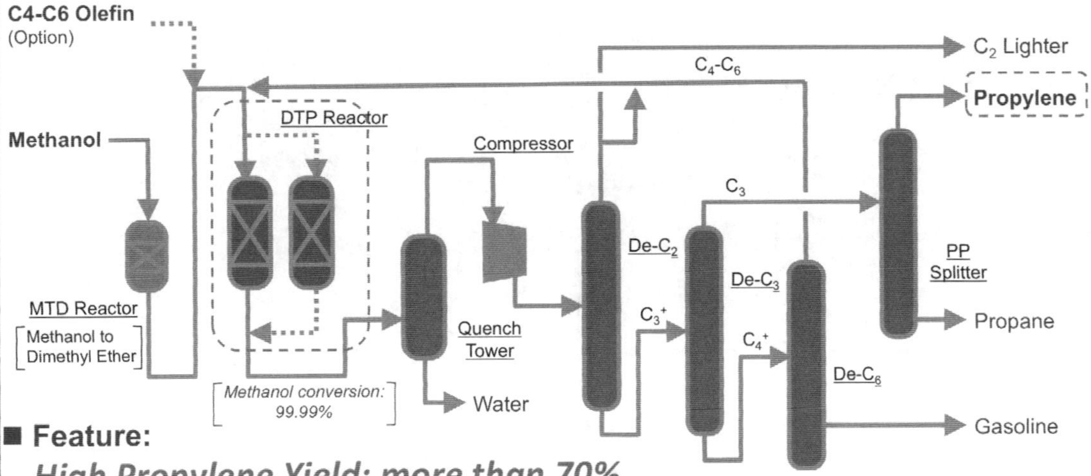

■ **Feature:**

- ***High Propylene Yield: more than 70%***
- ***Applied Existing Proven Technologies***

Process Flow (Integration Case)

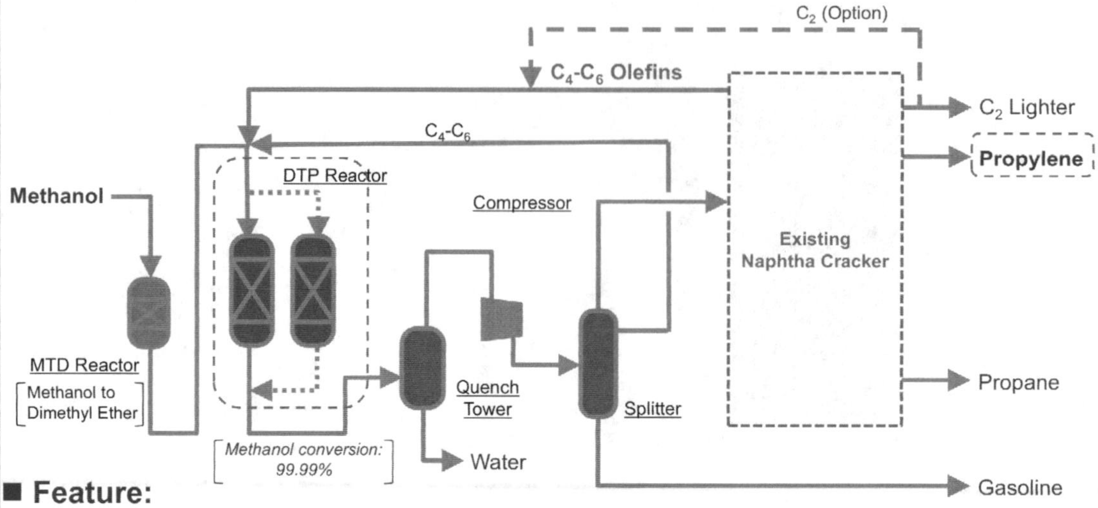

■ **Feature:**

- *High Propylene Yield: more than 70%*
- *Applied Existing Proven Technologies*
- *Low Investment Cost in case of Integration with Naphtha Cracker*

Application

(1) Naphtha Cracker Integration Case (Propylene Capacity: 20KTA – 200KTA)

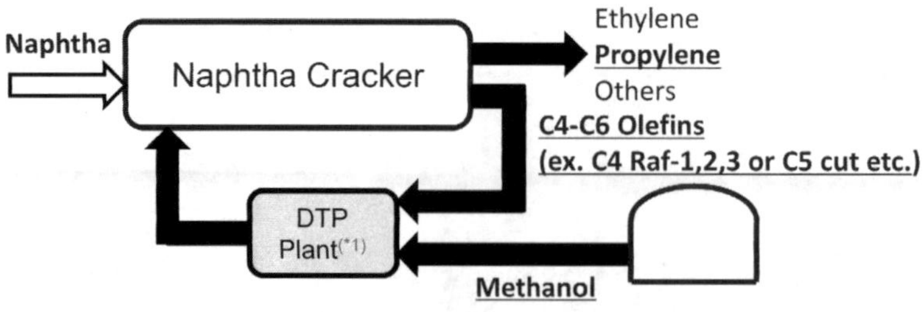

(*1) Utilizing facilities in Naphtha Cracker.

(2) Standalone Case (Propylene Capacity: 250KTA – 500KTA)

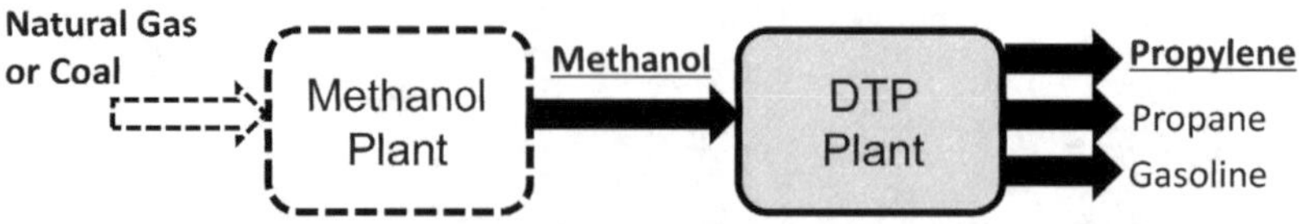

- *Two Cases available depending on Client requirements !!*

DTP is...

- *High Yield Propylene Production Process*
- *Flexible Feed Methanol to Olefin Process*
- *Process that increase P/E Ratio for Cracker*

Thank you very much !!

Contact:
Mitsuo Morita
Market Development Dept.
Global Marketing Division
2-3-1, Minato Mirai, Nishi-ku,
Yokohama 220-6001, Japan

Phone: 81-45-682-8371
FAX: 81-45-682-8183
E-Mail: morita.mitsuo@jgc.co.jp

352533 Self-Cleaning "Bernoulli" Type Filters Used in Onshore and Offshore Applications

Abstract:

Georg Schuenemann GmbH (SAB-BREMEN) has developed a new Filter Type F-480, dubbed the "Compact Giant" for applications in onshore water intake and cooling water systems, sea water desalination plants, and in offshore projects, such as FPSO's, Drillships, and Production Platforms.

This paper will describe the essential design features of the filters, based on the "Bernoulli principle" which result in a fully automated self-cleaning filter system, and choice of materials in a sea water service with high corrosion rates.

A case study will be presented in which stringent project requirements such as substantial flow rates, adaptibility to space limitations and pipe connections had to be observed. Reliability of the system, extended run times without the need for frequent replacement of filter elements, as well as ease of monitoring and operator involvement have been achieved.

352653 Effects of Support on Sulfur Tolerance and Regeneration of Pt Catalysts Measured By Ethylene Hydrogenation and EXAFS

Abstract:

We investigated the effect of support on sulfur tolerance and regenerability under reducing environments by measuring the rate of ethylene hydrogenation, combined with hydrogen chemisorption, and extended x-ray absorption fine structure (EXAFS). Alumina and titania (P25) supports were used to prepare Pt catalysts at 1 wt.%. The catalysts were tested after treatment in H_2S/H_2 at 250 °C followed by regeneration treatments in H_2 at 250 °C, 350 °C and 450 °C. The sulfided catalysts exhibited a 20-27 times decrease in the rate of ethylene hydrogenation and a 4-6 times drop in the Pt surface area. Regenerations up to 450 °C were unable to remove all the sulfur, as shown by the presence of Pt-S bonds by EXAFS. However, a partial recovery of the hydrogenation rate per mole of Pt was observed on sulfided Pt/Al_2O_3 after reduction at 450 °C, while the induction of strong metal support interactions (SMSI) at reduction temperature above 350 °C was observed on Pt/P25, regardless of the presence of sulfur. For Pt/P25, the reversal of the SMSI state together with sulfur removal by mild oxidation suggests that sequential reduction/oxidation treatments may be more effective in restoring the S-free state of TiO_2-supported catalysts.

352799 Improved Operational Efficiency and Reliability through Insulation Materials Selection

Abstract:

This paper will explore the implications of insulation materials selection on the energy efficiency and reliability for Ethylene production units. Considerations will include insulation properties, structural support, installation methods, installation and maintenance costs.

BOILER MACT COMPLIANCE: MERCURY IN FUEL GAS

PATRICK LAINE, WES YONGE, & VIC VICKERY
PEI, HOUSTON, TX 77043

Patrick Laine, Ph.D.
Chief Scientist
Houston, TX
850-866-6686
plaine@pei-tx.com

Wes Yonge, PE
Chief Engineer
Houston, TX
713-806-8396
wyonge@pei-tx.com

James 'Vic' Vickery
Chief Technology Officer
Houston, TX
281-536-0899
vvickery@pei-tx.com

1 Introduction & Objectives

For most ethylene producers the primary objective of this sampling and analysis effort will be to verify and "establish a mercury fuel specification through which gas-fired boilers that use a fuel other than natural gas or refinery gas may be considered Gas 1 unit" as identified in 40 CFR Part 63.7521.(f).

Measurement and Monitoring Requirements: Decision Rules

❑ If Hg concentration ≤ 50% of Gas 1 specification* then no further testing required

❑ If Hg concentration > 50% and ≤ 75% of the specification* then semi-annual sampling required

❑ If Hg concentration > 75% of the specification* then monthly sampling required

*Specification is **40 μg/Sm³**

The goal will be to execute a sampling and analysis campaign that will produce compliance quality data. The key to producing and validating compliance quality data is a well-designed and executed QA/QC protocol. At a minimum, a performance based QA/QC protocol is recommended that includes the following components:

- Sorbent Trap Breakthrough Bed:
  - ❑ Sorbent traps consisting of a minimum of two sorbent beds in series so that the trailing bed can be analyzed to verify that no breakthrough occurred from the leading bed.
- Duplicate Samples:
  - ❑ The sampling system should be configured with two trains to accommodate the collection of simultaneous duplicate sorbent trap samples, in parallel, providing a means of verifying precision in the sample collection process.
- NIST (National Institute of Standards and Technology) Traceable Spike:
  - ❑ Sorbent traps with a third bed (downstream) that is spiked with a NIST traceable spiked mass of mercury serves as a matrix spike and provides a means of verifying that components in the matrix (the fuel gas) are not affecting the trapping efficiency of the sorbent trap.

2 Methods

Sorbent Trap Mercury Sampling Methods Comparison

Method	Description	Method Type: S = Sampling, A = Analysis	Evaluation	Sample Collection Phase QA/QC
ISO 6978	Hg in Natural Gas by Thermal Desorption (CVAFS)	S & A	Method calls for low sample volumes can lead to non representative results.	No
EPA Method 30B	Determination of Gaseous Hg Emissions from Coal Fired Combustion Sources using Carbon Sorbent Traps	S	Requires modification (Modified M30B)	Yes
ASTM D6350	Hg S&A in Natural Gas (AFS)	S & A	Method calls for short term samples (i.e., low sample volume) which leads to poor accuracy.	No
ASTM D5954	Hg S&A in Natural Gas (AAS)	S & A	AAS is less sensitive than AFS.	No
EPA Method 1631	Measurement of Hg in Water	A	No	Yes

3 EPA Method 30B QA/QC Performance Criteria

QA/QC Test or Specification	Acceptance Criteria	Frequency
Analyzer calibration	Correlation Coefficient: 0.999	Daily
Transfer efficiency of analyzer	%Transfer Efficiency: 90%–110%	Once per sorbent trap
Initial calibration blank	< 0.5 ng	Once before each calibration
Sorbent Trap Section B breakthrough	≤ 10% of Section A Hg mass, for Hg concentrations > 1μg/dscm; ≤ 20% of Section A Hg mass, for Hg concentrations ≤ 1μg/dscm	At each sample point
Spike Recovery Test	Average recovery between 75% and 125 % for Hg_0	Once per sample point
Paired Sorbent Traps Agreement	≤ 10% RD for Hg concentration > 1μg/dscm; ≤ 20% RD or 0.2 μg/dscm absolute difference for Hg concentrations ≤ 1μg/dscm	At each sample point

4 Equipment

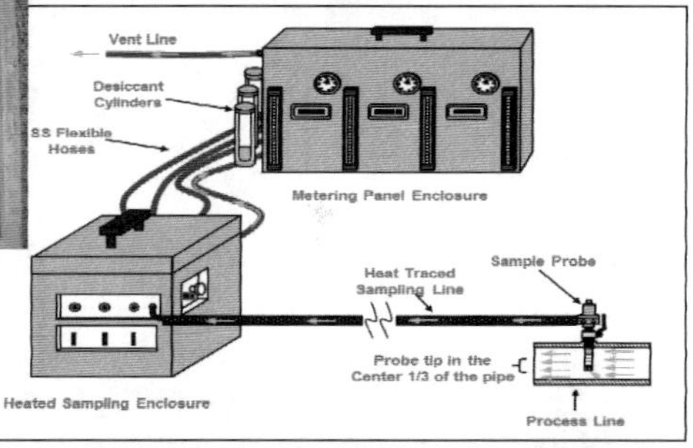

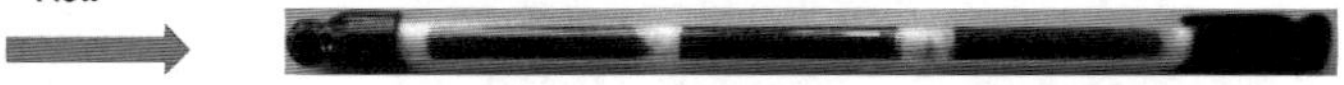

QA/QC Acceptance Criteria:	99% of Total Collected Hg	≤ 10% of Primary Mass	Recovery between 75% and 125%

Gold Trap:

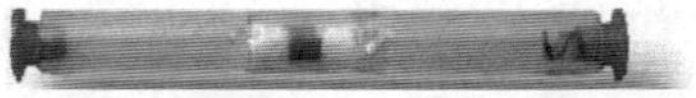

5 Conclusions/PEI Recomendations

❑ There are five methods accepted by the Boiler MACT rule. PEI has extensive experience and expertise employing **all five** of these methods.

❑ Operators can achieve compliance without submitting a Fuel Analysis Plan by using ASTM D6350, however, we have noticed variable results with this method.

❑ The **PEI recommendation** for obtaining the most accurate compliance quality data is to use Modified EPA Method 30B in conjunction with EPA 1631 where traps should be analyzed in a certified ISO 17025 laboratory. The recommended pathway includes a rigorous, performance based QA/QC protocol that validates accuracy and precision of all sample collection procedures.

351376 Young Professional Simulation Tutorial

Abstract:

18C00 Young Professional Simulation Tutorial

Authors: Naomi Hua and Mike Donahue, Invensys

Process simulation is a complex skill to master. Even with years of experience, a typical user is constantly facing new problems that need to be addressed. There are many issues that can lead to unexpected and unnecessary errors with the use of process simulation tools. Regardless of what simulation program you are using, these issues are common from one program to another. In this presentation, targeted for the young professional, we will discuss the "Most Common Pitfalls in Process Simulation". For each topic, we will discuss various techniques and methods these troublesome problems and prevent their occurrence.

Topics to be discussed include:

1. Erroneous data entry

2. Simulation defaults

3. Bad plant data

4. Thermodynamic method selection

5. Simulation tolerances and convergence

6. Recycle identification and handling

7. Model complexity

8. Difficult or impossible specifications

9. Assay pseudo-component development

10. Assay property blending

1. Erroneous data entry is the most common mistake made when building a process simulation model. This presentation will cover methods that can help ensure proper data entry in your process simulation model, such as checking the data you entered, particularly the units of measure, using "alternate forms" of data input such as copy/paste and data import, and reviewing all warning and error messages.

2. Defaults are methods, values, and conditions that are automatically selected for you by the program when you begin a new simulation model. These defaults will affect the results calculated by your model and every simulation program has them. Most defaults within simulation programs fall under categories such as convergence criteria, thermodynamic methods, tolerance and units of measure. The problem with defaults is unintended results from not fully understanding the effects a particular default has on the model. Take the time to understand the defaults of the program and how these defaults will affect the results. If you do not agree with the defaults selected by the program, most simulation programs will allow the user to change the default.

3. Plant data changes over time. The feed information and process conditions of the unit operations may be different. Not only does plant data change over time, but measurement information is often very inaccurate. This could be due to faulty or poorly calibrated measurement equipment, leaks in lines, sensor failures, and other issues. Because of these changes and inaccuracies over time, you always have to validate the information that is entered in your model to ensure that you are obtaining the proper results. To do so, you can utilize programs that can help with data validation and reconciliation like SimSci's ROMeo online optimization program. With it, you can connect your model to your data historian to constantly import current data measurements into your model and then reconcile these measurements to determine when problems arise within your plant and where those problems are occurring. You will also need to use your engineering judgment to determine if the data is valid. More details will be provided during the presentation.

4. The proper selection of thermodynamic methods for your simulation is the most important decision that you make when modeling a process. The method you select could have a drastic effect upon your results. Often, any number of thermodynamic methods is suitable for a given application. It is up to the user to determine the best method to use based upon knowledge, personal experience, and best judgment. Some important factors that need to be considered are the components being used and the operating conditions.

Oftentimes, the selected thermodynamic method still needs to have additional inputs such as binary interaction parameters to simulate the process accurately.

5. Many of the calculations in process simulation are iterative. When these calculations iterate, the program compares the results calculated from one iteration to the results calculated in the next iteration in order to determine

whether it is converged to a solution or needs to continue calculating. This method of comparing results is where tolerances come into play and simulation programs use tolerances to determine when a solution is reached. One of the major mistakes when modeling a process is not understanding or using the correct tolerance values. A looser tolerance may make the file easier and quicker to converge, but it also reduces the accuracy of the model. A tighter tolerance may be more accurate, but might make it impossible for the model to solve or increase the convergence time exponentially. A common practice is that when you have reached a converged solution with your model, tighten the tolerances within the model and solve it again. If you notice that your answers have changed significantly, you need to continue tightening the tolerance values until your results change little from one iteration to another. You will then know that you have reached an accurate solution for your model.

6. Recycles in process simulation occur when you return to a previously calculated unit operation to recalculate the unit due to a change made elsewhere in the model. They are very common and can often be difficult to converge. Some suggestions on troubleshooting recycle convergence include initially solving your model without the recycles to see how it behaves, providing initial estimates for recycle streams, using tighter tolerances on unit operations within recycle loops than the total recycle tolerances themselves, and ensuring that all components have a way in and out of a recycle by performing mass balances and using purge or makeup streams where necessary.

7. The general rule of thumb when modeling is to start simple and add complexity as you go. Each time you add a few streams or unit operations converge and save the model before moving on. Make sure you justify any additions that you make to the model. You can most likely model your system without including every unit from your PFD in the simulation. When modeling, think about the goal of your simulation and not about literally translating every real-world piece of equipment into the simulation model. Make the judgment call on what is most important and what the purpose is of every unit you add to your simulation. For example, if your real-life process shows a heat exchanger, a pump and a splitter, but all that is necessary is that the temperature, pressure, and separation are achieved properly, consider using a flash drum to accomplish all three.

8. Specifications within process simulation are everywhere and every stream and unit operation you place in your model contains specifications. In its simplest form, a specification is a constraint that you place on the model that the model needs to meet in order to successfully converge. For streams, these are usually temperatures, pressures, flowrates and compositions. For unit operations, they range from process conditions, like temperatures and pressures, to product conditions, like an overhead composition in a column. There are some pitfalls, however, that you can run into with specifications such as azeotropes in columns, high purity specifications, and the possibility of multiple solutions.

9. Petroleum assays are sets of distillation data that are used to characterize all of the components that reside in a given crude sample. When you enter distillation data into a process simulator, the program converts it into components that it can use when modeling your process. The components that are generated are called pseudo-components, and each pseudo-component represents all of the unknown components that would be within a given boiling point range in the crude. Some of the things you will want to ensure when properly developing your assay components are did you use the correct distillation curve basis, did you enter the data correctly, did you use enough cuts and are you using the correct characterization methods .

10. When you have more than 1 assay in a simulation model, you need to be concerned with assay blending and how it will affect the results of your file. Assay blending is the mixing of 2 or more groups of components (i.e. ¨C sets of cuts/pseudo-components) in order to calculate averaged component properties for cuts that are in the same boiling point range as one another. Once these new averaged component properties are determined, the program will use these averages from this point on in the simulation model any time it needs to access component

properties for a given pseudo-component. All component properties are blended, including normal boiling points, specific gravity, molecular weight and refinery inspection properties, and usually the blended component properties are calculated using a weighted average

These topics are based on experience and feedback from end-users to technical support representatives for process simulators. They often represent the most commonly asked questions or areas that most users find to be problematic. While the specific examples highlighted in each topic come from SimSci software, the intent of this tutorial presentation is to be broad-based and is particularly applicable to young professionals seeking to develop their experience with simulation.

Polymeric Sulfonic Acid Composites for Fuel Cell Applications: Effects of Casting Solvent and Sonication Time on Membrane Performance

*Jimoh K. Adewole[1], Abdullah S. Sultan[1, *], Amir Al-Ahmed[2], and S. M. Javid Zaidi[3]*

[1]Center for Petroleum & Minerals, King Fahd University of Petroleum & Minerals, Dhahran 31261, Saudi Arabia

[2]Center of Research Excellence in Renewable Energy, King Fahd University of Petroleum & Minerals, Dhahran 31261, Saudi Arabia

[3]School of Chemical Engineering, The University of Queensland, Australia

Abstract

Polymer blends comprising sulfonated poly (ether ether ketone) (SPEEK) and 3wt% (2-acrylamido-2-methyl-1-propanesulfonic acid) (PAMPS) were prepared by solution casting method. Two solvents (N,N-Dimethylacetamide (DMAc) and N-Methyl-2-pyrrolidone (NMP)) were uesd separately. Residual solvents in the membrane samples after drying were quantified using thermo gravimetric analysis. Effects of periods of sonication (0 to 90 minutes) and types of solvent on the membrane performance were evaluated interms of morphology and water uptake. An attempt was made to correlate membrane performance with the time of sonication in order to select the most suitable solvent and membrane casting conditions for subsequent membrane preparations. For both solvents, the water uptake was observed to first decrease at 10 minutes of sonication and then increased as the sonication time was increased. The minimum water uptake recorded for DMAc samples was 19.23% at 20 minutes sonication time while that of NMP sample was 24.24% at 10 minutes. However membrane morphology was found to deteriorate with increase in sonication time. No correlation was obtained between the membrane properties and the sonication time. In general, membrane performance was better for membranes samples using DMAc than NMP solvent. Therefore, DMAc was selected as the casting solvent for membrane preparation on which further studies will be carried out.

Keywords: morphology, water uptake, residual solvent, thermogravimetric analysis

Introduction

Water soluble polymers have received academic and industrial interests due to their potential applications in fuel cells and other electrochemical devices (Ye et al. 2012). One of the water soluble polymers that have attracted publication interest is poly (2-acrylamido-2-methyl-1-propanesulfonic acid) (PAMPS) (Florjańczyk et al. 2003). Generally, some of the major performance criteria for fuel cell materials include proton conductivity, water and methanol uptake, good mechanical properties, chemical stability, thermal stability and methanol permeability (Gil et al. 2004; Swier et al. 2005; Ye et al. 2012). PAMPS was reported to posses higher proton conductivity than partially hydrated Nafion. In addition, it has been found to be more tolerant to water fluctuation in high temperature operations. Some of the major drawbacks in the use of PAMPS are its poor mechanical properties, thermal stability, water and methanol uptake and solubility in many solvents (Florjańczyk et al. 2003). One of the approaches that have been proposed to solve these problems is development of polymer composites.

This research paper investigates the synergistic effect of combining water soluble PAMPS with sulfonated poly (ether ether ketone) (SPEEK) which is non - water soluble polymer to prepare composite membranes for application in fuel cells. Specifically, preliminary results on the performnance of the composite were presented. Performance were evaluated based on the composite morphology and water uptake. Effects of sonication time and types of casting solvent on membrane properties were also evaluated.

Material and Method

Material

Sulfonated poly (ether ether ketone) (SPEEK) purcahsed from FuMA-Tech, Germany, and poly (2-acrylamido-2-methyl-1-propanesulfonic acid) (PAMPS) purchased from Aldrich Chemical Co., are the two major polymers used in this research. The SPEEK has a DS of 0.533 and solubility of 99% in DMAc. SPEEK is derived by sulfonation of Poly (ether ether ketone) (PEEK). PEEK is a commercially successful polymer due to its excellent thermal stability and solvent resistance (Mohr et al. 1991). For sometimes, further application of PEEK have been stalled due this solubility issues. By sulfonation, PEEK can be made soluble in most commercial solvents. Importantly, the solubility of SPEEK also depends on its degree of sulfonation (DS). Since, this work involves the use of two polymers, a solvent which will disolve both the PAMPS and SPEEK must be selected. Literature review showed that PAMPS dissolves in N,N-Dimethylformamide (DMF), N,N-Dimethylacetamide (DMAc), propylene carbonate (PC), and water (Florjańczyk et al. 2003). Initial solubility studies that was carried out in our laboratory was done for N,N-Dimethylacetamide (DMAc) and N-Methyl-2-pyrrolidone (NMP). The choice of NMP was due its unique solvent properties and its structtural distiction from DMF and DMAc (Figure 1). Moreover, no literature was found where NMP was used as casting solvent for PAMPS. Thus, DMAc and NMP were used as solvents in this work. All materials were used as received from the manufacturers.

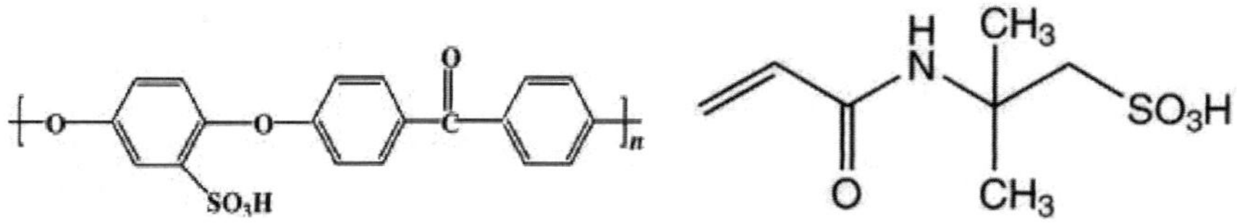

Figure 1 Structure of SPEEK and PAMPS

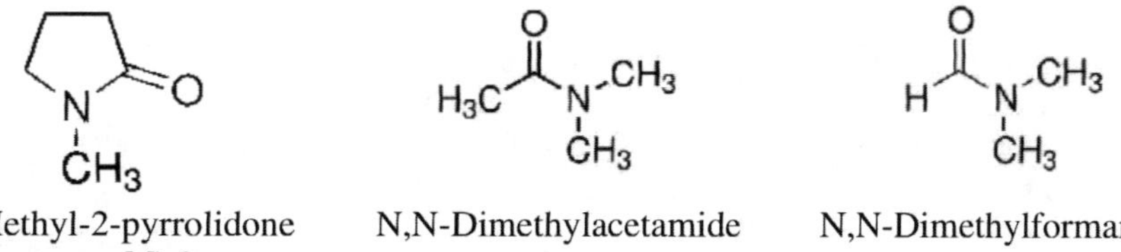

N-Methyl-2-pyrrolidone N,N-Dimethylacetamide N,N-Dimethylformamide

Figure 2 Structure of Solvents

Experimental Procedure

Membrane Preparation

SPEEK was first dissolved in solvents (DMAc and NMP) and stirred for 2hours at 70°C. PAMPS was then added and the blend was stirred for 24hours. The solution was sonicated for various periods of time ranging from 10 to 90minutes using an Ultrasonic bath. In untralsonic bath, the final temperature achieved for a given time of sonication depends on the tempearutre of the laboratory (summer or winter). It is therefoe genrally advisable that maximum temperature of the bath is determined and maintains (as the so - called equilibrium tempearture), when operating continuously under ambient conditions (Santos et al. 2009). During the sonication, equilibrium temperature was maitained by simply filling the bath with water that was preheated to sample temperature.

Membrane Characterization

Water uptake and morphological tests were performed. Water uptake was carried out by first vacuum drying all membrane samples at 100°C until constant weights were obtained. Dried samples were then immersed in deionized water for 24h at room temperature (25°C). Samples were taken out of the water after 24h, wipped to remove surface water and then weighed immediately. Water uptake was calculated using Eq. 1.

$$water - uptake = \frac{W_{wet} - W_{dry}}{W_{dry}} \qquad \text{Eq. 1}$$

Surface morphology of the membrane was done using scanning electron microscopy (SEM). Thermogravimetric analysis (TGA) was done from room temperature to 350°C using TA Q600 analyzer at heating rate of 20°C/min under nitrogen atmosphere with flow rate of 20mL/min using ~ 8 -10mg membranes samples.

Results and Discussion

Membrane Morphology by Scanning Electron Microspcopy (SEM)

Results of surface morphological analysis performed on membrane were shown in Figure 3 and 4. The overall pictures of this figure depicts a dense membrane with some surface defects observed for most of the sonicated samples. For example, Figure 3A showed some kind of layered arrangement with no visible pores at the surface. Also, the PAMPS seems to be well distributed throughout the SPEEK. The micrograph in Figure 3B can be observed to be rough and possesses some uneven ramdom distribution of PAMPS agglomerates indicating that the distribution of PAMPS have been disturbed by the 10minutes sonication. An increase in sonication time to 20 minutes was observed to improve the distribution of PAMPS. This was however accompanied by some cracks believed to be due to the slight increase in the viscousity of the blend solution. The vicousity increase was brought about by the excape of some fraction of the solvent during sonication. As the sonication time was increased, PAMPS was observed to become more evenly distributed and the cracks become more visible and larger in size (Figure 3D to F). SEM images of samples at sonication time of 10 and 30 minutes were reproduced in higher magnification of 10μm (Figure 4) in order to further revealed the effect of sonication on the morphology of the somposites. Similar morphological behaviour has been reported in literature, though for inorganic/organic clusters of nanoclay/epoxy composites membranes (Lam et al. 2005).

Water Uptake

Water uptake have been reported to be affected by the degree of sulfonation, membrane pretreatment, hydration, ambient relative humidity and water temperature (Xing et al. 2004; Wu et al. 2006). Figure 5 showed the effect of sonication time and solvent type on water uptake that was measured at room temperature (25°C). For both solvents, water uptake is same at zero sonication time. There was however a further decrease in the water uptake for NMP at 20 minutes sonication time before it beging to increase. Thus, the best sonication time for both solvent lied between 10 and 20 minutes with respect to water uptake. Interestingly, this result correspond to that reported by Lam (2005) on the effect of ultrasonication on microhardness of composites.

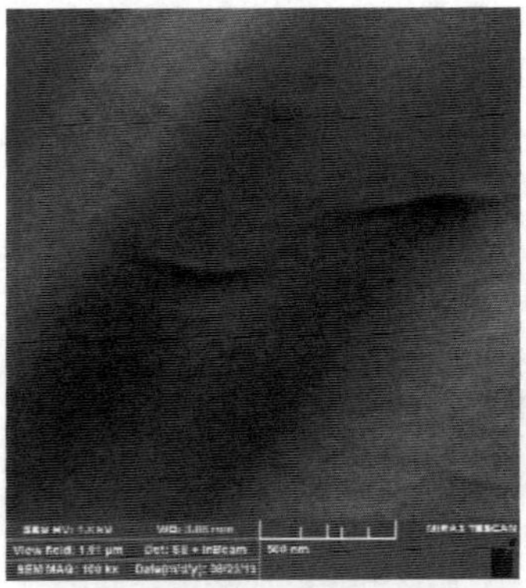

A) No sonication

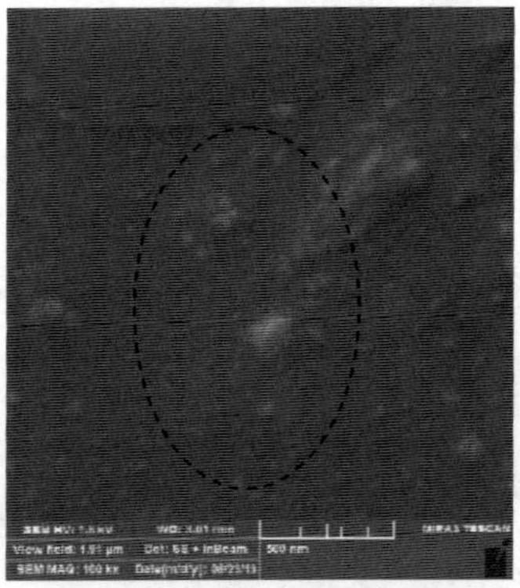

B) 10 minutes sonication

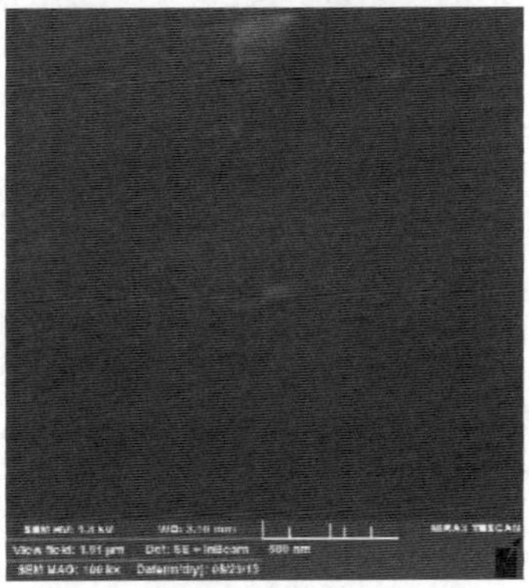

C) 20 minutes sonication

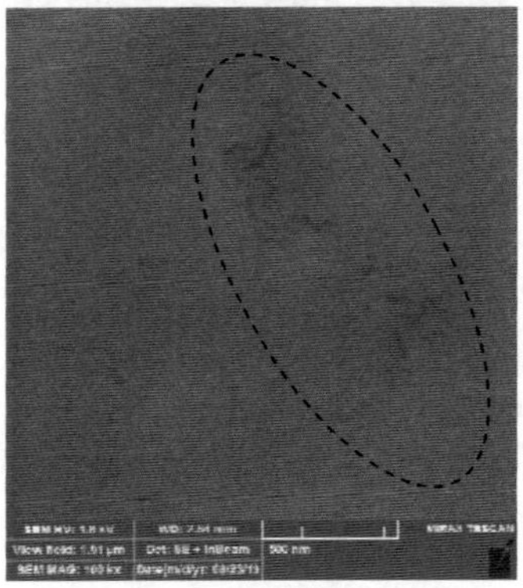

D) 30 minutes sonication

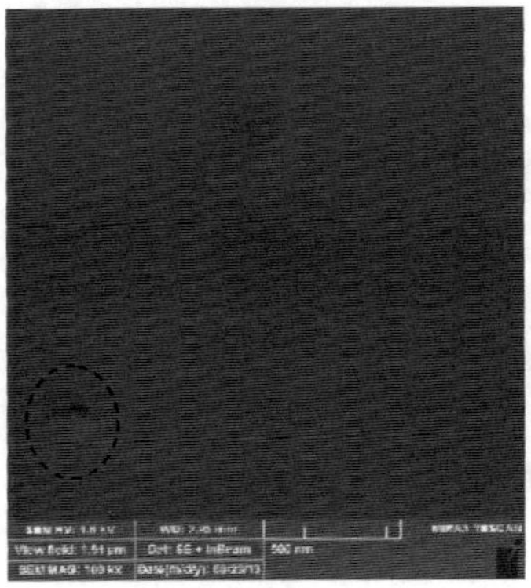

E) 60 minutes sonication

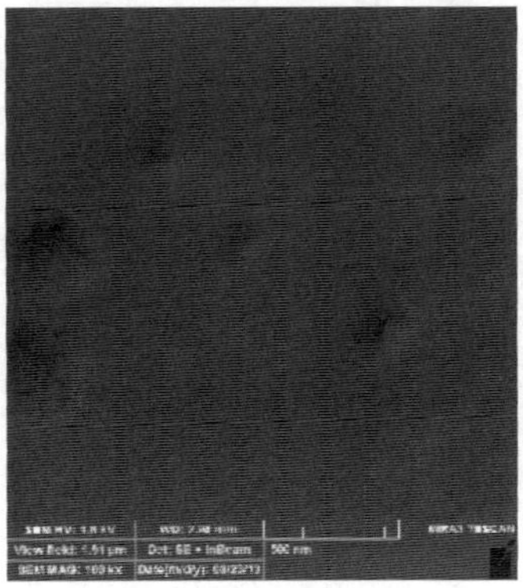

F) 90 minutes sonication

Figure 3 SEM micrograph of SPEEK-PAMS composite with various times of sonication with magnification of 500nm

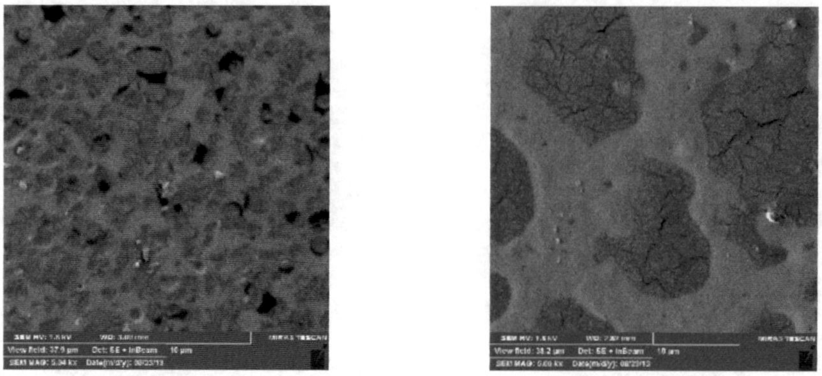

A) 10 minutes sonication B) 30 minutes sonication

Figure 4 SEM micrograph of SPEEK-PAMS composite with various times of sonication with magnification of 10μm

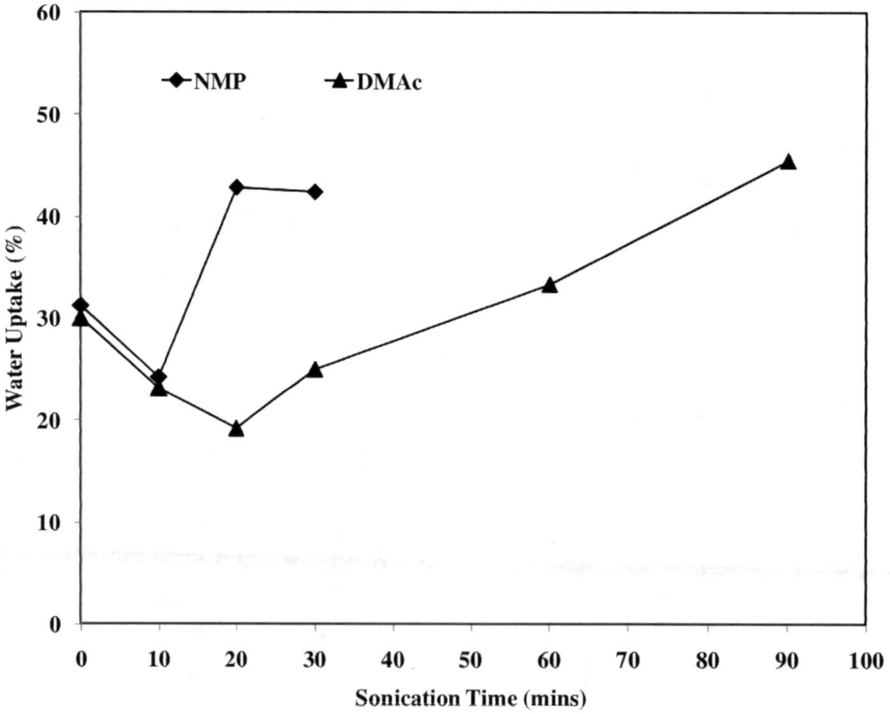

Figure 5 Water uptake of SPEEK-PAMS (3wt%) composite with variuos sonication times

Thermogravimetric analysis (TGA)

The difference in membrane performance resulting from the use different casting solvents have been attributed to the presence of residual solvents in the membrane samples after evaporation (Joly et al.

1999). TGA was thus used to quantify the residual solvent and evaluate its effect on the membrane performance as has been previously demonstrated in lieterature (Recio et al. 2007; Iqbal et al. 2008). Figure 6 shows the TGA curve for samples prepared from NMP and DMAc with different post vacuum heat treatment conditions. From this figure, weight loss can be attributed to evaporation of variety of substances, including solvents, from the samples (Rich et al. 1997). Table 2 contains the solvent and its boiling points, and weigth loss at the boiling point range that were extracted from TGA results. For sample heat treated for 24hrs, the overall percentage weight loss up to 204°C was about 8wt% of membrane sample for NMP while it is about 6.3wt% for DMAc. These will definitely include compnenents such as water, solvents and other unidentified substances. To quantify residual solvents, weight loss was estimated between 192 -214°C for SPEEK-PAMPS-NMP and between 154 and 176 for SPEEK-PAMPS-DMAc. The results obtained showed that the amount of the residual solvent in NMP is larger than that of DMAc by 1.25wt% (Table 1). To further investigate if this difference is responsible for higher value of water uptake for the NMP sample, membrane samples made from both solvents were further heat treated under vacuum for 72hrs. Residual solvent after this treatment was slightly higher in DMAc than NMP. Despite this, the value of water uptake of the NMP membrane is still higher than that of DMAc by approximately the same value as before the heat treatment (1.71wt%).

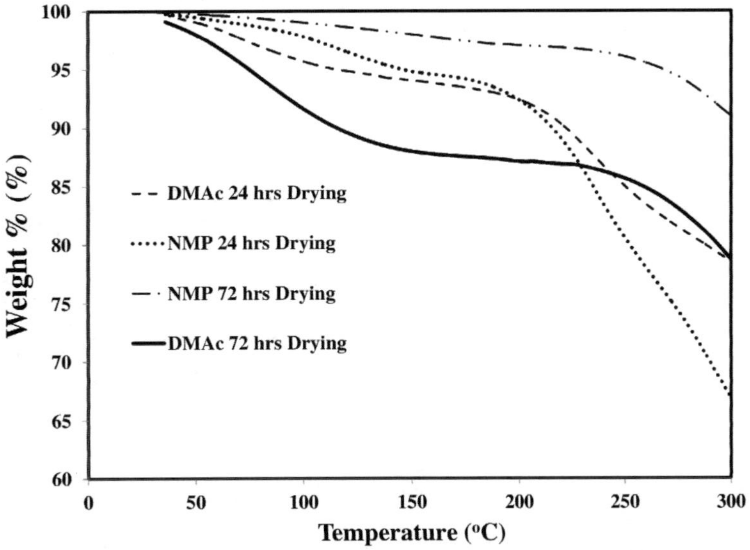

Figure 6 TGA of SPEEK-PAMPS-NMP and SPEEK-PAMPS-DMAc

Table 1 Residual solvent in SPEEK-PAMPS composites membrane samples

Membrane Sample	Solvent	B. Pt. (°C)*	Residual Solvent (wt%)	Water Uptake
DMAc 24 hrs Drying	DMAc	164.5 – 166	0.57	30.00
NMP 24 hrs Drying	NMP	202 – 204	2.85	31.25
DMAc 72 hrs Drying	DMAc	164.5 – 166	0.367	21.21
NMP 72 hrs Drying	NMP	202 – 204	0.217	22.92

*Boiling Points of Solvents

On a general note, results from Figire 1 – 5 showed that the best perfomance can be obtained for variety of sonication time depending the property of preference. Also, performance of DMAc solvent was observed to be better than that of NMP on the basis of water uptake. For this reason, DMAc was selected as the casting solvent for membrane preparation on which further studies will be carried out. Water uptake of DMAc membrane, prepared without sonication, was comapred with those reported in previuos literature. Table 2 illustrates the comparison between SPEEK-PAMPS and other relevant membranes previuosly reported in literature. In addition, this table serve to provide insight into the possible potentials of the composites. From this table, it was observed that the water uptake for SPEEK-PAMPS is within the range of values of other membranes on the table. Its water uptake is less than that of Nafion 112 which is a commercial membrane. The differences observed between the water uptake values of SPEEK – PAMPS and other membranes from the table can be explained based on three important factors - SO_3H group, the degree of sulfonation and experimental temperature (Gil et al. 2004; Shan et al. 2006; Wu et al. 2006). Water uptake of SPEEK – PAMPS is higher due to the presence of SO_3H group in the PAMPS molecules (Figure 2). SO_3H group have been reported to enhance membrane acidity and hydrophilic properties (Shan et al. 2006). Infact, the higher the SO_3H group the larger the water uptake. The degree of sulfonation has been reported to affect the water uptake of membranes. The higher the degree of sulfonation, the larger the value of water uptake (Gil et al. 2004). Thus, water uptake of SPEEK – PAMPS is lower than SPEEK -3 due their different degree of sulfonation, 53.3 and 72% respectively and the experimental temperature 24 and 60°C, respectively. The experimental temperature of SPEEK – 2 is close to that of the composite. However, the results cannot be compared in this case since the DS of SPEEK – 2 was not known. Above explanation is also true for the values of water uptake for SPEEK – 1.

Table 2 Comparison of water uptake values obtained with those from Literature (Xing et al. 2005)

Membrane	Water Uptake (wt%)	Experiment Temperature (°C)	**D S (%)	Reference
Nafion 112	22.85	N/A	-	(Xing et al. 2005)
SPEEK – 1*	16.00	30	45	(Ren et al. 2005)
SPEEK – 2*	44.90	25	N/A	(Zhong et al. 2008)
SPEEK – 3*	85.00	60	72	(Wu et al. 2006)
SPEEK– 4*	15.00	N/A	N/A	(Handayani and Dewi 2011)
SPEEK/PAMPS (3wt%)	21.21			This work

*The numbering have been included for explanation purposes.
**DS means Degree of Sulfonation

Conclusion and Recommendations

Blend of SPEEK and PAMPS was prepared by solvent casting method and its performance was evaluated with respect to its morphology and water uptake. Effect of sonication time (10 – 90minutes) and solvent types (DMAc and NMP) were also investigated. Performance of membrane prepared using DMAc was observed to be better than that of NMP. There was no correlation between the sonication time and membrane properties, but it was observed that the sonication deteriorate the membrane morphology. Membrane prepared without sonication shows better surface morphology, whereas the best water uptake was observed between 10 and 20 minutes. Thus, other properties need to be investigated in order decide the best sonication time. Therefore, further investigation such as thermal stability, methanol permeability, ion exchange capacity and gas (O_2 and H_2) permeability of the membrane are recommended for a complete evaluation of the membrane performance for direct methanol fuel cell (DMFC) application. All these investigations are in progress in our laboratory.

Acknowledgement

Authors would like to thank the National Science, Technology and Innovation Program (NSTIP) of KACST for funding this research under project number 10-ENE1374-04. Also, authors extend their appreciation to Center of Petroleum & Minearls, Center of Excellence in Renewable Energy, Mr. Dan Jerry D.Cortes of the Nanotechnology Laboratory, Chemical Engineering Department, King Fahd University of Petroeleum & Minerals, Saudi Arabis for their supports.

REFERENCE

Florjańczyk, Z., et al. (2003). "Proton conducting electrolytes based on poly(2-acrylamido-2-methyl-1-propanesulfonic acid)." Electrochimica Acta **48**(14–16): 2201-2206.

Gil, M., et al. (2004). "Direct synthesis of sulfonated aromatic poly(ether ether ketone) proton exchange membranes for fuel cell applications." Journal of Membrane Science **234**(1–2): 75-81.

Handayani, S. and E. L. Dewi (2011). "Influence of Silica/Sulfonated Polyether-Ether Ketone as Polymer Electrolyte Membrane for Hydrogen Fueled Proton Exchange Membrane Fuel Cells." Internat. J. of Sci. and Eng. **2**(2): 26-30.

Iqbal, M., et al. (2008). "Solvent effect on morphology and CO2/CH4 separation performance of asymmetric polycarbonate membranes." J. Membr. Sci. **318**(1–2): 167-175.

Joly, C., et al. (1999). "Residual solvent effect on the permeation properties of fluorinated polyimide films." Separation and Purification Technology **16**(1): 47-54.

Lam, C.-k., et al. (2005). "Effect of ultrasound sonication in nanoclay clusters of nanoclay/epoxy composites." Materials Letters **59**(11): 1369-1372.

Mohr, J. M., et al. (1991). "Gas transport properties of a series of poly(ether ketone) polymers." Polymer **32, 13**: 2387-2394

Recio, R., et al. (2007). "Gas separation of 6FDA–6FpDA membranes: Effect of the solvent on polymer surfaces and permselectivity." J. Membr. Sci. **293**(1–2): 22-28.

Ren, S., et al. (2005). "Surface modification of sulfonated poly(ether ether ketone) membranes using Nafion solution for direct methanol fuel cells." Journal of Membrane Science **247**(1–2): 59-63.

Rich, D. C., et al. (1997). "Effect of cure conditions on Probimide 32 polyamide-imide." Journal of Applied Polymer Science **63**(9): 1113-1126.

Santos, H. M., et al. (2009). The Power of Ultrasound. Ultrasound in Chemistry: Analytical Applications. J.-L. Capelo-Martinez. Weinheim, Wiley-VCH.

Shan, J., et al. (2006). "Sulfonated polyether ether ketone (PEEK-WC)/phosphotungstic acid composite: Preparation and characterization of the fuel cell membranes." Pure Appl. Chem **78**(9): 1781 - 1791.

Swier, S., et al. (2005). "Sulfonated poly(ether ketone ketone) ionomers as proton exchange membranes." Polymer Engineering & Science **45**(8): 1081-1091.

Wu, H.-L., et al. (2006). "Preparation and characterization of poly(ether sulfone)/sulfonated poly(ether ether ketone) blend membranes." European Polymer Journal **42**(7): 1688-1695.

Xing, D. M., et al. (2005). "Characterization of Sulfonated Poly(Ether Ether Ketone)/Polytetrafluoroethylene Composite Membranes for Fuel Cell Applications." Fuel Cells **5**(3): 406-411.

Xing, P., et al. (2004). "Synthesis and characterization of sulfonated poly(ether ether ketone) for proton exchange membranes." Journal of Membrane Science **229**(1–2): 95-106.

Ye, Y.-S., et al. (2012). "Water Soluble Polymers as Proton Exchange Membranes for Fuel Cells." Polymers **4**(2): 913-963.

Zhong, S., et al. (2008). "Modification of sulfonated poly(ether ether ketone) proton exchange membrane for reducing methanol crossover." Journal of Power Sources **180**(1): 23-28.

Fixed Plant Facility Siting – A Mega Project Lifecycle Framework

M.F. Haider
mech_haider@yahoo.com
mhaider@suncor.com

Prepared for Presentation at
American Institute of Chemical Engineers
2014 Spring Meeting
10th Global Congress on Process Safety
New Orleans, LA
March 30 – April 2, 2014

Fixed Plant Facility Siting – A Mega Project Lifecycle Framework

M.F. Haider
mhaider@suncor.com
mech_haider@yahoo.com

Keywords: Facility Siting, Project Management, Mega Project, Consequence Analysis (CA), Quantitative Risk Assessment (QRA), Adaptive, Change driven, Agile methodology

Abstract

Facility Siting based on tools such as Consequence Analysis (CA) and Quantitative Risk Assessment (QRA) is often times successfully applied for Brownfield operating unit projects, projects with relatively smaller footprints and for projects lasting relatively short time-periods. Challenge remains to apply and realize the full potential and value of Facility Siting for Mega-Projects. This is mostly due to project complexity and length – this paper will introduce a Management Framework to successfully apply CA and QRA as part of Facility Siting for Mega-Projects. The paper will also outline key strategies to ensure successful (i.e. value adding) implementation of facility siting, suggest a list of activities, associated sequence and pitfalls to avoid. Project Management team (e.g. Project Sponsor, Project Engineers, Managers) of both owner/operator and contractor/vendor may find this paper more useful as it provides a broad overview (i.e. when to initiate, how to plan, execute and close such assignment to maximize value) including real life examples of facility siting process over the project lifecycle.

1. Introduction

The driving force for process safety has been primarily based on catastrophic events [1]. With an increasing number of incidents, the process industry and governments started to take measures such as Facility Siting initiatives with hope to minimize loss of life and property. In the 1970s and 1980s, some of the world's most tragic industrial accidents took place. The Flixborough explosion in 1974 was reinforced by that of the Seveso tragedy in 1976 [2]. The Macondo blowout in the Gulf of Mexico on April 20, 2010, killed 11 employees and led to an uncontrolled oil spill lasting 87 days. This blowout was the most significant offshore incident in the US to-date. Finally, the Fukushima Daiichi nuclear plant incident in Japan in March 2011 drew the attention of the global process and power industries and encouraged them to incorporate natural disaster risks in a hazard analysis study [3]. As part of inherently safer design, many companies now-a-days apply CA and QRA as part of facility siting for their new projects. Due to the scarcity of information at the beginning, the studies in many cases start later in the project lifecycle for them to be effective and value adding. This paper will focus more on how to manage such initiatives for new Mega-Projects over the technical intricacies of the CA/QRA processes. The intention of the paper is to enhance the Body of Knowledge publicly available for management team in as much non technical languaga as possible to make their greenfiled Mega-Project initatives a success i.e. safer for people, environment and property.

So, what is different for industrial Mega-Projects?

- Industrial Megaprojects, as the name implies, are very large. To provide a simple and simply applied definition, we are defining a megaproject as any project with a total capital cost of more than $1 billion (U.S. dollars) as measured on January 1, 2003 [4].
- Most often engineering and construction activities run in parallel and not in sequence
- Multiple sub-projects under the 'Mega-Project umbrella' run in parallel
- Complex – design is almost 'never' complete in early stages – goes through many iterations
- Progressive elaboration is evident
- Scarcity of critical workforce with mega-project experience
- Scarcity of information – requires project team to make assumptions for CA/QRA
- Timeline: 2-5 years

2. Key Strategies

2.1 Facility Siting Charter

It is not only important or good to do but critical that the stakeholders need to identify, understand and agree on the purpose and value to conduct facility siting for their projects. The Author's personal experience over the years has been in order to maximize value of Facility Siting the project team has to start 'early' to develop project Charter and start with the end in mind. As can be seen from the following generalized Figure 1 that as the company starts to appraise opportunities, it is

Figure 1: Stages and Milestones to apply CA/QRA in a Mega-Project

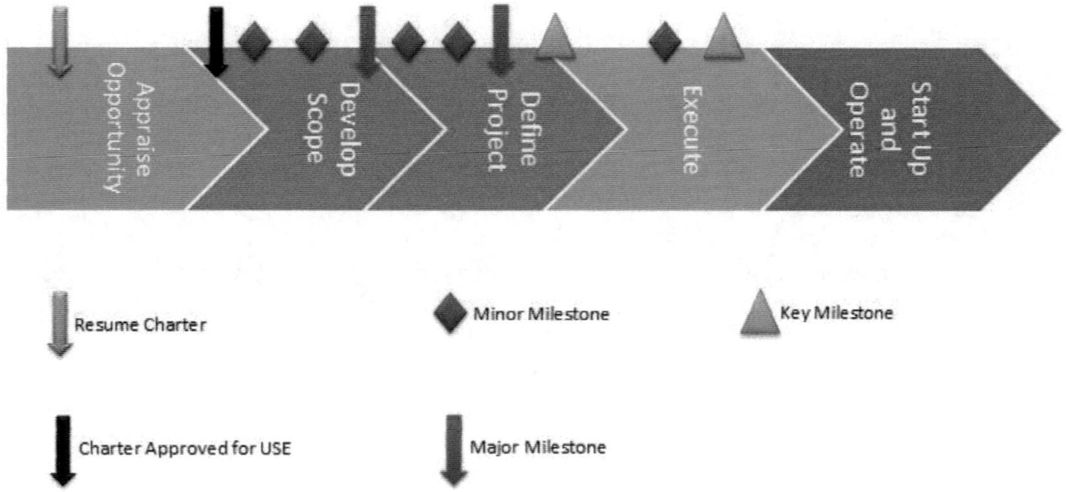

prudent to resume developing the Facility Siting Charter by bringing in the stakeholders together. Figure 1 shows a generalized project lifecycle for a typical Petro-Chemical Mega-Project. The

Figure also shows that the Key Deliverable is developed through a successive series of repeated cycles over the project life-cycle. The Milestones will be defined later in the paper. Part of starting with the end in mind is to clearly outlining ultimate goal/s, e.g. vision is to design a safer facility and utilize Mitigated QRA Report to formally document and communicate that facility is within the Risk Threshold or justification for accepting risk profile as is by Senior Management. The Vision in the Charter also provides opportunity to strike a balance between business goals and climbing the Risk Mountain – at each stage of increasing safety (decreasing risk). As part of striking a balance, the Project Engineer has to Identify options up-front if and/or when it becomes apparent that aggregate Residual Risk will not be within the Risk Threshold or acceptable so that team can put more effort in developing justification for either accepting or rejecting the risk profile by Management

Figure 2: Strike a balance between Business goal and Climbing the *expensive* Risk Mountain

2.2 Management Strategy

In order to successfully manage Facility Siting, CA and QRA for a Mega-Project, the stakeholders have to align on a management strategy. There are two primary management approaches for the purpose of managing such work:

1. Predictive or Plan driven
2. Adaptive or Change driven

The Author's past experience has been if the organization undertook very 'similar' projects in the near past (<5 years), critical experienced workforce is retained, Lessons Learned is readily available, and most importantly, the past project is still considered to be a success, it is advisable to use Plan driven approach. On the other hand, if the organization is introducing new technology, has limited past experience with similar project, involves relatively high number of stakeholders, capital cost > $1B, project life spans >2 years, it is recommended to utilize Change driven approach to maximize value. It is to be noted that it is not essential to meet all the suggested criteria mentioned above to pick one management strategy over the other.

2.2.1 Benefits of Adaptive or Change Driven Approach

In Adaptive/Change Driven Planning, project team intentionally repeats one or more project activities as the project team's understanding of the product increases ....Also known as ...agile methods .....intended to respond to high levels of change and ongoing stakeholder involvement [5]. The heart of Facility Siting involves making timely but informed decision. Some of the characteristics of this framework:

1. To add small incremental improvements that will deliver value to stakeholders
2. Milestones develop the product (e.g. Mitigated QRA Report) through a series of repeated cycles, while increments successively add to the functionality of the product
3. For each Milestone, a deliverable (read 'model') or set of deliverables will be completed. Future iterations may enhance those deliverables or create new ones. Each iteration incrementally builds the deliverables until the exit criteria for the phase are met, allowing the project team to incorporate feedback.

The biggest challenge with Mega-Project is scarcity of key information needed as inputs to develop scenarios and models. Some of the critical benefits, a project may gain by applying Change Driven approach:

1. The approach acts as Leading Indicator on Process Safety for Plant Layout - allows for multiple opportunities to make critical decisions on plant layout over the project life-cycle capitalizing on what is known at a given time. This lets the project team to use "what they know", to ensure "what they don't know" falls into the right place
2. Opportunity to make decisions over the project lifecycle not only works to save life but also helps with striking a balance between business needs and climbing the risk mountain
3. Aides design-development with progressive elaboration while producing final CA/QRA deliverable for the project

2.3 Execution Strategy

The main idea of Change Driven process starts with decomposing the known preliminary layout and process information into a set of requirements, work to be performed and their priority sequence, sometimes referred to as a *Product Backlog.*

It is advisable that the owner/operator forms a multi-use agreement (contract) with the chosen vendor. This way the plant areas under the Mega-Project umbrella can get their specific work done while the CA/QRA Project Team provides overall oversight and guidance.

At the beginning of an iteration, the team will work to determine how many of the highest priority items on the backlog list can be delivered within the next iteration. The time-length of each iteration should be less than 4 weeks. As shown in Figure 1, during "Develop Scope" and "Define Project", the length of each iteration could be even less than 2 weeks. This iterations (otherwise known as "Minor Milestone" as shown on Figure 1) will develop scenarios, assumptions and preliminary CA/QRA models. Assumptions are required due to lack of available quality information in the early stages of a project.

At the end of each iteration, the Minor Milestone/deliverable should be ready for review by the customers/stakeholders. This does not mean that the customer is required to accept delivery, just that the product should not include unfinished and/or incomplete features.

The sponsor and customer representatives should be continuously engaged with the other project stakeholders to provide feedback to Contractor on deliverables as they are created and to ensure that the *Product Backlog* reflects Sponsor/stakeholders' current needs. Below, you will find two examples using two different management strategies. For confidentiality purpose, the specific name/location is kept generic and few items were modified to suit.

2.3.1 Project X Facility Siting Using Change Driven Approach

A CA/QRA team was given the assignment to support design development during "Develop Scope" phase (Figure 1). The Sponsors already selected the site for the project. But the fixed plants were yet to be located on the site. The Vision for the initiative was for the QRA result to come under the organization Risk Threshold. Multiple EPC's were given to engineer different plant areas. The CA/QRA team selected a vendor to support them with defining the *Product Backlog* based on the preliminary Process Design Basis, intended product the facility is expected to sell, the feedstock necessary etc. The CA/QRA team had prioritized the Product Backlog based on hazard level of feedstock, preliminary PHA schedule and project schedule. This also involved obtaining feedback from different Process Engineering and Operations teams to collect necessary inputs to put together the assumptions and scenarios to be modeled within the 1st iteration.

Out of 25+ scenarios, only 7 perceived to be the most hazardous were modeled for CA that gave ample input (e.g. explosion, toxic and fire contours) for the CA/QRA team to work with the stakeholders to recommend location for specific process units on the given site. Note that, CA/QRA team integrated the potential consequence to understand local incident and global consequence events. At this time QRA was not performed. This was used as input for minor modification of preliminary plant layout. It is to be noted that the site plan was still preliminary after the first iteration but assisted different EPC accountable for their specific area to recognize where to locate the specific plants i.e. finalized key plans for the overall site plan.

With further design-development, more refined information became available from each EPC. This new information led to 130+ scenarios and was used for further CA/QRA iterations that created either new model and/or embellished the past ones. Each EPC was accountable for facility siting within their boundary while the CA/QRA team provided technical oversight and integration support for the overall project. The Sponsor/stakeholders were engaged on a regular basis to ensure that the Product Backlog reflects future modeling requirements to properly site facilities e.g. Fire Water Pump house, Tank Farms, site Medical Centre etc. Early engagement of CA/QRA team and higher management willingness to incorporate their feedback with the progressive elaboration of the project led to QRA result meeting the organization risk threshold and ultimately safer facility. Changes suggested based on CA/QRA work in earlier stage (i.e. Develop Scope, Define Project) were easier to incorporate into the design.

2.3.2 Project Y Facility Siting Using Plan Driven Approach

The sponsors for project Y selected the site for the project and decided to hold on to not perform CA/QRA work with the progressive elaboration of the project activities. This was mostly decided based on the fact that not all information was available to properly conduct the 'studies'. Fundamentally, the sponsors looked CA/QRA as deliverables requiring a more 'complete' plan. Once the Develop Scope (as referred in Figure 1) is almost complete, the sponsors decided to engage CA/QRA team and a vendor. By this time, the project had significantly high number of stakeholders. Many of the stakeholders were not very familiar with CA/QRA and the value it could bring to them. There was not much time to schedule and train the stakeholders as well to bring them up to speed. As a result, the stakeholders were unable to align on a Vision. The Vendor was asked to provide explosion, toxic and fire contour maps for specific scenarios and locations that were almost 'fixed' on the site plan at that point in time. The resultant contour maps showed significant consequence on nearby facility with over 700+ people. It was also noted that the plant area consuming the most hazardous substance would impact the nearby facilities significantly from consequence perspective. It should be noted that the result was only based on a selected few scenarios. By the time the project reached the "Execution" phase, more scenarios were added as part of the scope. But it was already too late to incorporate most of the recommended changes resulting from CA/QRA work. The project went through several area by area risk reduction workshops only to accept (by the company executives per organization risk matrix) the aggregate residual risk level which was also found to be not within the organization risk threshold.

2.3.3 Importance of Starting Early

Figure 3: Impact of variable based on project time [5]

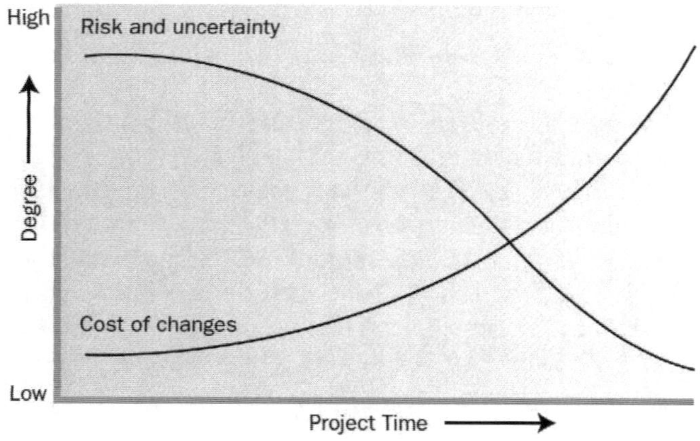

The above two examples highlight the importance of starting CA/QRA work early enough to realize and maximize the value CA/QRA brings to a Mega-Project. The high-level view provides common frame of reference for comparing projects for CA/QRA initiative, even if the projects were dissimilar. Change driven methodology allows for applying the known information at early stage in a way that supports laying out plant facilities to reduce risk and thus maximize potential value. While these characteristics remain present to some extent in almost all project life cycles, they are not always present to the same degree [5]. Change Driven framework, in particular, keeps stakeholder influences higher and the costs of changes lower especially in the early stages of the life cycle than in predictive/plan life cycles. The Figure 3 highlights a well-known fact of project management that even though risk and uncertainty remains high in the early stages of a project, cost of incorporating changes remains much lower. This translates into a key opportunity for project stakeholders to align on laying out the key areas on a site plan intelligently to reduce risk and maximize value by applying tools such as CA and QRA.

2.4 Key Decision Points

Facility Siting, from a management point of view, is all about bringing the stakeholders closer and find the 'right' spot for the facilities. Finding the 'right' spots requires making timely decisions. The challenge has always been to make timely decisions as the stakeholders more often find themselves forced to make those decisions based on information that are frequently of poor quality and at times uncertain, especially in the early stages of a project. At times, due to the scarcity of information, stakeholders rely on their past experience to make those decisions on plant layout. As a result, depending on the individual experience, two different teams for a project area may come up with two very different plant layout. CA/QRA work, if started early enough, will also depend on input that most often requires embellishment through assumptions to create needed scenarios and models. Even with this limitation, CA/QRA provides a key 'input' in allocating real estate for key facilities on a plant site at early stage of a project lifecycle. With each iteration and deliverable 'Change driven' methodology allows for multiple opportunities to make critical decisions on plant layout over the project life-cycle. It allows the stakeholders to decide when to 'sharpen the pencil' (e.g. PHAST to Computational Fluid Dynamics model), relocate facility or accept the consequence and/or risk, 'pull the plug' on further modeling/refinement and start developing justification, if required, to recommend management to accept/reject the aggregate risk profile.

2.5 Alignment with Process Hazard Analysis (PHA) Planning

CA/QRA results are a useful input to PHA (e.g. HAZOP, LOPA) activities. This is particularly true as projects move forward after establishing the plant layout on a high level and starts to conduct "Define Project" (Figure 1). PHA activities may also result in recommendations (i.e. new scenarios) that goes into the *Product Backlog* and requires project team to develop new models or refine old ones. Mega-projects have multiple sub-projects running in parallel most of the time, having an Integrated Schedule for all the reviews is a good strategy so that the right stakeholders are booked for the sessions. The integrated PHA schedule also supports prioritization of *Product Backlog* for CA/QRA activities over the lifecycle of the project.

2.6 Integration Strategy

It is critical for a Mega-Project involving multiple plant areas/silos to understand the BIG picture so that site risk can be dealt with on an aggregate basis. There is strong possibility of one area to impact other plants in the vicinity. It is recommended to review integration aspect and aggregated results on a regular basis. The following Figure 4 shows that the origin of an incident may reside in one particular plant area, while it may impact other areas in the vicinity.

Figure 4: Consequence of an incident in one area impacting multiple areas

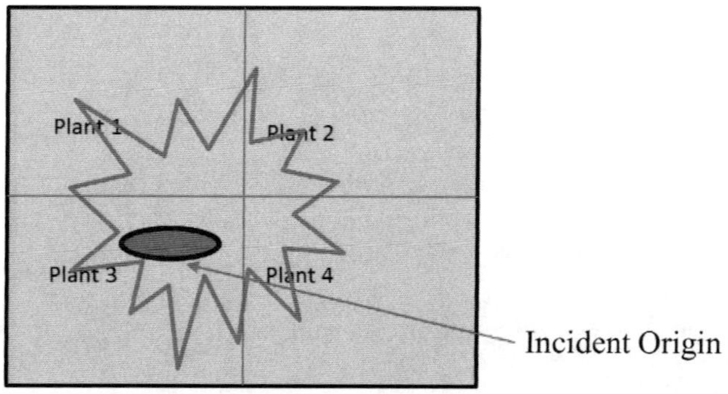

Recommended Strategy:

- Setup Strong Center - Task Force Single Point of Contact (SPOC)
- Collaborate – SPOC will promote Collaboration among Plants/Areas

3. List of Activities (suggested)

3.1 Overall Sequence

The CA/QRA work typically should start with the approved Charter as shown on Figure 1. A project Task Force i.e. SPOC is formed comprising of Project Engineer, Operations rep, Process rep and CA/QRA Technical Expert. The SPOC is ultimately accountable to produce the CA/QRA deliverable studies but also acts to support the design development and site plan finalization efforts in the early stages of the project.

The team collects 'input' such as Preliminary site Layout and design information, works with area process and operations team to identify scenarios, develop assumptions for known-unknowns and unknown-unknowns, develop modeling 'Product backlog', prioritize backlog based on PHA schedule and produce modeling output (i.e. deliverable) for stakeholder use. The stakeholder usually has the option of accepting or rejecting consequence. If the result of an iteration is acceptable within area causing the incident – communicate to outside stakeholders to ensure there is no cross boundary impact. If the result of the iteration is NOT acceptable to outside of the area – use the result as input to PHA for the areas to mitigate. Collaboration recommended between origin area and impacted area/s to identify appropriate mitigations. There is also a possibility that the consequence resulting from the iteration is not acceptable to the area causing the incident, in that case the area will initiate a PHA to mitigate the consequence. As part of developing mitigations, consider the following:

- Develop QRA if appropriate
- Perform Building Assessments for Critical Buildings
- Apply Cost-Benefit analysis to optimize risks/opportunities

- Determine Safe Haven locations and their characteristics

If Mitigation is not possible from Business perspective – perform As Low As Reasonably Practicable (ALARP) justification for Senior Management to formally accept the posed risk

- To apply mitigations on a site wide basis, review Milestone deliverables by taking on a BIG picture perspective
- Future iterations may enhance those deliverables or create new ones as new information is available. Each iteration incrementally builds the deliverables to produce the Mitigated CA/QRA report to support Facility Siting for the stage

The overall effort is expected to produce a maximum of two single reports for the overall site:

1. Mitigated version of QRA Report
2. and Consolidated CA Report

It is recommended to develop as-built, pre-start up CA/QRA reports prior to start-up. Table 1 and 2 outlines typical activities performed by SPOC and plant areas respectively once key process areas are established/finalized on Site Plan. Please note that the activities are project dependent and may warrant modifications to suit specific situation.

Table 1. SPOC Sequence of Activities to conduct Facility Siting

Initiative	Key Steps
Setup Facility Siting Team	Sponsor to support to setup a team (i.e. Project Engineer, Technical Expert, Operation, Process representative) to act as SPOC
Integration	Provide functional (technical and management) guidance to Areas
	Assess cross boundary impact/opportunity
	Communicate incremental outputs after each Area PHA/Risk Reduction
IFA versions (prior to "Define Project" stage)	Consolidated CA Report
	Mitigated and Integrated QRA Report
IFD CA/QRA (prior to "Execution" stage)	Re-validate IFA CA/QRA post PHA and vendor PHA
As-built versions	Post-oil version

3.2 Typical Area Activities

Table 2. Area (Typical) Sequence of Activities to conduct Facility Siting

Initiative	Key Steps
Plant 1 (Typical)	Perform Modeling and Calculations
	Conduct Risk Reduction Workshop (if appropriate) – Apply mitigations
	Incremental output for use based on reduced risk level
	Assess Buildings/facilities
	Identify Safe Haven/shelter Locations (if required)
	Estimate Area Project Scope impact
	Perform area scope updates/changes e.g. passive fire protection requirement
	Carry out QRA (if appropriate)
	Update Product Backlog
	Develop/update contract to support area specific activities

4. Pitfalls to Avoid

The following observations were noted by the Author during his time to manage Facility Siting scope:

1. Facility Siting does not replace good and common sense engineering practice e.g. locating a plant that uses high volume of hazardous material will significantly increase risks to other plants/areas (e.g. camp) within the vicinity
2. Evaluate effectiveness of 'Cost Reduction' philosophies as opposed to added risk e.g. different modularization techniques may decrease the plant footprint by reducing pipe lengths but makes the plant more compact and likely to increase severity of explosion consequence
3. Avoid developing models with hope to realize risk reduction
4. Sharpening the pencil for models most likely may not reduce risk
5. Avoid jumping into CFD or other similar methods without evaluating cost-benefit for undertaking such effort
6. Utilization of QRA results is much more controversial than the methodology. Watson [6] has suggested that QRA should be considered as an argument, rather than a declaration of truth. In this view – focus on accuracy over precision
7. CA/QRA Task Force should be part of Management of Change (MOC) process
8. Evaluate carefully before locating facilities e.g. tank farms close to main entrance/egress as part of Facility Siting

9. Keep the focus on VISION for the overall effort to avoid scope creep
10. Note the difference between Facility Siting as a Deliverable and tool for design-development activities

5. Conclusion

This paper presented Adaptive/Change driven framework to manage initiatives like facility siting/CA/QRA as part of Mega-Projects. Change driven framework does not mean "lack of planning or no planning". As demonstrated in the paper, this methodology starts with a vision and then develops the ultimate product both iteratively and incrementally when it is possible to define small incremental improvements that will deliver value to stakeholders. This allows making key decisions on laying out the facilities in early stages of a project rather than waiting for design to develop to a significant extent. Often times, project team wait too long for design to mature enough before applying CA/QRA to validate consequence impact and/or risk level. Most often, especially for Mega Projects, this leads to delays in executing the CA/QRA work properly at the right time to influence site layout. This delay in execution leads to scenarios when scope changes cannot be incorporated anymore. This translates into value erosion as project was unable to reap the full benefit of Facility Siting. The management framework recommended in this paper i.e. Change driven/adaptive framework allows for the stakeholders to start early in the project lifecycle to apply CA/QRA in laying out the plant facilities. The key is to start EARLY but START with the END in mind. Alignment among stakeholder on a VISION up-front is also important as this is expected to save life while striking a balance between business needs and climbing the Risk Mountain. Vision also allows avoiding scope creep while applying Adaptive/Change driven approach. It is highly recommended that stakeholders focus more on accuracy over precision. In the early stages of project lifecycle, quality information is scarce. Change driven/Adaptive framework gives you as the project stakeholder the opportunity to use "what you know" intelligently to ensure "what you don't know" falls into the right place.

6. References

[1] M. Mannan, A. Chowdhury and O. Reyes-Valdes, "A portrait of process safety: From its start to present day," College Station, 2012.

[2] M. Mannan, Lees' Loss Prevention in the Process Industries, Volumes 1–3 (3rd Edition), , 2005, Elsevier, 2005.

[3] K. McAndrews, "Consequences of Macondo: A Summary of Recently Proposed and Enacted Changes to US Offshore Drilling Safety and Environmental Regulation," in *Society of Petroleum Engineers, Americas E&P Health, Safety, Security and Environmental Conference*, 2012.

[4] E. W. Merrow, Industrial Megaprojects Concepts, Strategies and Practices for Success, New Jersey.: John Wiley & Sons, Inc., 2011.

[5] A Guide to the Project Management Body of Knowledge (PMBOK® Guide) – Fifth Edition, Newtown Square: Project Management Institute, Inc., 2013.

[6] S. Watson, "The Meaning of Probability in Probabilistic Risk Assessment," *Reliability Engineering and System Safety,* vol. 45, pp. 261-269, 1994.

348480 Global Energy and Transportation and Mobil Oil

Abstract:

Biofuels, Computational Methods, Engineering Economics, Polymers, Thermodynamics. Polymers-plastic, Thermodynamics- oil, Engineering Economics – the money it takes to finish the method of Global Engineering Research, Computational Methods – numerical input of analytical data, Biofuels – the transport fluids that utilized in more than one method from industrial usage to everyday living which enters the atmosphere. Exxon has completed a $600 million dollar deal for production of algae biofuel, amongst Mobil Oil has performed major research Global Energy and the effects of oil. Reading a most recent article on Mobil website.

348748 Characterization of Iron Phthalocyanine As the Active Material for Lithium Batteries

Abstract:

Results on the determination of the lithium-ion diffusion coefficient in the iron phthalocyanine (FePc) active material as a function of intercalation level, using a modified electrochemical impedance spectroscopy technique, will be presented. Also, developed thermodynamic functions such as the Gibbs free energy, entropy, and enthalpy changes for the lithium insertion reaction with FePc, derived from cell open-circuit voltage measurements as a function of intercalation level, are presented. Also presented are estimates of the maximum amount of lithium that can be intercalated into the FePc based on constant volume and temperature conditions.

349086 Design of a Free-Fall Reactor for Fast Pyrolysis of Waste Plastics

Abstract:

Thermal pyrolysis has earned significant reputation in the recovery of valuable chemicals, especially petrochemical based fuels from various materials, including biomass and plastics. Of many available technologies, *fast* pyrolysis was preferred in his study owing to its minimal heat and mass transfer limitations and better control on process conditions. The focus of this study is towards efficient conversion of waste plastics into gaseous products consisting primarily of alkanes, alkenes and aromatics. Lab scale experiments were performed using a micro reactor to determine favorable conditions, including reaction temperature and particle size that would result in maximum yield of ethylene. Using these results, a simple industrial size radiatively heated free-fall reactor was designed for the purpose of pyrolyzing waste plastics. The goal is to design a *fast* pyrolysis reactor that improves upon inefficiencies in traditional reactors while resulting in high yield of ethylene.

349732 Biosorptive Dehydration of Ethanol/Water Azeotropes Using Compound Starch-Based Adsorbent

Abstract:

Abstract: Recently, the interest in the production of bioethanol has greatly increased as a result of the increasing fossil fuel prices and associated environmental issues. Bioethanol is made from either grains or biomass and the broth contains 6 to 12 percent ethanol. In addition, water removal represents a serious problem as ethanol form azeotrope with water at atmospheric pressure. Distillation to water-free ethanol consumes 50 to 80 percent of energy used in the fermentation ethanol manufacturing process. Various techniques have been developed to break the azeotropic point, such as salting-out method, azeotropic distillation, extractive distillation, reactive distillation, adsorptive distillation and pervaporation. Among the methods, the adsorptive option is the most attractive, especially the biosorption, which has distinct advantages over conventional methods as it is environmentally friendly, highly selective, easily available, easy to operate, cost effective and reusable in repeated cycles in the treatment of aqueous organics.

In this work, a specially formulated compound starch-based adsorbent (CSA), which consists of corn, sweet potato and foaming agent, was developed. As an exploratory study, this research was focused on the investigation of CSA for biosorptive dehydration of aqueous ethanol and optimization of conditions for its maximum adsorption under various conditions of vapor feed flow rate, bed temperature and bed height through an orthogonal design method. Furthermore, to investigate the adsorption mechanism, dynamic saturated adsorbance of CSA was studied and characterization of the adsorbent before and after adsorption, as well as after regeneration, was performed.

In the first step, the net retention time and separation factor of ethanol and water were obtained using inverse gas chromatography (IGC), the results indicated that the net retention time of water on CSA is significantly longer than that of ethanol which attested the feasibility of this adsorptive separation purpose and low temperature could be propitious to the process which was consistent with the traditional theory of adsorption.

Then, through a batch adsorption experiment, the optimum adsorption condition was determined under vapor feed flow rate, bed temperature and bed height by orthogonal design method, which indicated that the optimization condition for ethanol was at vapor feed flow rate of 1.1g/min, bed temperature of 83 and bed height of 25cm. The packing density was 445kg/m^3. The residence time is determined by vapor feed flow rate and bed height. The residence time is more influenced than bed temperature. The longer the residence time, the more the output is. But it cannot be ignored that the adsorption selectivity may decreases. Low bed temperature is benefited to the adsorption process which agrees with the result of IGC. Five cycles experiments of adsorption-regeneration were carried out at the optimum adsorption and regeneration condition, the results showed that there is practically no change in the regeneration efficiency of CSA after the cyclic process adsorption–regeneration. It is due to the properly chosen regeneration conditions assuring complete desorption of the water adsorbed without changes in the CSA structure. This unchanged adsorption capacity after a number of cycles demonstrates reusability of the adsorbent.

To study the adsorption mechanism, experiment was conducted to study the dynamic saturated adsorbance of CSA. The water adsorption capacity of CSA has been found to be similar to some molecular sieves, which is about 0.15g/g. Besides, field emission scanning electron microscopy (FESEM) and mercury porosimetry (MP) were applied to investigate the change of surface morphology and microstructure of CSA before and after the adsorption, as well as after regeneration, as shown in Fig.1 and Fig.2. The FESEM images showed that compared with the particles before the adsorption, they became plump and the void spaces became smaller after adsorption, which indicated that the CSA was saturated with water. However, the CSA particles after regeneration became wizened like the nature particles, which demonstrated that almost all the water had been desorbed from CSA. The pore size distribution (PSD) indicated that compared with the PSD curve before adsorption, the peak at 6000nm disappeared and the peak at 17000nm became remarkably smaller after adsorption, which manifested that CSA is mainly composed of macropores which play a major role in the adsorption. However, the peaks at 6000 and 17000nm reappeared after regeneration, and the curve looked exactly the same with that before adsorption, which further demonstrated that CSA after regeneration can be totally regenerated and recovered to the original structure.

These results indicated that the starch-based adsorbent can be considered as an effective and low-cost bio-sorbent to dehydrate alcohol.

349781 Purification of 2-Amino-1-Phenylethanol Enantiomers By a New Technique Combining Distillation and Crystallization

Abstract:

Stripping crystallization (SC) is a new separation technique which combines distillation and crystallization. Basically, SC is operated at a triple-point condition, in which the liquid is simultaneously vaporized and crystallized due to the three-phase equilibrium. In essence, the process is continued until the liquid phase is completely eliminated and only the pure solid crystals remain in the feed. Thus, crystal washing is not required since only pure solid crystals remain in the feed and no impurities are adhered on the crystal surfaces at the end of SC. In the previous research, SC has been successfully applied to separate p-xylene from the mixed xylenes by Shiau et al. (2005, 2008).

Phenylethanolamine, also known as 2-amino-1-phenylethanol (2-APE), is an important basic skeleton in many adrenergic drugs and is often used as ingredients of cold medicines and anorectics. As the biological activity and pharmacokinetics of the enantiomers of a chiral drug can be different, the objective of this research is to apply SC to purify R-2-APE from the chiral 2-APE mixture. The experiments show that no crystallization occurs during SC without seeding due to subcooling of the liquid enantiomers. On the other hand, seeding can eliminate subcooling during SC, leading to the R-2-APE crystals gradually formed preferentially on the seeds. Thus, seeding with magnetic stirring can be successfully employed to purify R-2-APE from the 2-APE enantiomers.

1. Shiau, L. D., Wen, C. C. and Lin, B. S. Separation and purification of p-xylene from the mixture of m-xylene and p-xylene by distillative freezing, *Ind. Eng. Chem. Res*, **44(7)**, 2258-2265, 2005.

2. Shiau, L. D., Wen, C. C. and Lin, B. S. Separation of p-xylene from the multicomponent xylene system by stripping crystallization, AIChE Journal, 54(1), 337-342, 2008.

350841 Fabrication of Low Cost Insulating Material from Kaolin Clay for Construction Purposes

Abstract:

A new low cost, thermal insulator, construction mineral polymeric material was synthesized from local Saudi Arabia resources using Inorganic Mineral Polymerization technique (MiP). In this technique, the mineral polymer is produced from the following components: clay mineral rich in koalinite, sand, sodium hydroxide solution, and water. Inorganic polymerization experiments are conducted by mixing known amounts of clay minerals and sand in a variable speed stainless-steel mixer for some period of times until a homogenous solid mixture is obtained. Sodium hydroxide solution is added to the mixture and the polymerization reaction takes place at different mixing speeds and times. The produced mixture is divided into a number of samples of known weights. Each sample is moulded in a stainless-steel cylinder at a medium pressure. Then the moulded samples are cured in an oven for 24 hours. The curing temperature was 80°C. The mineral polymerization conditions such as the ratio of clay to sand filler, particle size of the clay, particle size of the sand, concentration and amount of sodium hydroxide, speed and time of mixing were optimized to obtain stable and hard mineral polymer. Compressive strength for different specimens of fabricated mineral polymers was tested under dry and wet conditions. Values of compressive strength with values ranging from 11.5 MPa to 27.9 MPa were obtained.

351141 Thermodynamic Study of Binary PAH (Anthracene + Phenanthrene) Solid Mixtures

Abstract:

Polycyclic aromatic hydrocarbons (PAHs) are common components of many materials, such as petroleum and various types of tars. They are generally present in mixtures, occurring both naturally and as byproducts of fuel processing operations. It is important to understand the thermodynamic properties of such mixtures in order to better understand and predict their behavior (i.e., fate and transport) in the environment and in industrial operations. The Knudsen effusion technique was combined with gas chromatography and differential scanning calorimetry to measure the vapor pressure, composition, and phase behavior of binary anthracene and phenanthrene mixtures at near ambient temperatures. Mixtures of anthracene and phenanthrene exhibited non-ideal mixture behavior: In mixtures with less than 80% phenanthrene, this more volatile component preferentially sublimed and the vapor pressures decreased as the mixtures became enriched in anthracene. The vapor pressure data do not follow Raoult's law (that is, ideal mixture) predictions. Mixtures with high phenanthrene content (> 80%) showed azeotrope like behavior, in that anthracene and phenanthrene sublimed at nearly constant vapor pressure and molecular ratio. These results confirm that the behavior of few-component PAH mixtures does not follow ideal mixture models. Hence, as tars and petroleum (and potentially other organic compound-containing mixtures) weather to few component mixtures over time, complicated, non-ideal phase behavior might be observed.

351160 How Confident Are You That Your Major Accident Risks Are Under Control?

Abstract:

Major accidents in the process industries including those at Texas City , Buncefield UK and in the Gulf of Mexico have highlighted the critical role played by senior managers and executives. Effective leadership is essential if companies are to develop a positive safety culture that remains constantly vigilant towards the risk of a major process safety incident. In the UK the HSE has made leadership a key priority for the high hazard industries and there is a cross-industry initiative to develop and deliver training focussed towards board members responsible for process safety performance.

This paper explores the leadership failings that contributed to recent major accidents, and the cost to both the company reputation and the individuals involved. Key principles are discussed that can help improve leadership in the major hazard industries. These include; attitudes and decisions of senior manager affecting the investment strategy and the safety culture of the organisation, reinforcing the importance of safety by personal example, thorough understanding of major accident hazards and key risk control systems, investigating process safety incident and near misses to find the underlying causes, developing world class safety management systems, and identifying weaknesses in these systems using targeted performance indicators.

The author bases this paper on the development of a training standard for 'Process Safety Leadership' involving a cross-industry 'expert panel'. Training of senior executives in compliance with this standard has been carried out from late 2011 and feedback from these courses and the key requirements for effective leadership will be discussed.

351236 Sil Determination: Shortcomings with the Use of LOPA

Abstract:

Proposed Topic Area: Managing Risks

For a number of years, industry has been trying to apply the principles outlined in IEC 61508. In the process sector, this has resulted in a focus on the requirements of IEC 61511 (2003) and ANSI/ISA 84.00.01 (2004). These are risk based standards. They involve the identification of potential hazardous events, the causes that could lead to each event and the necessary risk reduction for each event. Where the risk reduction involves an instrumented function (e.g. a trip or an interlock), the required performance of the instrumented function can described in terms of the average probability of failure on demand. In the standards, the performance for such instrumented functions is assigned to one of four Safety Integrity Levels. The assessment of what performance level is needed in respect to a specific hazardous event is referred to as SIL Determination. In the field of SIL Determination, IEC 61511 provides examples of a number of different approaches – these are shown in IEC 61511 Part 3. These include Event Trees, Risk Graphs, Safety Matrices, Fault Tree Analysis and Layer of Protection Analysis (LOPA).

In the first few years, following the publication of IEC 61508 Edition 1 in 2000, industry generally embraced the use of Risk Graphs as the method of choice. However, over the subsequent few years, many companies using this method found that it was (a) giving higher safety integrity level targets than really needed and (b) was less flexible than other methods and consequently more difficult to use in many real situations. This has resulted in a general move to other methods and in particular to the use of Layer of Protection Analysis. Indeed, LOPA may now be the predominant "Method of Choice" for SIL Determination. As a method, LOPA gives the impression of simplicity and appears quite straightforward. It almost invites anyone with a minimum of understanding of the method to be a practitioner.

However, there are a number of shortcomings in the use of the LOPA method that have become evident through its use across the globe.

This paper will review the application of LOPA and provide insights into what may be seen as the proper use of the method and its core principles. It will also highlight various shortcomings in use that have been found during expert reviews carried out of a number of existing studies. The paper will be of benefit both to those who are new to the LOPA method, but also of value to those who have been using the method for some time.

Shale Oil and Gas – Wellhead to Fuel

Dr. Lawrence Kremer - Baker Hughes
AIChE - Hilton New Orleans Riverside
April 1, 2014

14AIChE Spring Meeting, New Orleans, LA

BAKER HUGHES

Outline Unconventional Shale Oil

- Properties of shale oils
- Problems
  - Transportation
  - Refining

14AIChE Spring Meeting, New Orleans, LA

BAKER HUGHES

General Characteristics of Shale Oils

- API gravity typically varies 30-55+
- Very paraffinic
- Low sulfur
- Can have hydrogen sulfide and mercaptans
- Low TAN
- Very low asphaltenes
- Light crude with very little residual material
- Many samples show great quality variation

Factor	Bakken	WCS
API	39.7	21.6
S%	0.25	3.44
Light ends	2.6%	
LSR	3.5%	3%
Light Nap	12.4	3.0
Heavy Nap	9.7	5.0
Jet	10.5	10.0
LFO	27.6	12.0
Gasoil	27.2	37.0
VacTar	6.7	29.8

14AIChE
Spring Meeting, New Orleans, LA

BAKER HUGHES

Shale Oil Property Comparison

Parameter	Eagle Ford	Bakken
API	52	40.8
TAN (g KOH/g)	< 0.05	0.09
Sulfur (wt %)	<0.2	0.304
Asphaltene (wt %)	0.1	0.41
Resin (wt %)	1.6	4.95
Filterable Solids (PTB)	225	76

14AIChE
Spring Meeting, New Orleans, LA

BAKER HUGHES

Utica Shale Crude - Three Phase

14 AIChE Spring Meeting, New Orleans, LA

BAKER HUGHES

Eagle Ford Crude Samples

14 AIChE Spring Meeting, New Orleans, LA

BAKER HUGHES

Analytical Breakdown

Value	Yellow	Red	Black
API	55.0	44.6	52.3
TAN	<0.05	0.07	<0.05
Sulfur	<0.20%	<0.2%	<0.2%
Na	1.0 ppm	1.6 ppm	1.6 ppm
K	0.3 ppm	0.4 ppm	0.5 ppm
Mg	3.4 ppm	2.9 ppm	3.0 ppm
Ca	2.6 ppm	2.8 ppm	3.8 ppm
Asphaltene	0%	0%	0.1%
Resin	0.5%	3.2%	1.6%
Filterable solids	176 PTB	295 PTB	225 PTB
P	5.3 ppm	4.6 ppm	5.0 ppm

14 AIChE
Spring Meeting, New Orleans, LA

BAKER HUGHES

Micrograph of Sludge

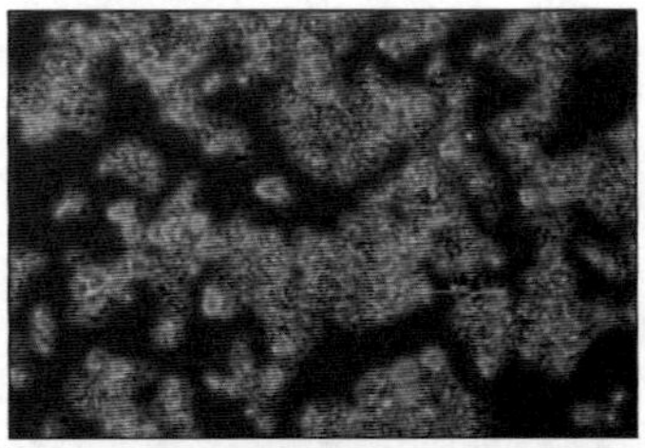

100 X

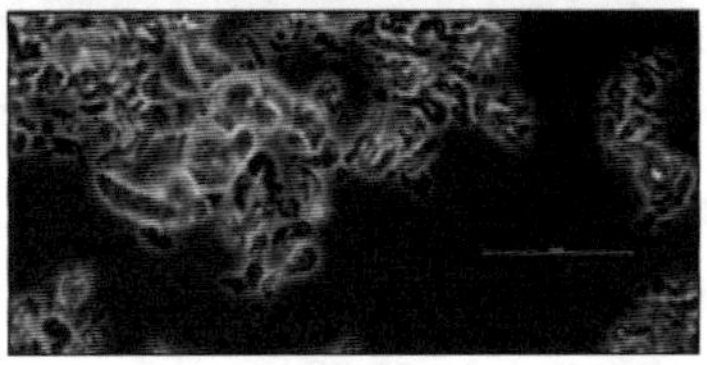

400 X

14 AIChE
Spring Meeting, New Orleans, LA

BAKER HUGHES

Paraffin Analysis of Sample

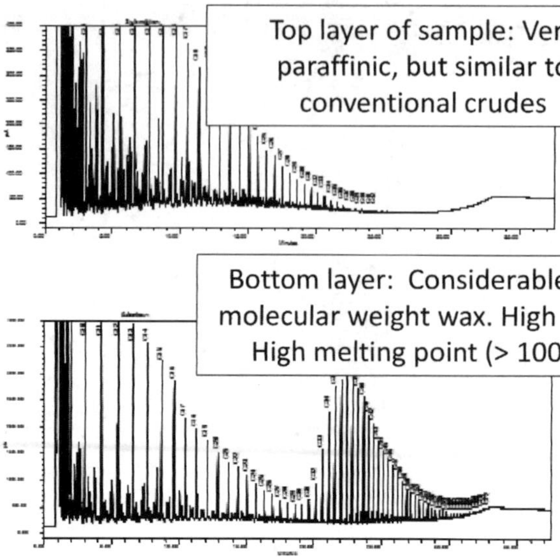

Top layer of sample: Very paraffinic, but similar to conventional crudes

Bottom layer: Considerable high molecular weight wax. High MW = High melting point (> 100 °C)

14AIChE Spring Meeting, New Orleans, LA

BAKER HUGHES

Transport of Shale Production

- Infrastructure does not exist to meet needs
  - Most shale oil moved by rail or trucks
  - Limited pipelines expanding to meet transportation needs
- Unit trains to East Coast, West Coast, and Gulf Coast
- Problems:
  - Tankers and barges with heavy wax deposits ROB
  - Tight emulsions that look like oil or white condensate but high BS&W
  - Cars not uniformly treated for H_2S
  - Corrosion in rail cars

14AIChE Spring Meeting, New Orleans, LA

BAKER HUGHES

Viscosity of ROB Before Treatment

ROB from untreated compartment

14AIChE Spring Meeting, New Orleans, LA

BAKER HUGHES

Eagle Ford Shale Oil Wax Dispersion Testing

Several products were effective in dispersing the paraffin wax in an Eagle Ford shale oil sample at ambient temperature for over a month.

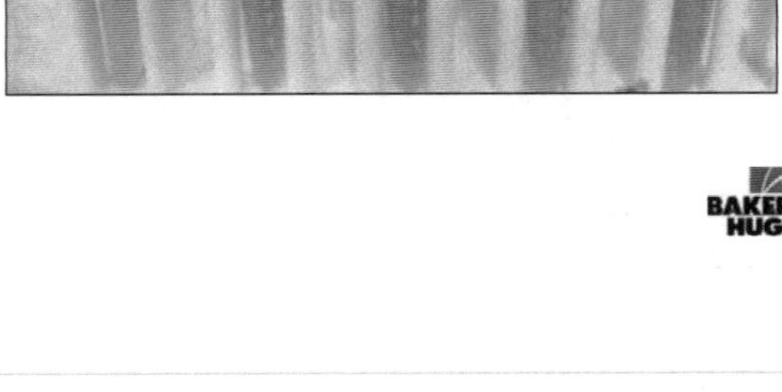

14AIChE Spring Meeting, New Orleans, LA

BAKER HUGHES

Product 'A' Effect on Wax Deposition (Utica)

Blank **200 ppm** **400 ppm**

14AIChE
Spring Meeting, New Orleans, LA

BAKER HUGHES

Refinery Issues

- Crude Blending
- Fouling
- Corrosion
- Product quality

14AIChE
Spring Meeting, New Orleans, LA

BAKER HUGHES

Crude Unit Fouling

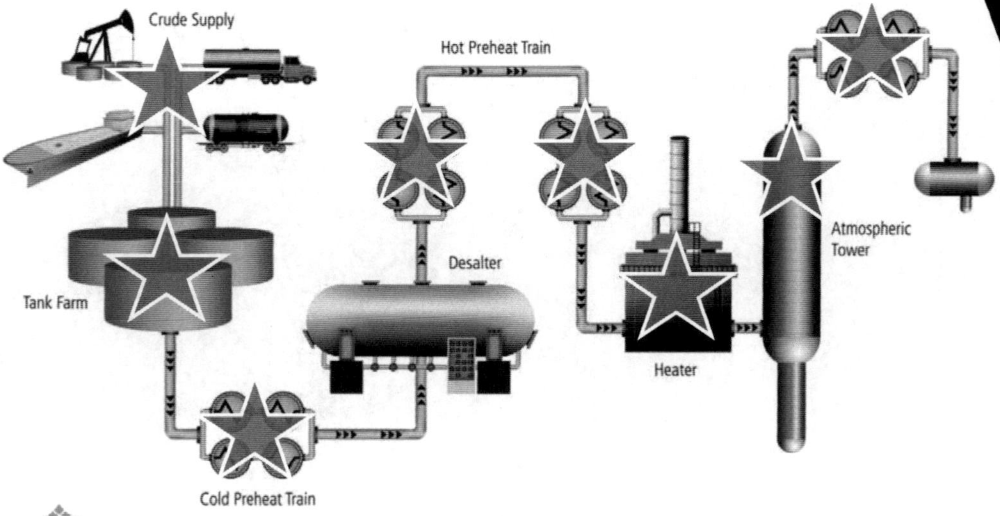

★ = Susceptible to Fouling

14AIChE Spring Meeting, New Orleans, LA

BAKER HUGHES

Asphaltene Precipitation

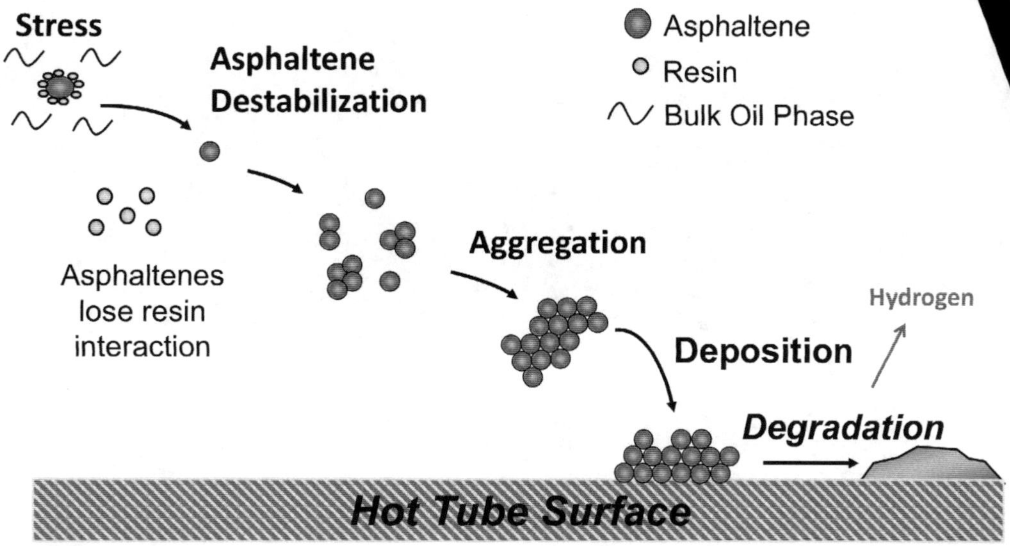

14AIChE Spring Meeting, New Orleans, LA

BAKER HUGHES

Asphaltene Stability Index Test

Field ASIT services™ technology measures the onset of the flocculation of the asphaltenes with high accuracy by inducing the asphaltene precipitation via titration with a paraffinic solvent.

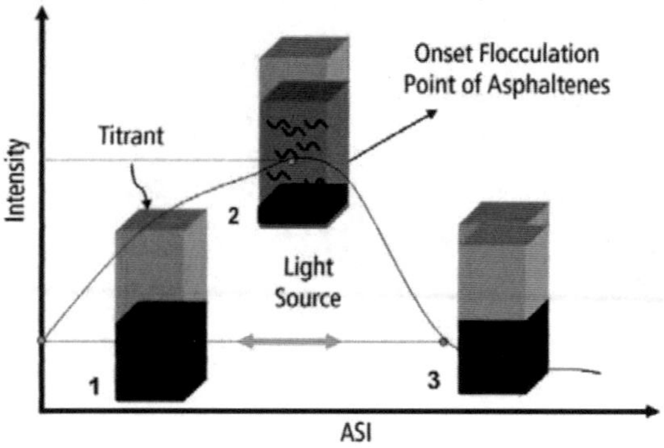

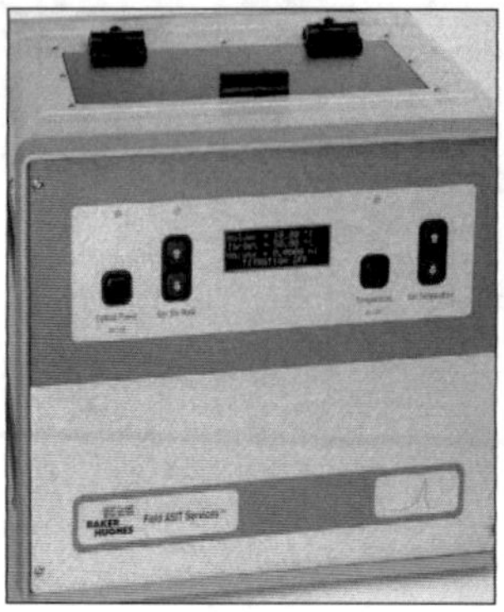

14AIChE® Spring Meeting, New Orleans, LA

Field ASIT services is a trademark of Baker Hughes Incorporated

BAKER HUGHES

ASI Shows Eagle Ford Can Destabilize Crude

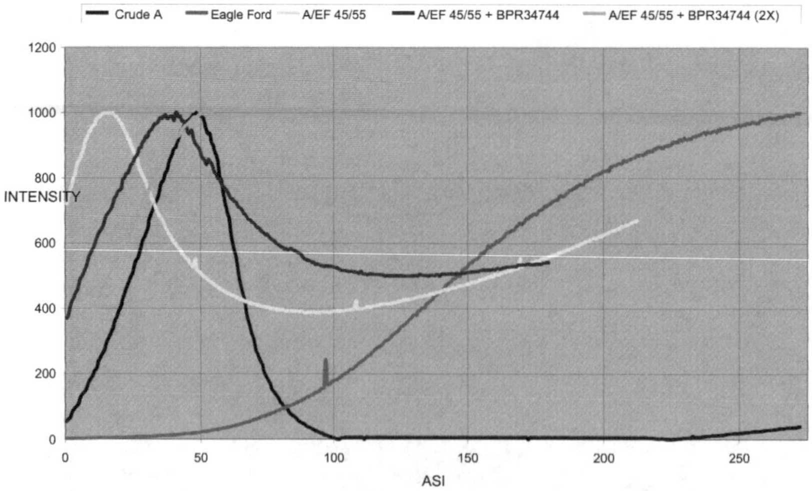

14AIChE® Spring Meeting, New Orleans, LA

BAKER HUGHES

Tower Tray Fouling - Phosphorus

- Seen by refineries running Bakken, Eagle Ford, and WTI
- Operationally manifested as increasing tower dP
  - Typically in the kero/jet zone
- Tower inspection shows deposits on trays in kero/jet zone
  - Some deposition may also show a few tray levels above and/or below kero/jet zone
  - Deposit scale typically tightly bonded to tray surface restricting flow and valve function

Above kero/jet zone

In kero/jet zone

Below kero/jet zone

14 AIChE Spring Meeting, New Orleans, LA

BAKER HUGHES

Shale Oil Impact – MEA Contamination

Corrosion and Fouling

- Shale oils transported by truck, rail, barge.
- Many shipments are treated with H_2S scavengers:

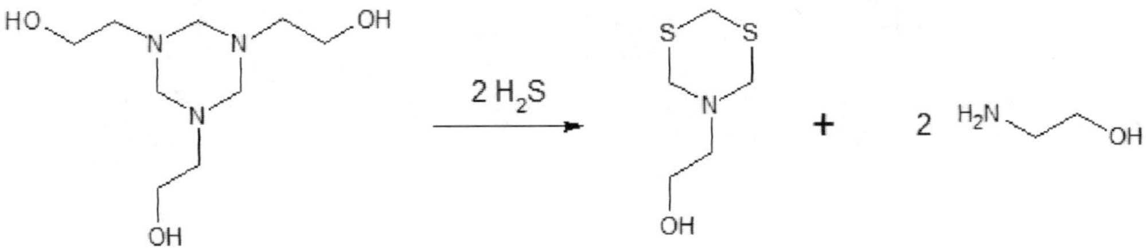

- Potential Impacts:
  - H_2S scavengers leave byproduct amine in crude oil (MEA, MA).
  - Tramp amines distill to OH system, create amine-HCl salts:
  - Corrosive to system metallurgy.

14 AIChE Spring Meeting, New Orleans, LA

BAKER HUGHES

Impact of Tramp MEA on OH Water

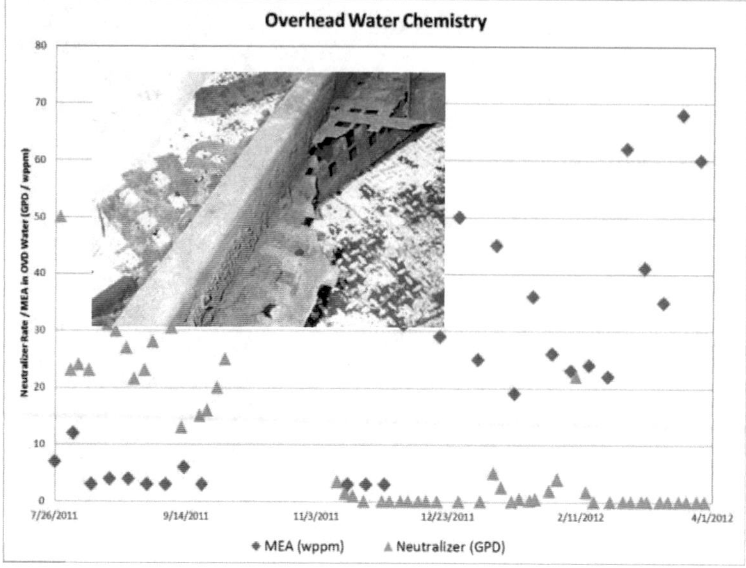

14 AIChE Spring Meeting, New Orleans, LA

BAKER HUGHES

Salt Risk w/MEA Concentration

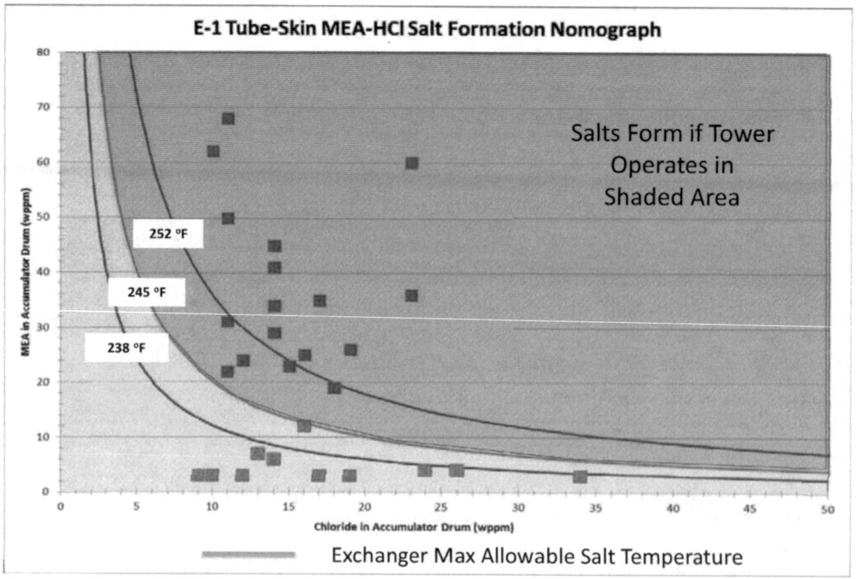

14 AIChE Spring Meeting, New Orleans, LA

BAKER HUGHES

Shale Oil – Refinery Process Concerns and Monitoring

BAKER HUGHES

CHARACTERISTIC OF SHALE OIL	CONCERN IN REFINERY	PROCESS DATA	SAMPLES AND ANALYSES Samples	Analysis	OUTPUT DATA	FREQUENCY	BHI PROGRAMS
Crystalline Wax	Tank bottom sludge	• Tank Sludge Depth	• Tank bottom	• Visual sludge • GC analysis	• Tank Sludge Depth • Crystal Wax Presence	• Monthly	• XERIC™ Pretreatment
	Cold Preheat Fouling	• Desalter Temperature			• Temp vs. Time	• Update 2x/mo.	• LIFESPAN™ Wax Dispersants
	Desalter emulsions	• Emulsion layer growth	• Triline emulsions	• Visual wax • GC analysis	• Crystal Wax Presence	• After Shale Oil begins • When emulsion grows	• XERIC Demulsifiers
Asphaltene Stability in Blends	Desalter emulsions	• Emulsion layer growth	• Triline emulsions	• Precipitated Asphaltenes	• Precipitated Asphaltenes	• After Shale Oil beings • When emulsion grows	• XERIC Asphaltene Stabilizers • Field ASIT services™
	ColdPreheat Fouling	• Desalter temperature	• Crude blend	• ASIT	• Temp vs. Time • ASIT #	• Pre-Shale Oil • Update 2x/mo.	• LIFESPAN Antifoulants • Field ASIT services
	Hot Preheat Fouling	• Furnace inlet temp • Furnace tube temps • Furnace fuel usage • Preheat DP	• Crude blend	• ASIT	• Temp vs. Time • Fuel use vs. Time • ASIT #	• Pre-Shale Oil • Update 2x/mo.	• LIFESPAN Antifoulants • MILESTONE™ Furnace Fouling Control • Field ASIT services
Paraffinic	Furnace Coking	• Furnace tube temps • Furnace fuel usage	• Crude blend	• ULPS	• Coking tendency	• Pre-Shale Oil • W/Shale Oil	• MILESTONE Furnace Fouling Control
Tramp MEA	Tower/OH corrosion (Early Warning)	• OH Neutralizer rate • EXCAL usage rate	• OH water • Desalter brine	• OH pH • Brine pH	• Early Sign of Amine • Early sign of Amine	• Pre-Shale Oil • W/Shale Oil	• TOPGUARD™ Overhead Corrosion Control • EXCALIBUR™ Contaminant Removal Technology
	Tower/OH corrosion (Quantitative)	• Inputs for CRM	• OH water • Desalter ww • Desalter brine	• IC Amines	• Confirm MEA in Crude • OH MEA ppm • Salt Risk Tower • Salt Risk OH System	• Pre-Shale Oil • W/Shale Oil • Update 2x/mo.	• TOPGUARD Corrosion Control • EXCALIBUR Technology • TOPGUARD Corrosion Risk Monitor (CRM)
Phosphorus	Tower P Fouling	• Tower DP			• Tower DP Increase	• 2x/mo.	• LIFESPAN Antifoulants Dispersants
			• Crude	• ICP for P	• P in Crude	• 2x/mo.	
			• Kerosene	• Filt Solids • XRF Filter P • ICP for P	• Incr. Kero Solids • P in Solids • P in Kerosene	• 2x/mo.	
			• Kerosene • Diesel	• ASTM D86	• Incr. in Diesel IPB – Kero EP overlap	• 2x/mo.	

Technology Validation Process for Offshore Fields: A Winding Road

Ankur Jariwala

Senior Product Manager – Gas Technology
Cameron Process Systems, Houston

Gary W. Sams

Managing Director R&D
Cameron Process Systems, Houston

The technology qualification process is usually a winding road to commercialization. In this paper we will discuss Cameron's experience in working with companies to qualify gas technology for their offshore and onshore commercial operations. In last 30 years, we have commercialized over 18+ gas technologies and have gone through significant internal and external technology qualification steps starting from concept validation, bench scale testing, commercial approach validation, pilot testing, engineering, design and installation. The process can vary significantly from onshore to offshore. We have a wide range of technology validation experiences with technologies targeted for offshore, mainly Cynara CO_2 Membranes, MEG Reclamation and Regeneration technology, electrostatic oil dehydrators and high efficiency phase separation internals. Although the external technology validation process may vary, there are several common aspects to this process. Common aspects are operational history in a similar environment, mean time between failures, cause and effect chart, mechanical and structural integrity, process variations, utility consumption, ongoing research and development and aftermarket support.

The technology validation process also requires integrating different Original Equipment Manufacturer (OEM) and associated technology validations to align with parent technology. This is also important to identify, define and incorporate into an overall technology validation strategy for the technology success. In this presentation, we will also cover several challenges and hurdles we faced with OEMs and their associated technology and how we overcame those risks via a technology validation process.

Technology evolution is a process where technology runs through a maturity phase via prototype, bench scale, pilot unit and a semi or a full commercial unit operation. Each of these steps provides valuable data to main commercial unit and there by minimizes the overall risk from laboratory model to commercial scale unit. Especially Oil and Gas patch, where technology has to be robust due to potential safety and failure risks which can drive significant revenue loss. Key measures are put in place to validate the technology before applying to the main field.

For offshore oil and gas production, the technology validation process is more rigorous compared to onshore applications. Cameron has developed several oil and gas processing technologies for the offshore which has gone through technology validation process. The four main technology validation process experiences listed here shows different sets of processes than any standardized procedure. However, they all have one thing in common, long term offshore testing to analyze technology uptime and its operational success under a tough upstream environment.

Technology ideation starts with a problem statement or a pain point. The problem statement creates multiple options or ideas for the solution. Ideas can be easy to implement when it's a proven technology in the respective field. The available technology can be in the parallel field and may be required to go through a qualification process for the required field. The technology qualification process is identified as technical as well as commercial. The technical qualification can run through different stage gate process which includes an assessment of the Technology Readiness Level (TRL). The TRL levels usually run from 1-9 and depending upon how the technology is proven and tested, it's ranked accordingly. Most of the times, companies/clients would like to investigate technology once it has reached to level 3-4. Again, it may vary depending upon how difficult the problem it's going to solve and what is the biggest impact the technology can have, even if it is at TRL level 1 or 2.

The commercial qualification is usually evaluated once the technology has reached TRL level 4-5. The commercial validation process really focuses on Return on Investment (ROI), technology risk, potential market penetration versus cost and time to market. Each of these factors is weighted for the given technology to get a combined commercial qualification score.

The process of technology qualification can be application specific, which can be completely developed and proven by the technology owner. However, the technology owner may choose to take the technology to where the industry or end users have enough data and comfort level to accept technology for the required problem statement. In some cases, end users have more interest and may take the full responsibility of the technology validation or testing of the technology for the required application.

Technology qualification can also be done through a third party. For example, a technology owner can choose to use innovative technology from the third party as a part of their main technology. This process is also more like Original Equipment Manufacturer (OEM) where the component of the main technology is unique and may require the same qualification process as main technology. In this case, technology owner can work with third party or OEM provider to run in a same TRL or stage gate process which is parallel to main technology qualification.

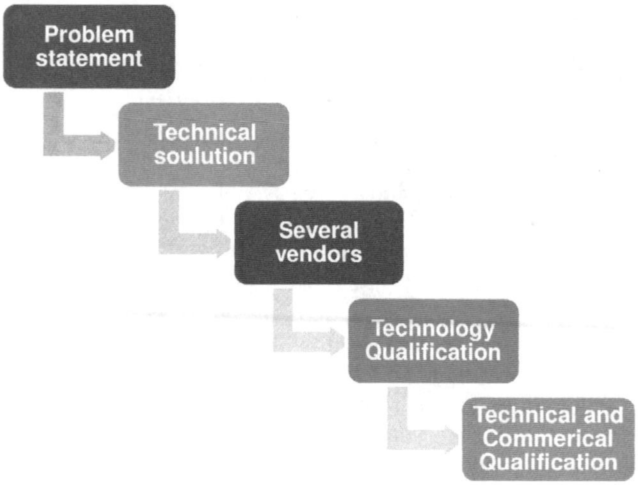

Figure 1.0 – Technology Qualification Process and Various steps

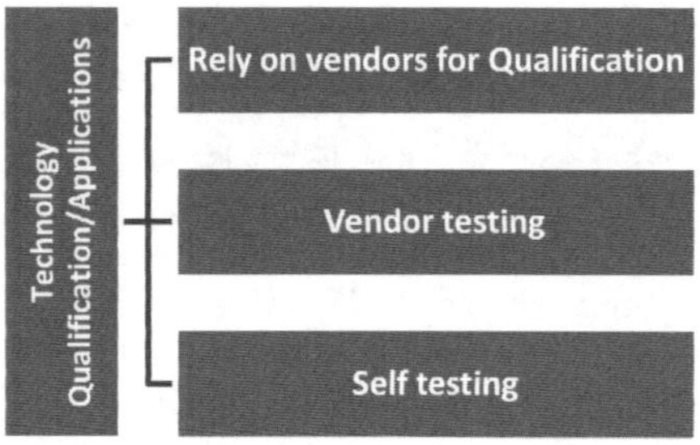

Figure 2.0 – Technology application qualification options

The four cases below expand on the Technology Qualification methods used by Cameron to qualify technology for a range of oil and gas processing applications.

1) Cynara Membrane System – Technology validation process

Problem:

Membrane for separating CO_2 from natural gas is a technology which is very well proven for onshore gas processing. For the offshore, the system footprint has to be small and should be able to address the change in inlet feed gas composition in the dynamic manner. The onshore CO_2 removal using membrane systems have limited control system in past and cannot go full automation which is required by the remote offshore gas fields. For the offshore fields there were several challenges to address like mean time between the failures, higher uptime reliability, membrane change out, sparing philosophy and turn down capability.

Most of the offshore units are designed with the highest design capacity as per life of the field to plan overall space and weight on the platforms or Floating Production, Storage and Offloading (FPSO) vessels. During the first year, the production may not see the full capacity and therefore need a design to address that turndown.

Cameron's recent projects on FPSO require compact footprint and higher automation to minimize operator attention. These challenges require different problem solving approach.

Technology Solutions:

Cameron has proven membrane technology, Cynara membranes, with 28 years for operational experience onshore. Cameron's extensive experience in operating some challenging Enhance Oil Recovery (EOR) gas fields with high CO_2 at medium to high pressure on an ongoing basis provides significant operational data for hollow fiber membranes. Cameron's Cynara membranes use Cellulose Tri Acetate (CTA) polymer based hollow fiber membrane which are well proven. For small footprint requirements offshore, Cameron never changed their core fiber and stuck to CTA material based membrane while improving thee overall packaging method to reduce the overall size and weight of the skid.

Cameron started with 5" diameter membrane module using the same CTA membrane fibers and developed a 16" membrane diameter element which can replace 18 of the 5" membrane elements. As the offshore fields are getting larger with higher CO_2, the approach is to reduce this footprint even further by developing a 30" diameter membrane module. Each 30" membrane module can replace 4 of the 16" diameter membrane elements, significantly reducing space, weight and system cost.

Figure 3.0 – Different Cynara Membrane Module Sizes.

The 30" diameter membrane provides significant footprint reduction with higher uptime.

Technology Validation Process:

The 30" diameter membrane element is about 72" tall and has more capacity to withstand a higher pressure with higher hydrocarbon retention rate in the product gas.

These membrane modules were not tested offshore and our client was a bit concerned over the performance of these modules, because they were going to put them in a main service where they play critical role in their revenue chain. The first approach is to test this in a real process pilot unit at SACROC; Cameron's operated membrane based CO_2 removal facility in West Texas. Cameron developed the data set range for the client who was interested in testing this offshore at their facility. The process data for the 30" membrane element showed predictable performance under desired conditions for 3 years of testing at SACROC. After reviewing these data, the client agreed to work with Cameron in testing this 30" membrane module at their offshore facility. They started the offshore test and ran for another 3 years to understand its performance under variable feed gas conditions.

During this period, Cameron had developed and improved their control system to minimize any downtime and developed better sparing philosophy by working with the client.

After 3 years of testing offshore, the client qualified this technology for their further projects for onshore, as well as offshore.

2) MEG Reclamation and Regeneration Technology

Problem:

The technology was very well qualified for the given field, however the client's application was unique and required testing and qualification for the major capital investment decision.

One of the Cameron clients approached them regarding their existing problem at their gas terminal in UK. The facility was developed long time ago and it is currently processing gas more than their original design conditions. This was a major revamp which the customer was looking to do in order for them to operate this facility efficiently.

In their facility, monoethylene glycol is injected into the gas production pipelines to prevent formation of gas hydrates in the raw gas at pipeline conditions (high pressure and low temperature). The facility includes MEG Regenerators to remove the produced water from the incoming Wet MEG feed (50 wt% MEG) to yield Dry MEG by removing water.

Technology Solution:

A Desalination Package was installed during the earlier phase of production to remove salt from the Dry or Wet MEG. Both monovalent salts (predominantly sodium chloride) and divalent salts (calcium chloride) were present in the feed. The divalent salts are removed as insoluble carbonates in a Pre-Treatment Package and the monovalent salts are precipitated from the Dry/Wet MEG in the Desalination Package using flash vaporisation.

The Pre-Treatment and Desalination Packages were to be replaced as part of the Rejuvenation Project. Client's production projection for that gas fields indicated increased water and salt production rates.

Technology Validation:

In 2010, the Client engaged Cameron to carry out a FEED study for the Desalination Package with a new feed design basis.

As part of the FEED study, pilot plant trials were carried out at Cameron's PureMEG™ pilot plant at the University of Manchester in 2012 using Dry MEG supplied by Client from the existing Desalination Package from their field. The pilot plant trials were carried out to demonstrate Cameron's proprietary salt clean-up technology to Client's process engineering, projects and operations personnel. In Cameron's PureMEG™ process, MEG is washed from the salt crystals as they settle under gravity through a column of saturated sodium chloride brine. The result is reduced MEG losses with the salt waste stream. The PureMEG™ Pilot Plant comprises a 600mm ID x 1.6m high flash separator above a 6 m high, 4" diameter column fabricated from borosilicate glass which allows observation of the salt crystals falling through the brine phase.

Figure 4.0 – Cameron's PureMEG Pilot Unit at University of Manchester, UK

The pilot plant trials were also used to demonstrate (1) first fill of the system with MEG, (2) control of the brine column during salt processing and (3) the robustness of Cameron's flash separator design to the presence of Flow Assurance Chemicals (particularly corrosion inhibitors) which cause foaming problems on the existing unit.

The pilot plant trials were carried out over a period of 5 days and were witnessed by the client.

In addition to the Flash Separator trials at the University of Manchester, trials were carried out with filter press and centrifuge vendors to support the Client's Rejuvenation Project FEED study.

After several testing and trial runs, the Client was comfortable with the technology and awarded the contract for the design of the Desalination Package of the Rejuvenation Project to Cameron Construction / fabrication and commissioning is due to begin in 2014.

Again, the technology was very well proven; however, it still does not meet the required criteria from the Client unless they test this technology and get the detailed performance data. The approach is to really minimize the risk with the technology by investing more time upfront by qualifying any technology at their cost.

3) Dual Frequency Electrostatic Dehydrators

Problem:

When applying electrostatic processes to a crude oil for the purpose of oil dehydration, it became apparent that the electrical conductivity was compromising the process performance. The problem statement posed the question of determining a method that would permit a high voltage electrostatic field to be sustained even in the presence of a highly conductive crude oil. The existing technology was being compromised by the excessive current demand on the power supply and subsequent loss in voltage. Clients were experiencing poor dehydration due to the reduce voltage and loss of dehydration efficiency.

Technology Solution:

The technology solution was found by investigating the use of medium frequencies that could successfully sustain the voltage field. The application of a medium frequency power unit proved to be a valid method to achieve the desired dehydration from highly conductive oils. During the development of this solution it was discovered the conductivity resulted primarily from the blending of two oils that chemically reacted causing the resulting conductivity to spike.

The function and features of the medium frequency power supply were clearly defined to facilitate the work scope.

Technology Validation:

Using the function and features documents, the power unit was designed and fabricated for pilot testing on crude oil samples supplied by potential end users. Numerous oils were tested using the pilot power unit to demonstrate the benefit of this new technology and to validate and confirm the functionality. Following several months of pilot trails and a few design changes to the power unit a commercial power unit was design and built. The commercial unit was tested on a pilot plant before going for a field trial.

Figure 5.0 – Electrostatic treater in the field

Several oils were tested and a few features were altered in the power unit before looking for a field site.

The first commercial qualification by a customer was conducted at a large production facility operating five oil dehydrators in parallel. One dehydrator was converted to the new power unit and the field trials were started. Obviously, in a commercial facility the ability to alter performance parameters is typically limited to reduce the risk of loss production. For these trials, the oil capacity was the primary test variable. The rate was increased in several steps and the dehydration performance of the equipment was recorded. Once the rate tests were completed, the unit was permitted to operate without intervention to permit the equipment to handle the daily upsets experienced by the process. After 30 days of operation, the power unit was removed and shipped to the manufacturer for examination. These field trials resulted in a few final alterations to the design.

These field trials provided the needed testing to satisfy a TRL of 5 for this technology making it readily acceptable by many operators.

4) High Efficiency Separator Internals:

Problem:

Oil, gas, produced water and several other phases are separated in a separator via gravity separation. These phase separators are sometimes not efficient enough to meet the separation requirements due to inadequate distribution of phases inside the separator or due to a change in the inlet flow conditions.

Technology Solution:

High efficiency Separation internals are used in phase separation/separators where oil/gas/produced water and solids are separated to the highest efficiency and thereby protecting downstream equipment from any type of two phase handling. These internals are developed to either retrofit an existing separator to increase the overall throughput while maintaining efficiency, or they are installed in a new separator to meet highest level of separation performance.

The technology was initially developed as per the required standards from offshore projects. Cameron developed these cyclonic separator internal designs and developed some prototypes in the beginning based on the concept and computational fluid dynamics study.

For the offshore operational environment, these CFD analyses are not sufficient enough to qualify the technology. Our client, who is a highly technology oriented company, requires certain testing protocol before they can qualify these internals for their commercial separators offshore.

Technology Validation:

The protocol called for an extensive CFD analysis with various flow conditions and required certain changes in the physical structure of these internals. After rigorous analysis and satisfactory preliminary results, these internals were tested on a bench scale unit in-house to understand the basic performance in an air/water separation regime. Some limitations on the bench scale were the test pressure and actual fluid conditions which are different from oil and gas field. After successfully testing and understanding the operational envelope for these internals, the internals were tested at very high pressure and actual field simulated conditions at the client's laboratory. These tests provided complete real time performance and reduced the overall performance risk on new technology. This also provides technology limitations and operational efficiency envelope under different process flow conditions.

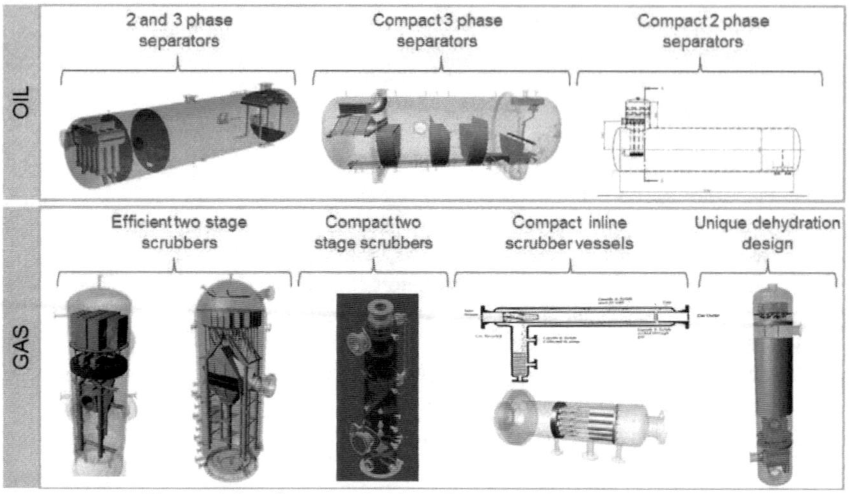

Figure 6.0 – Cameron's High Efficiency Separator Internals Suite

Cameron has successfully developed 5 different types of internals which are targeted for different gas loading factors. All of these internals were tested and approved by the Client. One of the internal designs was patented by the client and Cameron was exclusively authorized for licensing. The Client has full access to the internals for their existing or future projects.

By working with a client on technology development and qualification process, the technology accessibility becomes much easier for their key technical team and it also reduces the overall technology risk for the client.

Conclusion:

As offshore oil and gas production fields are getting more complex, they require more sophisticated technologies to produce cost effectively. The technology validation process is a key aspect to address safety and reliability. The technology validation process may vary significantly depending on the type of the application. The overall process requires close cooperation between technology user and vendor. It also requires strong research and development capability to validate technology from laboratory scale to fully commercial technology.

Cameron has developed extensive and successful experience in validating new technology for large complex offshore fields with significant research and development capability in house.

Technology Qualification – Two Subsea Applications

Martha Viteri – DNV-GL
Austen Tyler – DNV-GL

14 AIChE Spring Meeting, New Orleans, LA

DNV·GL

Technology Qualification

- Qualification is the process of providing the **evidence** that technology will **function within specified limits** with an acceptable level of **confidence**.

14 AIChE Spring Meeting, New Orleans, LA

DNV·GL

Why Consider Technology Qualification?

- Lack of evidence in the technology
- Internal or customer requirements
- Technology selection
- Expert evaluation
- Market competitive advantage

14 AIChE
Spring Meeting, New Orleans, LA

DNV·GL

Technology Qualification Process

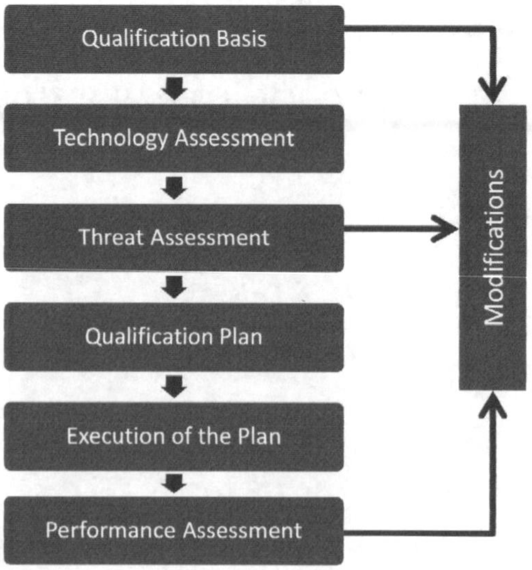

14 AIChE
Spring Meeting, New Orleans, LA

DNV·GL

Qualification Basis

- Common set of criteria to assess qualification activities
- Describe the following:
  - Intended use
  - Environment
  - Required functions
  - Acceptance criteria
  - Performance expectations

14 AIChE
Spring Meeting, New Orleans, LA

DNV·GL

Technology Qualification Process

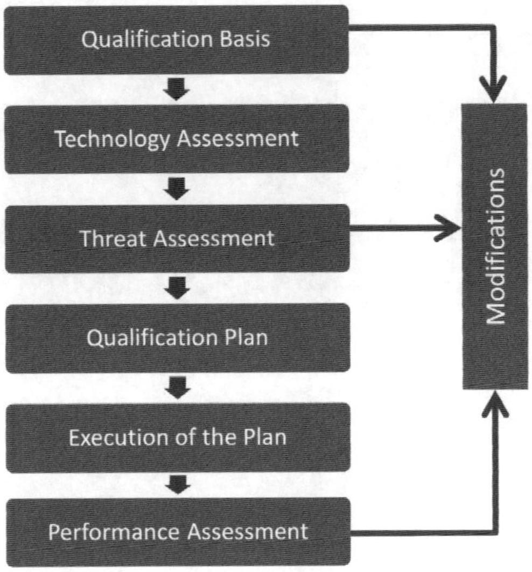

14 AIChE
Spring Meeting, New Orleans, LA

DNV·GL

Technology Assessment

- Identify which elements require technology qualification
- Identify key challenges and uncertainties.

14 AIChE
Spring Meeting, New Orleans, LA

DNV·GL

Technology Assessment

- Consists of the following steps:
  - Breakdown the technology
  - Assess technology novelty
  - Identify key challenges and uncertainties

14 AIChE
Spring Meeting, New Orleans, LA

DNV·GL

Technology Assessment

Application Area	Novelty of the Technology		
	Proven	Limited Field History	New or Unproven
Known	1	2	3
Limited Knowledge	2	3	4
New	3	4	4

- The classification categories:
  1. No new technical uncertainties
  2. New technical uncertainties
  3. New technical challenges
  4. Demanding new technical challenges

14 AIChE
Spring Meeting, New Orleans, LA

DNV·GL

Technology Assessment

- Statement of Feasibility:
  - Document issued by DNV-GL
  - Affirms technology feasibility
  - Based on DNV-RP-A203

DNV

DET NORSKE VERITAS

STATEMENT OF FEASIBILITY

<Technology Name>

14 AIChE
Spring Meeting, New Orleans, LA

DNV·GL

Technology Qualification Process

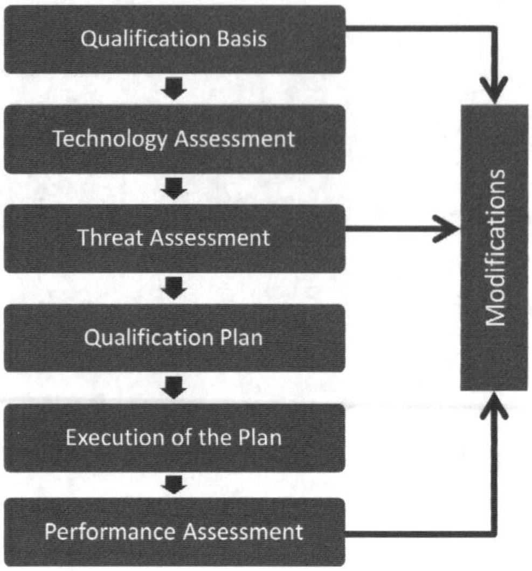

14 AIChE® Spring Meeting, New Orleans, LA

DNV·GL

Threat Assessment

- Identify failure modes / mechanisms
- Assess the associated risk
- Most common methodology chosen is FMECA

Failure Mode	Failure Mechanism	Failure Effect	Failure Detection	Safeguards	Risk Ranking
Mode 1	Mechanism 1	Effect 1			**High**
	Mechanism 2	Effect 1			**Low**
Mode 2	…				

14 AIChE® Spring Meeting, New Orleans, LA

DNV·GL

Threat Assessment

- Risk ranking performed using a risk matrix

		Increasing Likelihood				
		1	2	3	4	5
Increasing Severity	1	L	L	L	M	M
	2	L	L	M	M	M
	3	L	M	M	M	H
	4	M	M	M	H	H
	5	M	M	H	H	H

14 AIChE
Spring Meeting, New Orleans, LA

DNV·GL

Technology Qualification Process

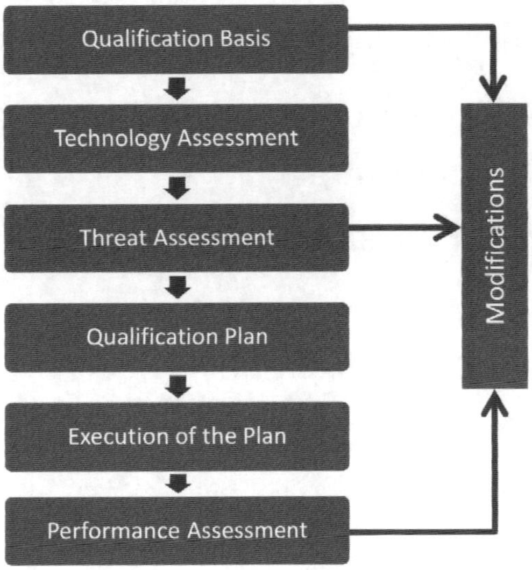

14 AIChE
Spring Meeting, New Orleans, LA

DNV·GL

Qualification Plan

- Evidence needed to manage the critical failure modes
- Describe the following:
  - Evidence linked to failure mode
  - Qualification activity linked to Qualification Basis
  - Success criteria

14 AIChE
Spring Meeting, New Orleans, LA

DNV·GL

Qualification Plan

- Potential qualification methods include:
  - Engineer judgment
  - Analytical methods
  - Numerical methods
  - Testing
  - Procedure development

14 AIChE
Spring Meeting, New Orleans, LA

DNV·GL

Qualification Plan

- Endorsement of Qualification Plan:
  - Document issued by DNV-GL
  - Affirms ability to qualify technology through the planned activities

DET NORSKE VERITAS

ENDORSEMENT OF QUALIFICATION PLAN

<Technology Name>

14 AIChE Spring Meeting, New Orleans, LA

DNV·GL

Technology Qualification Process

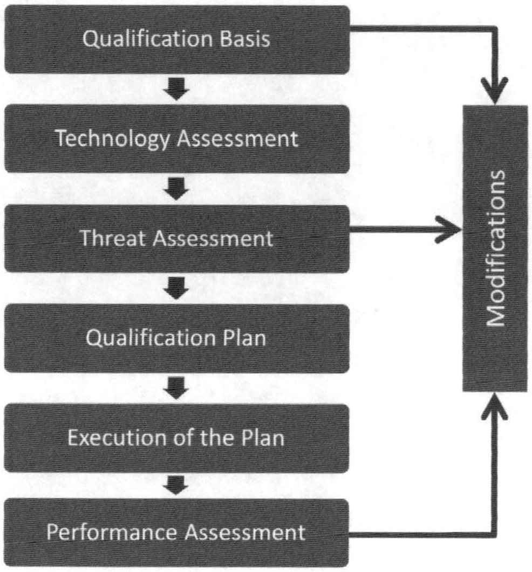

14 AIChE Spring Meeting, New Orleans, LA

DNV·GL

Qualification Plan Execution

- Involves the following steps:
  - Execution of qualification plan
  - Collecting and documenting data generated
  - Ensuring traceability of the captured information
  - Determining the performance margin

14 AIChE® Spring Meeting, New Orleans, LA

DNV·GL

Technology Qualification Process

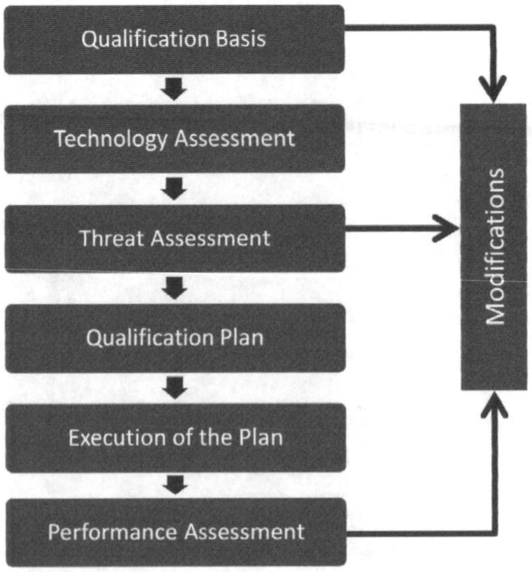

14 AIChE® Spring Meeting, New Orleans, LA

DNV·GL

Performance Assessment

- Confirmation that requirement have been met
- Risk and uncertainty reduced to acceptable levels

Parameter	Acceptance Criteria		Failure ID	Qualification Activity	Margin	
	Min	Max			Min	Max
Number of Cycles	100		2.5	Five samples will be cycled until failure	50	
...						

14 AIChE
Spring Meeting, New Orleans, LA

DNV·GL

Performance Assessment

- Technology Certificate:
  - Document issued by DNV-GL
  - Based on Qualification Basis and relevant standards
  - Affirms qualification according to DNV-RP-A203

14 AIChE
Spring Meeting, New Orleans, LA

DNV·GL

Qualification Progress by Development Phase

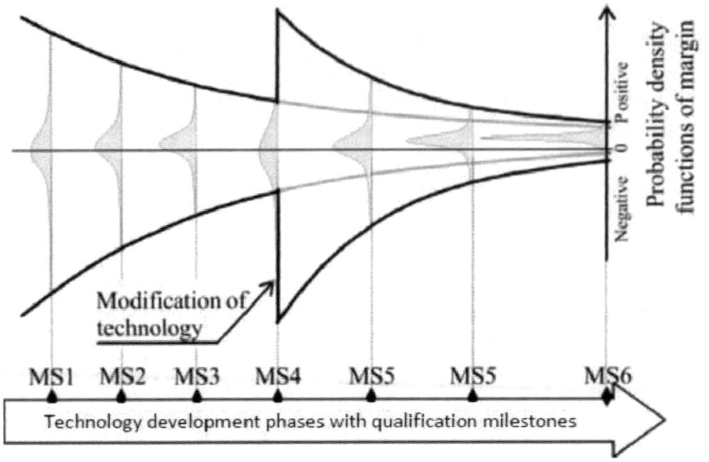

14 AIChE Spring Meeting, New Orleans, LA

DNV·GL

Enhanced Thermal Performance Reeled Pipe-in-Pipe System

- Challenge:
  - Maintain the appropriate product temperature
  - Avoid the formation of hydrates or waxing

14 AIChE Spring Meeting, New Orleans, LA

DNV·GL

Enhanced Thermal Performance Reeled Pipe-in-Pipe System

- Developed by Subsea 7 and ITP InTerPipe
- Reeled pipe in pipe system which can include electrical trace heating
- Stepped approach to qualification

14AIChE Spring Meeting, New Orleans, LA

DNV·GL

Reelable BuBi® Lined Pipe System

- Challenge:
  - Transporting corrosive media
  - Develop lower cost pipeline systems

14AIChE Spring Meeting, New Orleans, LA

DNV·GL

Reelable BuBi® Lined Pipe System

- Developed by Subsea 7 and Butting
- Without the occurrence of any wrinkling or other degradation of the corrosion-resistant liner, via reel-lay method.

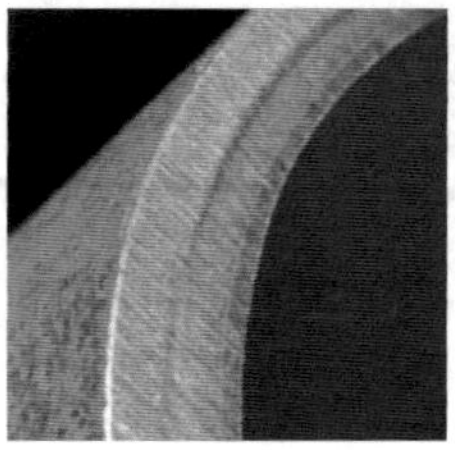

14 AIChE Spring Meeting, New Orleans, LA

DNV·GL

Outcome of the TQ Projects?

- Evidence / confidence in the technology
- Understand risk profile and failures
- Compliance with requirements
- Independent evaluation of the technology
- Market competitive advantage

14 AIChE Spring Meeting, New Orleans, LA

DNV·GL

Closing

- The authors would like to thank Subsea 7, ITP and Butting for their permission to publish
- austen.tyler@dnvgl.com
- Any Questions?

14AIChE
Spring Meeting, New Orleans, LA

DNV·GL

Improving Design of Hot oil unit in LNG plants using Dynamic Simulation

Amudha valli Narasimhan
anarasim@bechtel.com

Jaleel Valappil
jvalappi@bechtel.com

Sanjay Ganjam
sviswana@ bechtel.com

Bechtel Oil, Gas & Chemicals, Inc.
3000 Post Oak Blvd., Houston, TX

Introduction

In recent years, we see an increased use of dynamic simulation as an effective engineering tool during the engineering phase of a project in addition to its use during plant startup, shutdown, revamp and troubleshooting. This is due to increased awareness in the engineering community, especially among process engineers to use dynamic simulation for understanding the transient nature of processes. Also the capability of dynamic simulation to predict the design limitations within different operating envelops helps to validate the design. Usually different units are designed based on the maximum operating flow or a worst case temperature and so on. But they are not studied for different possible combination of these operating cases. Dynamic simulation is an appropriate tool to validate such designs. Furthermore, identification of any design change during early engineering phase can provide significant savings for the plant over long term.

System Description

The hot oil is an integrated closed loop system with heat producers like heaters and waste recovery units (WHRU) and different users all over the plant. Availability of reliable hot oil is very essential for continuous and steady operation of a LNG plant. It is the primary source of heat for column reboilers in the feed inlet facilities, amine treatment unit and heavies removal unit.

The hot oil utilizes the turbine exhaust flue gas in its waste heat recovery unit (WHRU). WHRUs extracts heat from the exhaust gas and distributes it to the process reboilers and other heat exchangers. After giving away the heat, the hot oil returns back to the WHRUs through the hot oil surge drum, for extracting heat and the cycle continues.

In this study, such an integrated heat recovery unit has been analyzed extensively for design verification, process control validation and hot oil unit robustness during various process upset scenarios. In addition, the dynamic response of the system was analyzed during a turbine trip leading to a plant turndown or a shutdown. The study also reflected the plant's integrity and transient behavior during trips and upsets. The study identified the need for an operator action required during any of these upset scenarios.

Figure 1, represents a simplified sketch of two integrated hot oil trains each with 6 WHRUs in parallel and 10 hot oil users.

Figure 1 :Hot Oil Circulation Schematics

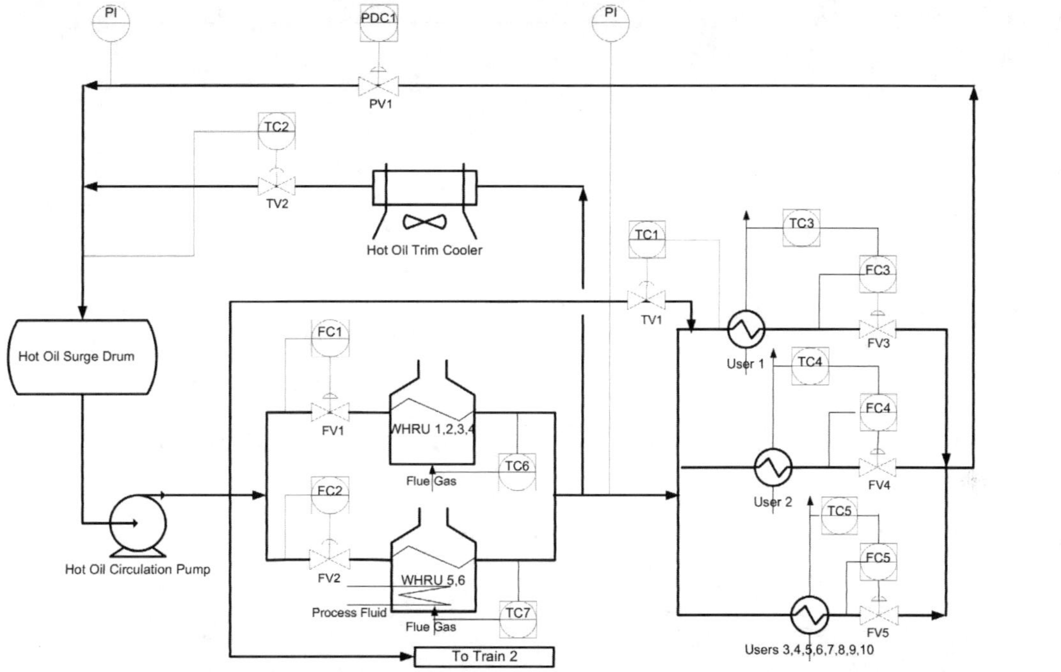

The hot oil surge drum collects and distributes the hot oil to the WHRUs and the users. The flow control valves FC1 and FC2 control the hot oil flow to the WHRUs. WHRUs 1,2,3,4 have only hot oil coils and WHRUs 5,6 shares its coil with a process fluid and hence its capacity is less than that of WHRUs 1,2,3,4. Temperature controllers, TC6 and TC7 maintain the hot oil outlet temperature by controlling the flue gas flow. User 1 is the amine reboiler and TC3 controls the amine outlet temperature.

TC1 tempers the hot oil inlet temperature to User 1. User 2 reboiler is the biggest consumer of hot oil and TC4 maintains its process outlet temperature. Users 3 to 10 are the other heat exchangers and reboilers in the LNG facility. The differential pressure controller, PDC1 maintains the differential pressure between the supply and the discharge header. The hot oil trim cooler cools down the return oil. Temperature controller, TC2 maintains the hot oil return temperature.

Dynamic model Development

A detailed dynamic model of such a complex, integrated network was developed and employed to study the plant disturbances. The model was developed in commercially available dynamic simulation software. Piping details were provided in the dynamic model with length, elevation, volume, and pipe fitting information from isometrics. The hot oil surge drum was modeled based on the vendor datasheet. The hot oil pump was modeled by providing vendor curve. The control valve characteristics and Cv's were given per vendor datasheet. WHRUs and the users namely reboilers, heat exchangers and hot oil cooler were modeled using vendor data. The controllers were configured and tuned to respond to disturbances.

Among a number of scenarios studied in the hot oil system, this article highlights two major cases that resulted in the design changes. The following Case 1 and Case 2 explain the two scenarios in detail.

Case 1: Prevent Possible Amine Degradation due to High Hot oil Return Temperature

Case 1 simulation predicted a high hot oil temperature to the amine reboiler (User1) leading to possible amine degradation.

As the hot oil circulates through a number of reboilers in different units, an upset in any one of them impacts the others. This was observed during an upset in User 2 reboiler resulting in its trip. User 2 is the biggest consumer of hot oil and when User 2 trips, TC4 action leads to FV4 going wide open.

Figure 2 shows the dynamic response of hot oil return temperature controller TC2. The reboiler trips at about 20 minutes of simulation time. Due to excess recirculation of hot oil, the return temperature of hot oil is much higher than the set point. A high return temperature, in turn increases the hot oil supply temperature to User 1. Figure 3 shows an increased hot oil supply temperature to User 1. TC1 remains wide open unable to maintain the hot oil supply temperature.

Such a high hot oil inlet temperature to the amine reboiler in User 1 overheats the amine. A prolonged operation with such high hot oil temperature could result in amine degradation. Such possible amine degradation has been avoided by increasing the size of the hot oil trim cooler valve (TV2), enabling more cooling. This resulted in a reduction in the hot oil return temperature to a value lower than amine degradation temperature. Re-sizing the hot oil trim cooler valve during the engineering phase avoided amine degradation. This saved the amine replacement cost, plant down time for maintenance without affecting plant production.

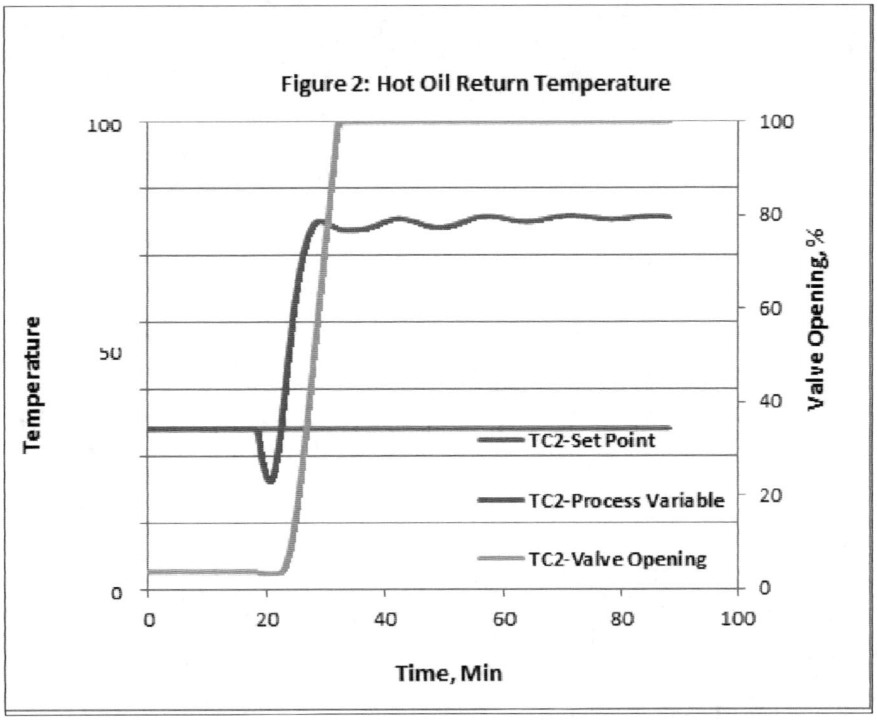

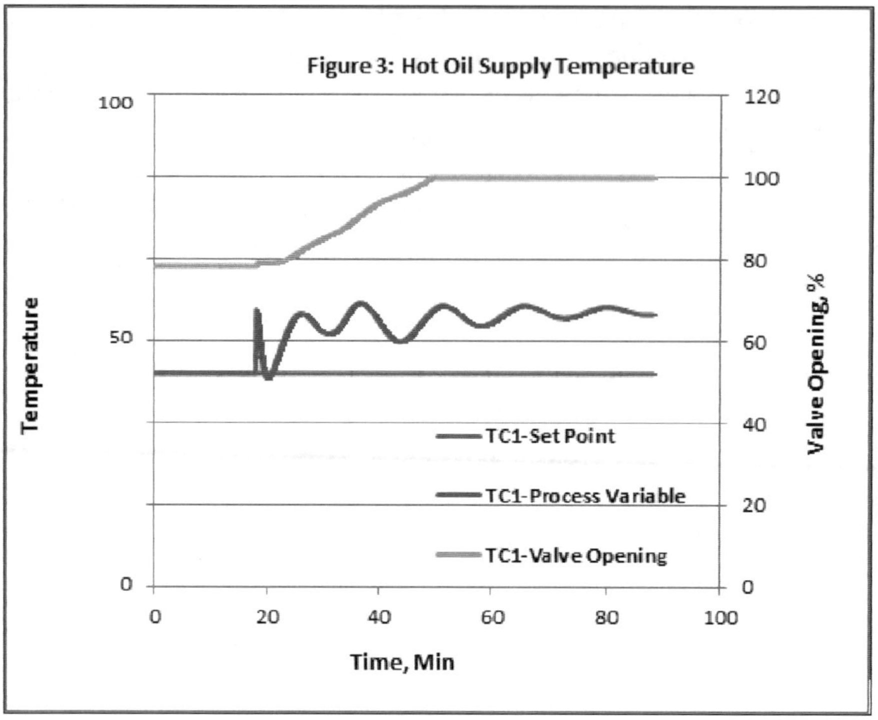

Case 2: Prevent WHRU Low Flow and its possible Trip

Case 2 simulations demonstrated a low hot oil flow in the smaller WHRU (5, 6) coils as a consequence of a major user upset, which could potentially trip the WHRUs. Such a situation occurred when one of the major users (User 3), of hot oil in the inlet facility was shutdown, cutting down hot oil flow through FV5. This reduced the overall hot oil circulation.

Figure 4 shows the dynamic response of User 3 flow controller FC5. User 3 trips at about 20 minutes of simulation time. User 3 shuts down cutting down the flow to zero. A major drop in the hot oil circulation flow causes low hot oil circulation through the WHRUs. Figure 5 shows the dynamic response of the smaller WHRU (5, 6) flow controller FC2. Unable to meet the minimum set point, the valve opens wide. The hot oil flow (process variable) drops less than its minimum value (set point).

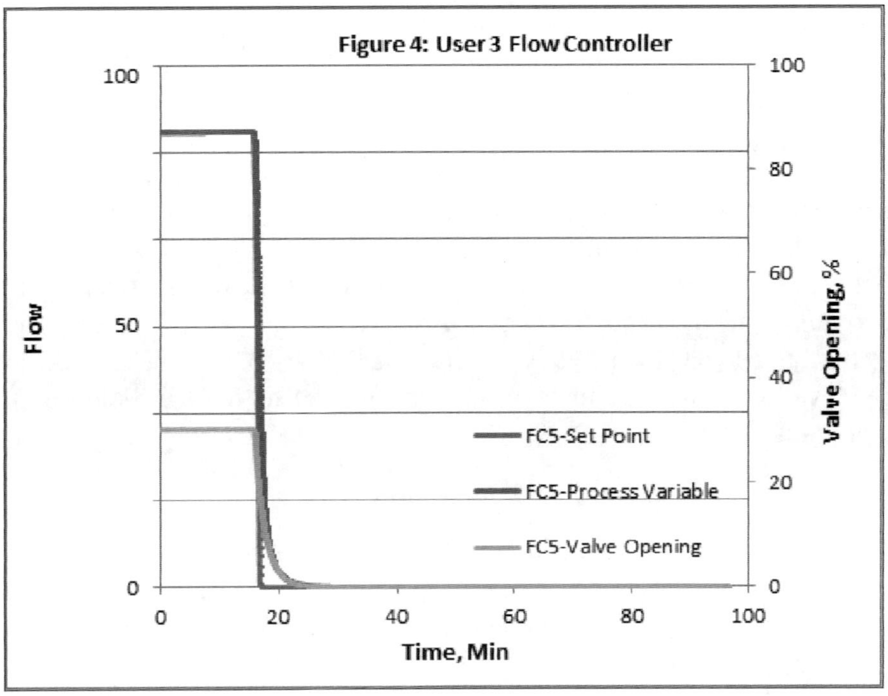
Figure 4: User 3 Flow Controller
100
50
0
Flow
100
80
60
40
20
0
Valve Opening, %
FC5-Set Point
FC5-Process Variable
FC5-Valve Opening
0
20
40
60
80
100
Time, Min

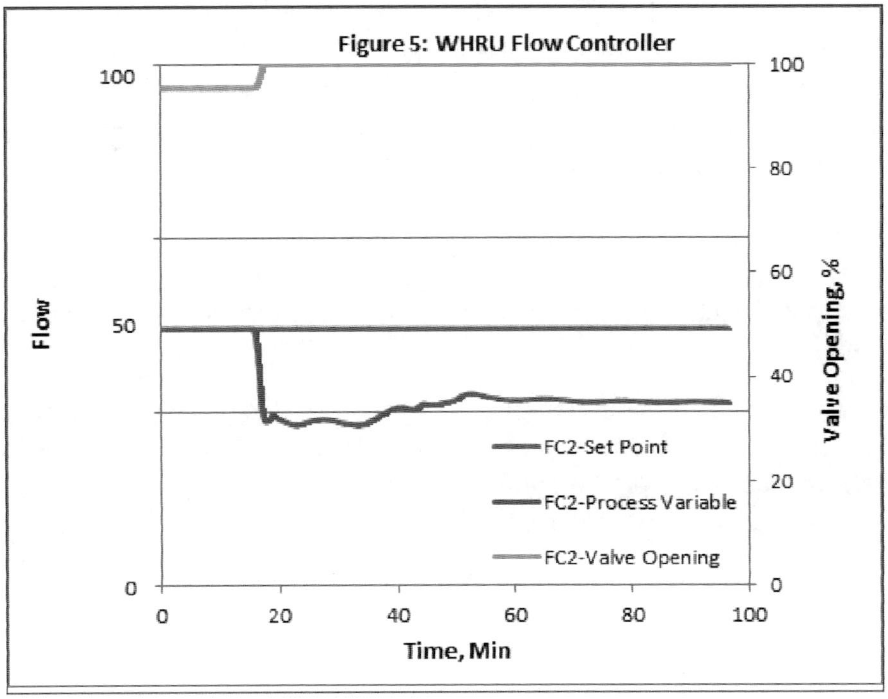
Figure 5: WHRU Flow Controller
100
50
0
Flow
100
80
60
40
20
0
Valve Opening, %
FC2-Set Point
FC2-Process Variable
FC2-Valve Opening
0
20
40
60
80
100
Time, Min

A low flow in the WHRU coils will overheat the hot oil resulting in coking inside the coils and can lead to tube rupture or damage. Among the WHRUs (WHRU 1,2,3,4) arranged in parallel, the flow through the smaller WHRUs (WHRU 5,6) decreases keeping enough flow through the others. The smaller WHRU accommodates two different set of coils, one for heating the hot oil and the other for heating a process fluid. A low flow would cause the smaller WHRUs to trip. The trip would lead to WHRU shutdown in order to protect and avoid damage to hot oil coils. In addition, this would also cut down the process heat and affect the associated process unit. However, tripping the WHRU for this scenario was uncalled for and unexpected. The low flow was prevented by increasing the hot oil valve size (FV2) for the smaller WHRUs (WHRU 5, 6). Increasing the hot oil valve size ensured sufficient flow to the WHRU coils avoiding unnecessary trip and a process upset.

Conclusion

The study avoided a plant shutdown due to an unforeseen dynamic condition. It assisted in improving an existing design enabling uninterrupted plant operation averting possible equipment damage and downtime for maintenance, thereby avoiding production losses.

Few other dynamic studies in the hot oil simulation (not discussed here), enabled the design engineers and operators understand the duration and recovery time of an upset. This in turn guided the engineers to specify certain operator instructions to be incorporated in the operating manual. This hot oil study was very instrumental in determining the PDC1 Controller set points at different operating conditions which proved to be a challenge during design phase. The study also benefited in the verification of process control and validation of equipment and valve sizing. The study aided in determining the plant's limitations and also its robustness. It also assisted in validating various design cases, operating conditions and also transitioning between design cases.

These design verification and improvements in the design could not have been achieved by conventional calculation or steady state simulation. This is due to the fact that the above scenarios are integrated transient effects of an upset which were predicted well using dynamic simulation analysis.

Mol. Sieve Dehydration unit potential problems for LNG plant.

AICHE spring meeting, New Orleans USA, April 1 2014

By J C Kuo, Ph.D. program, P.E. and Professor/Director K H Wang. Ph.D.

University of Houston, Houston Texas USA

1. Dehydration unit for LNG plant.

Treated gas from the CO_2 removal unit is chilled against high level Propane refrigerant in the Propane Feed Chillier to reduce the moisture content in the inlet gas feed to the gas dryers, thus reducing the size of the dehydration unit. The wet feed gas undergoes separation of condensed hydrocarbons and water which is accomplished with the Dryer Inlet Separator arranged in series some time with a Dryer Inlet Filter Coalescer prior to entering the Molecular Sieve Dehydrators.
The dehydration unit consists of multiple molecular sieve beds, any number of which will be in the adsorption mode with the other bed in regeneration or on stand-by.
Feed gas enters Molecular Sieve Dehydrators which are on adsorption cycle. The water vapour is removed from the feed gas and is retained within the molecular sieve during the whole adsorption cycle. A down-flow path is used to avoid fluidizing the beds.
Dryers are regenerated by back flowing clean, dry effluent gas from the Regeneration Gas Heater. The hot regeneration gas passes up through the molecular sieve bed currently off-line in the regeneration mode. The adsorbed water is stripped off of the bed together with some CO_2 and heavy hydrocarbons, restoring the adsorption capacity of sieves.
The hot, wet regeneration gas leaving the dehydrator is air cooled in the Regeneration Gas Cooler and passes to the Regeneration Gas K.O. Drum. Condensed water is Separated in this vessel. The regeneration gas on flow control is then re-circulated back to the front-end and combined with the feed gas via the Regeneration Gas Compressor. This ensures an adequate regeneration flow circulation through the bed as needed. The regeneration and dryer switching is controlled via programmable controller and switching valves. During the heating cycle and cooling cycle, the regeneration gas flow will be in the up-flow pattern. (Figure 1, typical dehydration unit)

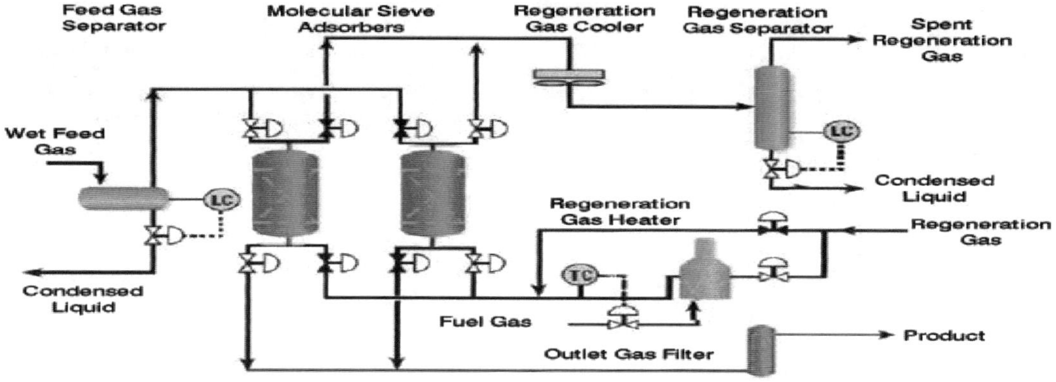

Table1. Typical LNG plant Dehydration specification.

		Process specification		
	units	**Treated gas from AGRU**	**Treated Gas from Dehy and Hg Removal**	**LNG**
CO2	**ppm**	100 (vol) Maximum		
H2O	**ppmv**	saturated	<0.1	
H2S	**ppm**	<3.0 (vol) Maximum	3.58 max	
Dew point	**°C**		-184	
CH3OH	**ppmv**		<1.0	
Hg	**µg/Nm3**		<0.01	
High Heating Value	**Btu/scf**			
Methane	**mol%**			88.0 - 96.0 max
Nitrogen	**mol%**			1.0 max
Ethane	**mol%**			8.6 max
Propane	**mol%**			2.9 max
Isobutane	**mol%**			0.70 max
N-butane	**mol%**			0.45 max
C5+	**mol%**			0.075 max
Mercaptan Sulfur	**g/Nm3**			0.0051
Total sulfur content	**ppm**			19.9 max
LNG Density	**kg/m3**			445.3-466.1
LNG Wobbe Index	**kcal/Nm3**			12,980-13,300
Vapor pressure	**bara**			
Corrossion				
Reid Vapor pressure	**bara**			

2. Potential problems

Many dehydration units in LNG based load plant have not been able to operate per design and premature moisture breakthrough has been occurring. From RCA (Root Cause Analysis) indicate the followings potential issues.

If liquid enters the molecular sieve, then the mass transfer zone length can be extended by a factor of 2-4 times that expected. Hence what is perceived as a capacity problem is in fact a breakthrough problem. Even when the mass transfer zone is not extended by this much, the extra liquids can cause extra duty for the regeneration part of the process, and prevent the regeneration from being done properly. One of the reasons to have hydrocarbon liquid is due to retrograde condensation. Liquid hydrocarbons can damage the sieve in the dehydration beds resulting in poor performance.

2.1 Retrograde Condensation

In one of the project which feed gas is significantly richer than feed gas for other LNG projects. This fact, coupled with the fact that feed gas pressure is higher than most LNG facilities, is causing retrograde condensation of hydrocarbons (HC) in the piping and equipment. (Figure 2. Phase diagram)

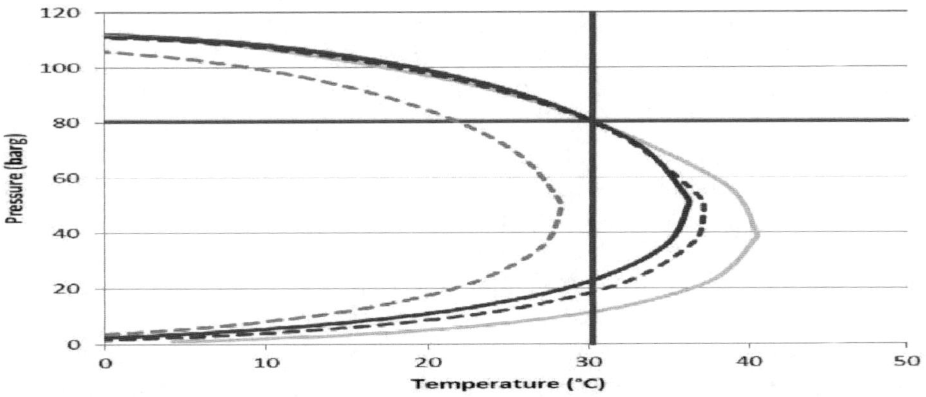

Retrograde condensation downstream of inlet separator occurs with any pressure drop at the design operating conditions for the average feed gas case as well as the rich feed gas case. The estimated hydrocarbon liquid condensed is more than 1000 kg/hour for the average feed gas at 10 psi pressure drop. Due to the pressure drop of the coalescer cartridges and the inlet piping, hydrocarbon liquids will be constantly introduced to the dryers even if separator has perfect separation. Furthermore, additional pressure drop in the equipment and piping downstream of the mol. sieve beds (mol. sieve post-filter, MRU bed, and MRU post-filter) will cause further retrograde condensation. Introduction of hydrocarbon liquids to the MRU bed will limit the MRU effectiveness and potentially cause mercury breakthrough and introduction the cold plant. The solution is simple, increase the temperature of the feed by, say, 3.0 °C to bring it outside the area of retrograde condensation. But this is expensive due to the refrigeration cost to cool down the temperature.

In some cases, if the gas is not so rich (MW < 18.0) and the pressure is not so high (<940 Psig) ,the dehydration unit is designed such that the heat of adsorption on the Mol Sieve is capable of vaporizing the retrograde condensation of the design case.

2.2 Piping Design with pocket between separator and dryer beds

Piping pocket is not allowed from the inlet separator to the dehydrator.

In one case there is a pocket has no permanent drain connections. Pockets can allow for liquid and accumulation and eventual introduction to downstream equipment. The 30 inch outlet process piping of separator to the three dryers is configured such that there is a common header from which the wet feed gas goes to Dryer A first, followed by Dryer B, and then Dryer C last. With liquid carryover from separator; when the dryers are in adsorption, the bulk of the liquid will tend to accumulate in the inlet pocket and also at the inlet piping of the dryer C. Dryer C has more chance to receive any accumulated hydrocarbon liquid in the pocket or inlet piping header.

2.3 Inadequate performance of separator vessels

These separators are often still under-designed. Entrained hydrocarbon droplets will tend to coat the surface of the beads adding a diffusion resistance layer which in turn causes the mass transfer zone length to increase by a factor of 2 to 4. If the droplets are entrained water, then this will substantially increase the adsorber load, both for adsorption and regeneration. If the droplets contain salt, then the damage is drastic and permanent. Any liquid droplets into the plant they can weaken the molecular sieve to a point where normal bed movement will cause serious erosion or attrition. A massive failure of the inlet separator can allow large slugs of liquid to hit the bed. This in turn will cause a hydraulic "hammer" which can crush the adsorbent.

In one case, the dehydration inlet separator vessel is not adequately removing hydrocarbon liquids. Testing results demonstrated liquid carryover downstream of the dehydration filter separator. SME review has confirmed that due to the high operating pressure and low surface tension, the proprietary vane pack design will not be effective for hydrocarbon separation. Also the inlet distributor is inadequate for proper flow distribution across the vane pack. Trending of separator level valve positions are further indicators of liquid carryover. Significant amounts of hydrocarbons have been sampled in regeneration gas separator during bed regeneration. All of the above factors indicate that separator is not performing as designed.

2.4 Inadequate regeneration gas flow rate

The design regeneration gas flow rate is very critical. The most significant factor affecting outlet specification is the regeneration, providing the unit has been correctly designed. Very often the regeneration gas flows are designed to operate at a rate which is turbulent, but which is close to the laminar region. A drift in the instrument can cause the flow regime to move into the laminar regime, and this will seriously affect the heat transfer characteristics in the column.

The outlet gas stream will be effectively in equilibrium with the very last layer of the bed at the outlet of the adsorption stream. If the specification in terms of dew point or impurity outlet specification suddenly is not reached, then the regeneration will often be the cause of the problem.

Monitor the regeneration heating outlet temperature. A lower temperature indicates a higher impurity outlet concentration, because it shifts the position of equilibrium. It also affects the kinetics of the desorption. A lower temperature also means a lower gas velocity. Lowering the pressure of regeneration may compensate here, if possible.

2.5 Regeneration Reflux

One cause of poor adsorber performance such as high pressure drop, dusting, shortened adsorbing time, or poor outlet dew-point can be traced to an unusual phenomena which we refer to as “refluxing". This is a situation where water is being stripped from the bed during regeneration at such a high rate that it condenses on the top head of the adsorber and subsequently "rains" back down into the bed.

When refluxing occurs, the liquid water phase moves down the bed until it contacts the heat zone moving up the bed. At this point the water will boil forming a reflux. This rolling boil will grind the molecular sieve into a powder which subsequently becomes a wet cake (mud) which is then baked by the rising hot gas. In extreme cases we have found a lump weighing several hundred pounds formed into an annulus or donut shape with a center opening of less than one foot. Even in less severe cases this rolling boil will destroy the molecular sieve so some design or operating changes are necessary to extend the adsorbent replacement periods.

Refluxing is most common when the regeneration gas is at high pressure (> 600 psig) because the volumetric heat content of high pressure gas is so high and the water capacity is low because of the high pressure. The following steps are suggested to minimize the reflux effect.

- Slow down the rate of heat input. Normally the regeneration gas rate is near the laminar flow region so the rate cannot be reduced but the rate of input heat can be reduced by ramping up the inlet gas temperature. If time is available the regeneration steps should start with an ambient temperature purge for one hour at the design regeneration gas flow rate. The temperature should be increased by 100 °F increments each hour up to the design inlet temperature. Normally this heating should continue at full flow for one hour after the outlet temperature has leveled out (approaches the inlet temperature by 50 -75 ° F.)
- Try to reduce the heat loss through the top end of the bed by adding extra insulation and even consider steam tracing.
- Change the heating gas flow direction to flow from the top to the bottom of the bed. This will push out any condensed water eliminating the refluxing boil. There is a cost to this because co-current regeneration requires more gas to thoroughly

strip the bed. This can be recalculated by the adsorbent vendor. Also the downward flow will push heavy liquid contaminants on down through the bed possibly increasing the fouling rate. Co-current regeneration was unthinkable originally but it has been proven workable in several olefin plants. It appears that a slow hydrocarbon fouling is preferable to sieve powdering.

- An alternate scheme is to reverse all flows and adsorb upward and heat down flow if the adsorbing flow rate is below the bed fluidization (lifting) velocity. The adsorbent vendors can easily check this.

In one case, the following charts were developed from plant historian data. Trends in the middle bed temperature measurements indicate refluxing due to insufficient heat distribution during regeneration. Data for bed show spikes in the middle bed temperature. This indicates that water is being driven off the sieve by hot regeneration gas and then cooled and condensed by the cooler material in the upper sections of the vessel (internal refluxing, Figure 3 and 4).

.

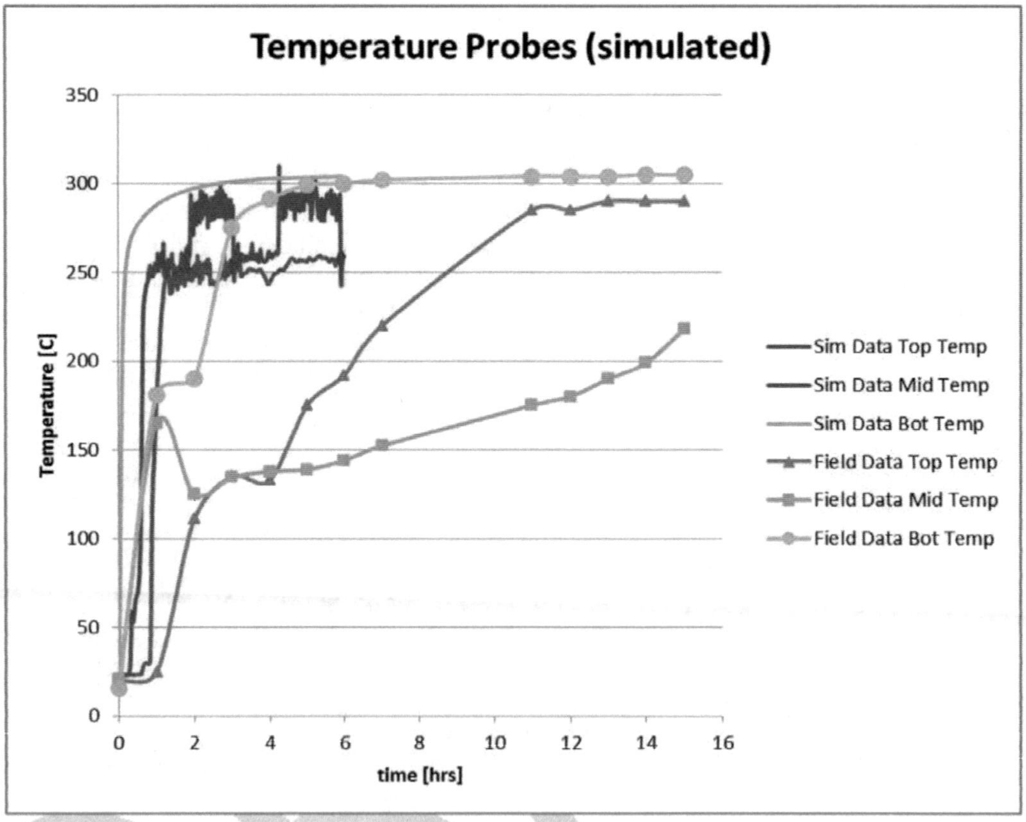

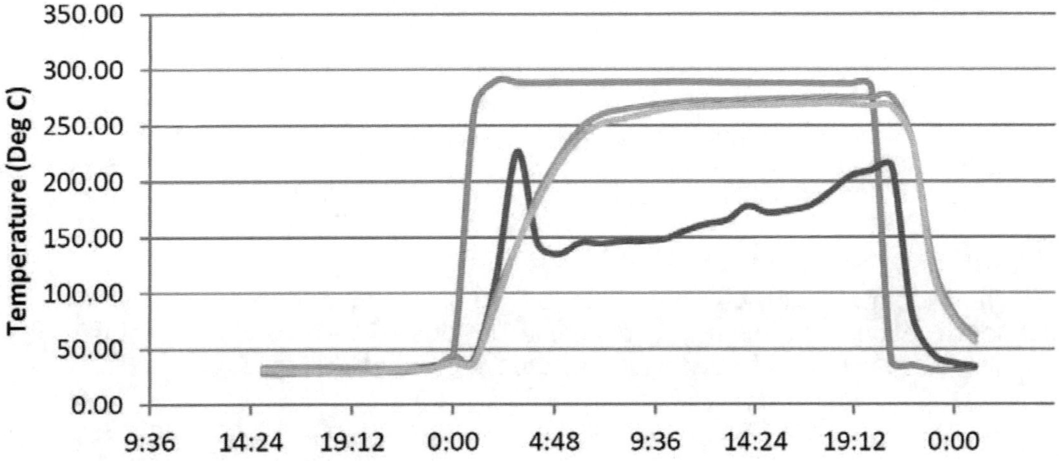

Blue : Regen Inlet Temperature
Red : Middle Bed A Temperature
Green :Bed A Outlet Temperature
Orang: Regen Outlet Temperature

2.6 Regeneration channeling

The regeneration gas must be adequately distributed throughout the cross section of the bed (by the specification of inlet distributors and support balls); to ensure there are no areas of poorly regenerated bed. A good indication that the flow is turbulent is that the pressure drop across the bed should be greater than 2.4 mbar per meter bed height. If it is 15-20% below this value, then there is a probability of channeling. The channeling may be due to a low flow rate, or may also be due to a buildup of coke on the bed, entrained salts, liquid carry over, or heavy ends in a gas stream. If channeling is confirmed on a bed which was working correctly, then the bed should be dumped and screened. During this time the distributors and bottom supports should be thoroughly checked for damage, as this area is a common cause of many channeling problems. There is a possibility that the bottom screen has lifted, may have warped during service, or that the screen is leaking due to a hole.

2.7 Oxygen in gas

Occasionally oxygen can get into the feed and regeneration gas streams. This can be for a variety of reasons, but may be due to poor condition of compressor or valve seals. If the concentration reaches above 20 ppm, then there is a danger that it will react or "light off' in the regeneration heater, where the combination of the temperature above 200°C and the iron oxide coating on the internal walls which acts as a catalyst, will cause the formation of C02 and water.

The regeneration gas is now wet and this will adversely affect the outlet dew point available. While the obvious cure is to find the cause of the oxygen leak and to repair it. If this cannot be done, then a second remedy may be to change from a 4A molecular sieve to a 3A in the case of dehydration, as 3A molecular sieve requires a lower regeneration temperature. If the molecular sieve grade cannot be changed, then the regeneration temperature must be lowered, and the penalties as above allowed for.

Some corrosion inhibitors can chemically attack the molecular sieve causing permanent damage to the adsorption properties. Sulphur compound can also react with oxygen at regeneration temperatures to form elemental sulphur. This will then deposit on the molecular sieve and block the pores slowing or stopping adsorption.

2.8 Switching valve pass or leak

It is surprising how small an amount of feed by-pass is required to drastically affect the outlet specifications. The remedy is to look for obvious sources of leaks such as leaking or passing valves. An ultrasonic leak detector is of great value here.

Also look for by-passes through instrument lines such as a delta P gauge. A pinhole leak in the diaphragm can be the culprit.

This liner is often a castable stucco-like material, and can crack with repeated heating and cooling cycles. This crack can by-pass a significant amount of wet gas during adsorption, and can produce confusing results. The bed will work satisfactorily until the crack which may be part of the way down the column is reached. Then a significant amount of inlet gas is passed through via the crack to the outlet bringing the outlet out of specification, but not by a large amount The second manifestation of a cracked liner is the hot spots on the outside of the column, and the subsequent increase in regeneration heat requirements.
The only solution is to dump the contents, screen if they are to be re-used, and then reload. Normally a 12 mesh screen is sufficient for large beads, and a 16 mesh for small beads.

2.9 Faulty moisture analyzer reading

This can be an embarrassing and expensive problem, but is also one of the most common problems. If early break-through is suspected, then confirm it by an increased pressure drop in the cold sections of the plant before calling in a task force to search for a water problem.
It pays to always have a back-up probe available, particularly as they can be easily fouled up with glycol and other chemicals. Also certain polar compounds such as methanol can "fool" the hygrometer even the hygrometer manufacturers argue this will not happen.

2.10 Regeneration Gas temperature too high.

Some base load plant could use 315C high temperature in order to minimize the regeneration gas flow. However for some impurities include heavy hydrocarbon, methanol or Amine solution; lower temperature at 240 C ranges may be required. Higher temperature tend to coke and damage the sieve.

3. References

1. Zeochem;
   Dehydration Bed trouble shooting guide

2. UOP;
   Dehydration Bed design

3. Chevron;
   Gas plant dehydration bed operating data.

353066 Use Exergy Analysis to Increase Efficiencies of Mid-Scale LNG Processes

Abstract:

Mid-scale LNG processes are widely used all around the world, and it has the potential to compete with pipeline transportation for small and medium shale gas fields.

To compete with pipeline transportation, cost is essential for mid-scale LNG processes. In a typical LNG value chain, the liquefaction section consumes 30%-45% of the total operating costs. If the operating cost of an LNG process could be reduced, the total cost of LNG value chain will also be reduced.

LNG processes are basically refrigeration systems. Exergy is a very useful thermodynamic property to analysis refrigeration systems. Exergy means the maximum useful work possible during a process that brings the system into equilibrium with environment. From exergy point of view, LNG processes get exergy from mechanical work and transfer it to natural gas to make it far from equilibrium with environment.

In this paper, a new exergy analysis method was developed for LNG processes. In this method, not only the overall exergy of the whole process, but also exergy transfer between facilities in the process are considered. Exergy is classified as three categories: temperature related exergy, pressure related exergy and phase-change related exergy. The three types of exergy could change to each other, and every facility can only process certain types of exergy. Exergy loss for each facility is evaluated. Process improvement methods are used to reduce exergy loss and increase process efficiency.

A C3MR process was used as case study. Two improvements were made for the original process. The first improvement only changes operation conditions, and the second improvement changes process itself. Results show that the overall operating cost could be reduced by 15% for the second improvement, and device investment for the second improvement is relatively small.

THERMODYNAMIK

14 AIChE Spring Meeting, New Orleans, LA

RUB

Advanced Thermodynamic Property Models for Natural Gas, LNG and Related Mixtures

AIChE Spring Meeting

Roland Span

April 2014

RUHR-UNIVERSITÄT BOCHUM
Fakultät Maschinenbau

Different Kinds of Property Models

14 AIChE Spring Meeting, New Orleans, LA

RUB

○ Limited suitability / accuracy ◉ Accurate but with limitations ● Accurate	NG			LNG			CCS			
	Processing	Transport Engin.	Custody Transfer	Liquefaction Engin.	Transport Engin.	Custody Transfer	Pre/Post Capture	Oxyfuel Capture	Transport	Storage (Geol.)
Engineering (SRK, PR, LKP)	○	○		○	○		○	○	○	○
Empirical, Simplified Analysis (S-GERG, AGA8-SC 92)			◉							
Empirical, Full Analysis (AGA8-DC 92)		●	●							
Empirical, Simplified Analysis (Revised Klosek McKinley)						◉				
Helmholtz Energy Mixture Model (GERG-2008, EOS-CG)		●	●	●	●	◉		●	●	

- Very different property models are used depending on the application

Different Kinds of Property Models

14 AIChE Spring Meeting, New Orleans, LA | RUB

○ Limited suitability / accuracy ◉ Accurate but with limitations ● Accurate	NG: Processing	NG: Transport Engin.	NG: Custody Transfer	LNG: Liquefaction Engin.	LNG: Transport Engin.	LNG: Custody Transfer	CCS: Pre/Post Capture	CCS: Oxyfuel Capture	CCS: Transport	CCS: Storage (Geol.)
Engineering (SRK, PR, LKP)	○	○		○	○		○	○	○	○
Empirical, Simplified Analysis (S-GERG, AGA8-SC 92)			◉							
Empirical, Full Analysis (AGA8-DC 92)		●	●							
Empirical, Simplified Analysis (Revised Klosek McKinley)						◉				
Helmholtz Energy Mixture Model (GERG-2008, EOS-CG)		●	●	●	●	◉		●	●	

- Very different property models are used depending on the application

Cubic EOS as Typical Engineering Models

14 AIChE Spring Meeting, New Orleans, LA | RUB

Cubic equations of state …

- are available for pure substances and mixtures in all kinds of tools,
- yield pretty reliable results for phase equilibria,
- have been combined with other models (e.g. electrolytes, hydrates),
- are a rather inaccurate source of densities and caloric properties.

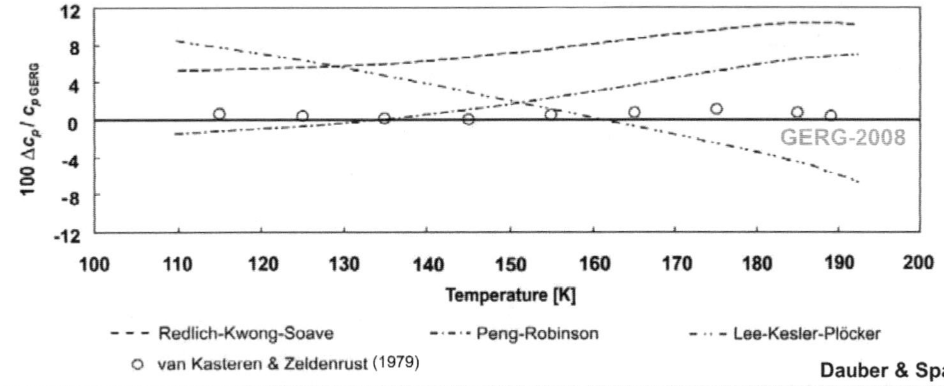

Dauber & Span (2011)

Cubic EOS and Liquefaction Engineering

- The main cryogenic heat exchanger is the key component
- Accurate data for natural gas and coolant flows were made available by Statoil / Linde

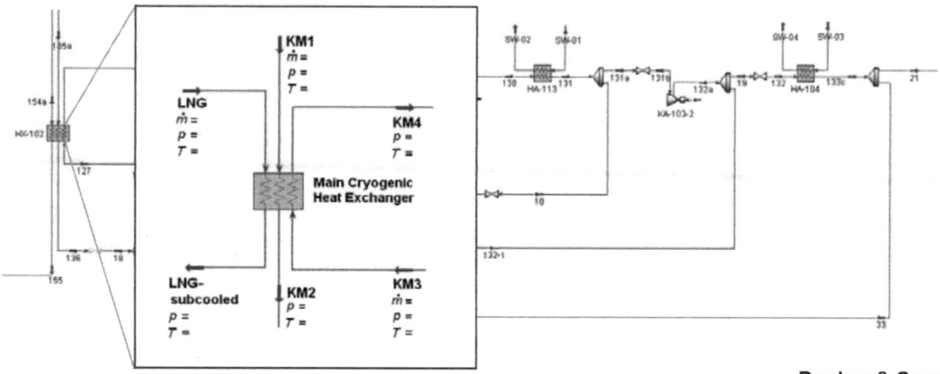

Dauber & Span (2011)

Cubic EOS and Liquefaction Engineering

Results of the energy balance

Property Model	$\Delta Q / Q_{cold}$ [%]
GERG-2008 PropertyPackage	-0.12
Peng-Robinson	2.48
Redlich-Kwong-Soave	2.73
Lee-Kesler-Plöcker	7.01

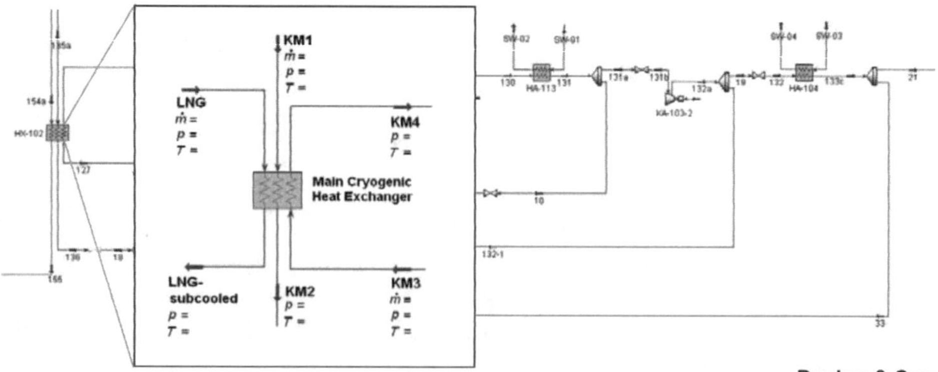

Dauber & Span (2011)

Models Based on a Simplified Analysis

14 AIChE Spring Meeting, New Orleans, LA | RUB

- The **S-GERG 88 / AGA8-SC 92** EOS was commonly used for density calculation in custody transfer for pipeline gases
- Z as a function of T, p, $H_{o,n}$, ρ_n, x_{CO2}, x_{H2} ⇨ no full analysis required
- At pipelining conditions usually accurate to ±0.1 %

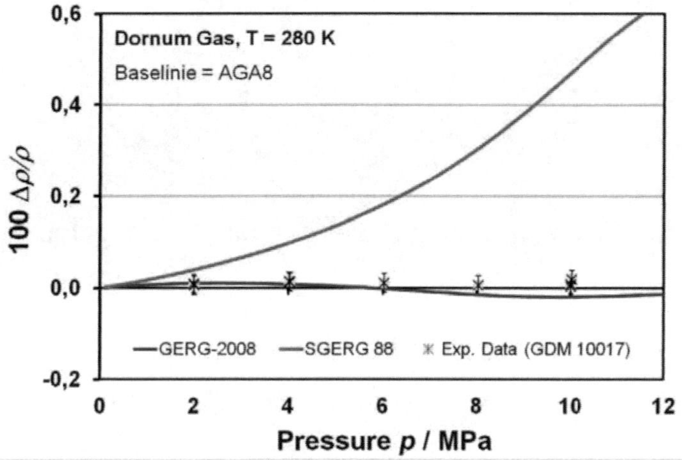

Composition	
CH4	87,7299
N2	0,8584
CO2	1,9950
C2H6	9,0234
C3H8	0,3476
n-C4H10	0,0196
i-C4H10	0,0136

Schley (2001)

Models Based on a Simplified Analysis

14 AIChE Spring Meeting, New Orleans, LA | RUB

- The **Revised Klosek McKinley** equation is commonly used for density calculation in custody transfer for LNG

$$v_{mix} = \sum_i x_i v_{o,i} - \left(k_1 + (k_2 - k_1) \cdot \frac{x_{N2}}{0.0425} \right) \cdot x_{CH4} \quad with \quad k_1, k_2 = f(T, M_{mix})$$

$v_{o,i}$ for CH_4, C_2H_6, C_3H_8, n-C_4H_{10}, i-C_4H_{10}, n-C_5H_{12}, i-C_5H_{12}, N_2

- k_1 and k_2 were fitted to results of a corresponding states model

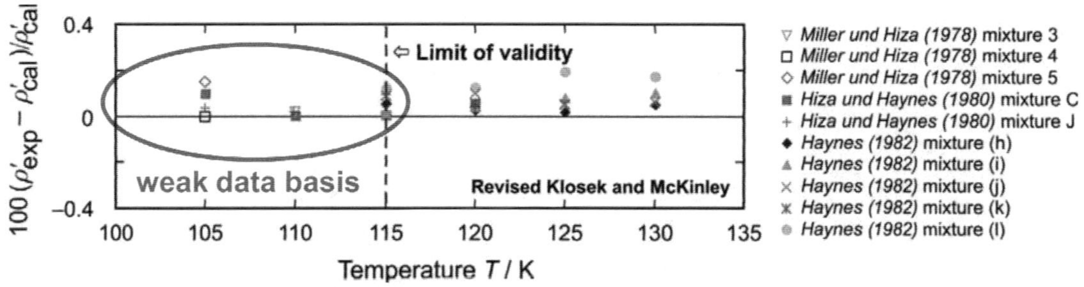

Models Based on a Simplified Analysis

14AIChE Spring Meeting, New Orleans, LA RUB

- The **Revised Klosek McKinley** equation is commonly used for density calculation in custody transfer for LNG

$$v_{\text{mix}} = \sum_i x_i v_{\text{o},i} - \left(k_1 + (k_2 - k_1) \cdot \frac{x_{\text{N2}}}{0.0425} \right) \cdot x_{\text{CH4}} \quad \text{with} \quad k_1, k_2 = f(T, M_{\text{mix}})$$

$v_{\text{o},i}$ for CH_4, C_2H_6, C_3H_8, n-C_4H_{10}, i-C_4H_{10}, n-C_5H_{12}, i-C_5H_{12}, N_2

- k_1 and k_2 were fitted to results of a corresponding states model

PRO:

- New data for LNG densities are represented within claimed uncertainty

CON:

- Substantially **more accurate** $v_{\text{o},i}$ available for relevant pure components
- Replacing the $v_{\text{o},i}$ results in a worth representation of LNG densities

Helmholtz Mixture Models

14AIChE Spring Meeting, New Orleans, LA RUB

- GERG-2004 / -2008 by Kunz and Wagner as new standard
- Introduced independently by Lemmon & Tillner-Roth in mid 90's

$$\alpha(\delta,\tau,\bar{x}) = \sum_{i=1}^{N} x_i \left[\alpha_{oi}^{0}(\rho,T) + \ln x_i \right] + \sum_{i=1}^{N} x_i \alpha_{oi}^{\text{r}}(\delta_{\text{m}},\tau_{\text{m}}) + \sum_{i=1}^{N-1} \sum_{j=i+1}^{N} x_i x_j F_{ij} \alpha_{ij}^{\text{r}}(\delta_{\text{m}},\tau_{\text{m}})$$

Helmholtz Mixture Models

- GERG-2004 / -2008 by Kunz and Wagner as new standard
- Introduced independently by Lemmon & Tillner-Roth in mid 90's
  - Pure fluid equations of state (EOS)

$$\alpha(\delta,\tau,\bar{x})=\sum_{i=1}^{N}x_i\left[\alpha_{oi}^{0}(\rho,T)+\ln x_i\right]+\sum_{i=1}^{N}x_i\alpha_{oi}^{\mathrm{r}}(\delta_{\mathrm{m}},\tau_{\mathrm{m}})+\sum_{i=1}^{N-1}\sum_{j=i+1}^{N}x_ix_jF_{ij}\alpha_{ij}^{\mathrm{r}}(\delta_{\mathrm{m}},\tau_{\mathrm{m}})$$

Helmholtz Mixture Models

14 AIChE Spring Meeting, New Orleans, LA RUB

Cubic EOS | Simplified Analysis | Helmholtz Mixture Models | Challenges

- GERG-2004 / -2008 by Kunz and Wagner as new standard
- Introduced independently by Lemmon & Tillner-Roth in mid 90's
  - Pure fluid equations of state (EOS)
  - Mixing rules for reduced input parameters δ_{m} and τ_{m}

$$\alpha(\delta,\tau,\bar{x})=\sum_{i=1}^{N}x_i\left[\alpha_{oi}^{0}(\rho,T)+\ln x_i\right]+\sum_{i=1}^{N}x_i\alpha_{oi}^{\mathrm{r}}(\delta_{\mathrm{m}},\tau_{\mathrm{m}})+\sum_{i=1}^{N-1}\sum_{j=i+1}^{N}x_ix_jF_{ij}\alpha_{ij}^{\mathrm{r}}(\delta_{\mathrm{m}},\tau_{\mathrm{m}})$$

- Two options:
  - Mixing rules with four adjustable, binary specific parameters

$$\tau_{\mathrm{m}}=\frac{T_{\mathrm{r}}(\mathbf{x})}{T}\quad\text{with}\quad T_{\mathrm{r}}(\mathbf{x})=\sum_{i=1}^{N}\sum_{j=1}^{N}x_ix_j\beta_{T,ij}\gamma_{T,ij}\frac{x_i+x_j}{\beta_{T,ij}^2x_i+x_j}\left(T_{c,i}T_{c,j}\right)^{0.5}$$

$$\delta_{\mathrm{m}}=\frac{\rho}{\rho_{\mathrm{r}}(\mathbf{x})}\quad\text{with}\quad \frac{1}{\rho_{\mathrm{r}}(\mathbf{x})}=\sum_{i=1}^{N}\sum_{j=1}^{N}x_ix_j\beta_{\rho,ij}\gamma_{\rho,ij}\frac{x_i+x_j}{\beta_{\rho,ij}^2x_i+x_j}\frac{1}{8}\left(\frac{1}{\rho_{c,i}^{1/3}}+\frac{1}{\rho_{c,j}^{1/3}}\right)^3$$

  - Combination rules

Helmholtz Mixture Models

14AIChE Spring Meeting, New Orleans, LA RUB

- GERG-2004 / -2008 by Kunz and Wagner as new standard
- Introduced independently by Lemmon & Tillner-Roth in mid 90's
  - Pure fluid equations of state (EOS)
  - Mixing rules for reduced input parameters δ_m and τ_m

$$\alpha(\delta,\tau,\bar{x}) = \sum_{i=1}^{N} x_i\left[\alpha_{oi}^{0}(\rho,T) + \ln x_i\right] + \sum_{i=1}^{N} x_i\alpha_{oi}^{r}(\delta_m,\tau_m) + \sum_{i=1}^{N-1}\sum_{j=i+1}^{N} x_i x_j F_{ij}\alpha_{ij}^{r}(\delta_m,\tau_m)$$

  - Binary excess functions

$$\alpha_{ij}(\delta_m,\tau_m) = \sum_{k=1}^{K_{Pol,ij}} n_k \delta_m^{d_k}\tau_m^{t_k} + \sum_{k=K_{Pol,ij}+1}^{K_{Pol,ij}+K_{Exp,ij}} n_k \delta_m^{d_k}\tau_m^{t_k} \exp\left[-\eta_k(\delta_m-\varepsilon_k)^2 - \beta_k(\delta_m-\gamma_k)\right]$$

- Two options:
  - Binary specific excess function – $F_{ij} = 1$, parameter in α_{ij} fitted
  - Generalized excess function – F_{ij} fitted, parameter in α_{ij} generalized

Cubic EOS | Simplified Analysis | Helmholtz Mixture Models | Challenges

Roland Span | AIChE Spring Meeting | April 2014 13

The GERG-2008 Model by Kunz and Wagner

14AIChE Spring Meeting, New Orleans, LA RUB

- GERG-2004 / -2008 by Kunz and Wagner as new standard
- Introduced independently by Lemmon & Tillner-Roth in mid 90's
  - Pure fluid equations of state (EOS)
  - Mixing rules for reduced input parameters δ_m and τ_m

$$\alpha(\delta,\tau,\bar{x}) = \sum_{i=1}^{N} x_i\left[\alpha_{oi}^{0}(\rho,T) + \ln x_i\right] + \sum_{i=1}^{N} x_i\alpha_{oi}^{r}(\delta_m,\tau_m) + \sum_{i=1}^{N-1}\sum_{j=i+1}^{N} x_i x_j F_{ij}\alpha_{ij}^{r}(\delta_m,\tau_m)$$

- Four levels of accuracy, depending on available experimental data and relevance of binary subsystem

UNIQUE!

  - Only combination rules for T_r, ρ_r
  - Mixing rules with four adjustable parameters for T_r, ρ_r
  - Adjusted mixing rules & generalized excess function
  - Adjusted mixing rules & binary specific excess function

Cubic EOS | Simplified Analysis | Helmholtz Mixture Models | Challenges

Roland Span | AIChE Spring Meeting | April 2014

The GERG-2008 Model by Kunz and Wagner

14AIChE Spring Meeting, New Orleans, LA | RUB

- GERG-2004 / -2008 by Kunz and Wagner as new standard
- Introduced independently by Lemmon & Tillner-Roth in mid 90's
  - Pure fluid equations of state (EOS)
  - Mixing rules for reduced input parameters δ_m and τ_m

$$\alpha(\delta,\tau,\bar{x}) = \sum_{i=1}^{N} x_i\left[\alpha_{oi}^{0}(\rho,T) + \ln x_i\right] + \sum_{i=1}^{N} x_i \alpha_{oi}^{r}(\delta_m,\tau_m) + \sum_{i=1}^{N-1}\sum_{j=i+1}^{N} x_i x_j F_{ij}\alpha_{ij}^{r}(\delta_m,\tau_m)$$

Methane (CH_4)	n-Pentane (n-C_5H_{12})	Hydrogen (H_2)
Nitrogen (N_2)	Isopentan (i-C_5H_{12})	Carbon monoxide (CO)
Carbon dioxide (CO_2)	n-Hexane (n-C_6H_{14})	Hydrogen sulphide (H_2S)
Ethane (C_2H_6)	n-Heptane (n-C_7H_{16})	Water (H_2O)
Propane (C_3H_8)	n-Octane (n-C_8H_{18})	Oxygen (O_2)
n-Butane (n-C_4H_{10})	n-Nonane (n-C_9H_{20})	Argon (Ar)
Isobutane (i-C_4H_{10})	n-Decane (n-$C_{10}H_{22}$)	Helium (He)

The GERG-2008 Model by Kunz and Wagner

14AIChE Spring Meeting, New Orleans, LA | RUB

- Uncertainty in density at pipelining conditions 0.1%
- Uncertainty in density at liquid states 0.1% - 0.5%

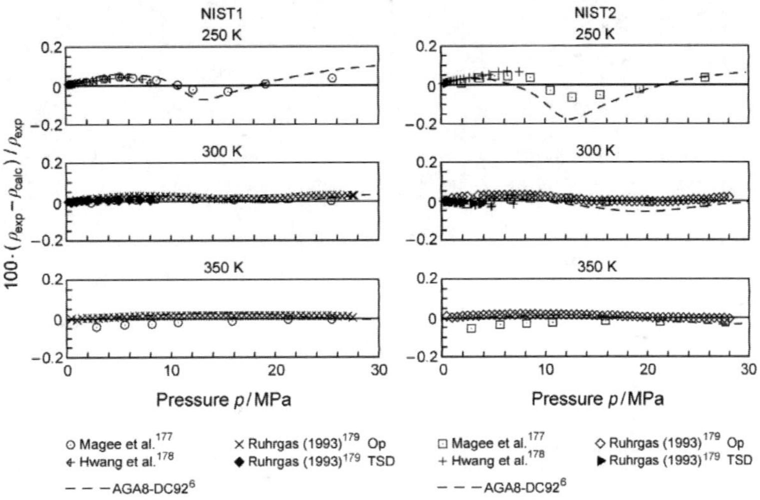

Natural gas at pipelining conditions

Kunz & Wagner (2012)

The GERG-2008 Model by Kunz and Wagner

- At least as accurate as AGA8-DC92 at typical pipelining conditions
- Superior for caloric properties, at uncommon compositions and at $T \leq \approx 270$ K
- Allows for calculations for arbitrary mixtures and down to liquid states

$$\alpha(\delta,\tau,\bar{x}) = \sum_{i=1}^{N} x_i \left[\alpha_{oi}^{0}(\rho,T) + \ln x_i\right] + \sum_{i=1}^{N} x_i \alpha_{oi}^{r}(\delta_m,\tau_m) + \sum_{i=1}^{N-1}\sum_{j=i+1}^{N} x_i x_j F_{ij} \alpha_{ij}^{r}(\delta_m,\tau_m)$$

Methane (CH_4)	n-Pentane (n-C_5H_{12})	Hydrogen (H_2)
Nitrogen (N_2)	Isopentan (i-C_5H_{12})	Carbon monoxide (CO)
Carbon dioxide (CO_2)	n-Hexane (n-C_6H_{14})	Hydrogen sulphide (H_2S)
Ethane (C_2H_6)	n-Heptane (n-C_7H_{16})	Water (H_2O)
Propane (C_3H_8)	n-Octane (n-C_8H_{18})	Oxygen (O_2)
n-Butane (n-C_4H_{10})	n-Nonane (n-C_9H_{20})	Argon (Ar)
Isobutane (i-C_4H_{10})	n-Decane (n-$C_{10}H_{22}$)	Helium (He)

Improving GERG-2008 for CO_2-Rich Mixtures

Status EOS-CG for 15 binary Systems

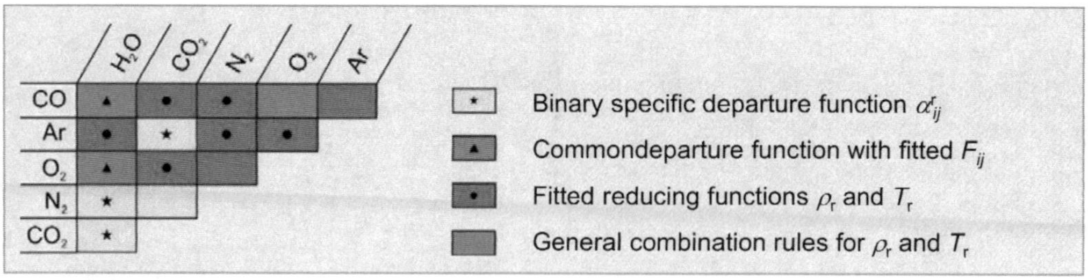

Database EOS-CG: ca. 350 data sets, > 23000 data points

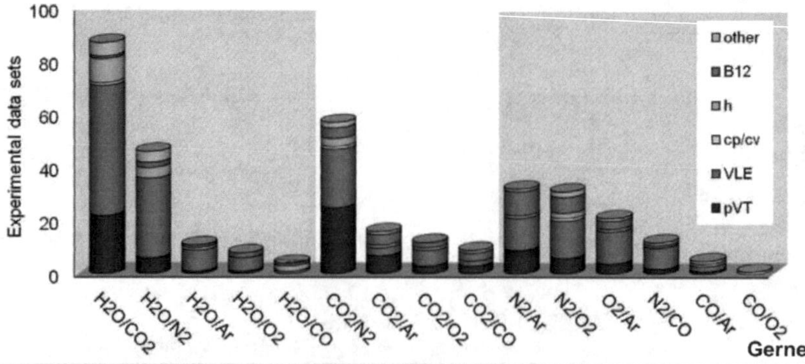

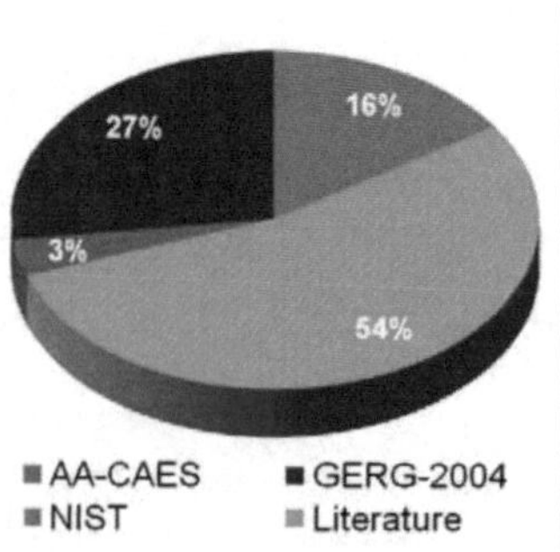

Gernert (2013), Gernert & Span (2014)

Improving GERG-2008 for CO_2-Rich Mixtures

14AIChE Spring Meeting, New Orleans, LA RUB

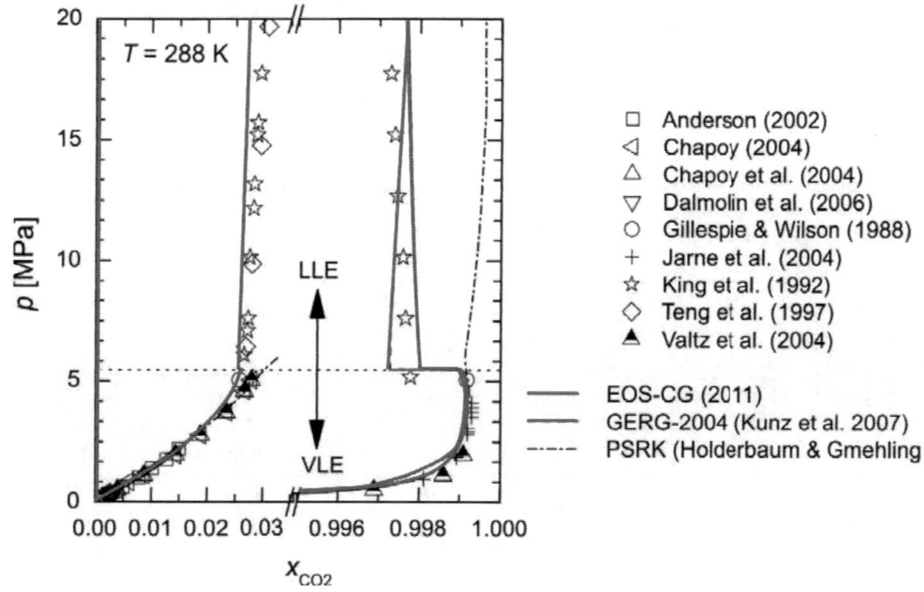

- Improved description of phase equilibria in the system CO_2/H_2O
- In general improved description of solubility of "non condensable gases"

Gernert (2013), Gernert & Span (2014)

Challenges

14AIChE Spring Meeting, New Orleans, LA RUB

○ Limited suitability / accuracy

◎ Accurate but with limitations

● Accurate

	NG			LNG			CCS			
	Processing	Transport Engin.	Custody Transfer	Liquefaction Engin.	Transport Engin.	Custody Transfer	Pre/Post Capture	Oxyfuel Capture	Transport	Storage (Geol.)
Engineering (SRK, PR, LKP)	○	○		○	○		○	○	○	○
Empirical, Simplified Analysis (S-GERG, AGA8-SC 92)			◎							
Empirical, Full Analysis (AGA8-DC 92)		●	●							
Empirical, Simplified Analysis (Revised Klosek McKinley)						◎				
Helmholtz Energy Mixture Model (GERG-2008, EOS-CG)	✗	●	●	●	●	◎	✗	●	●	✗

Custody Transfer for LNG

14AIChE Spring Meeting, New Orleans, LA — RUB

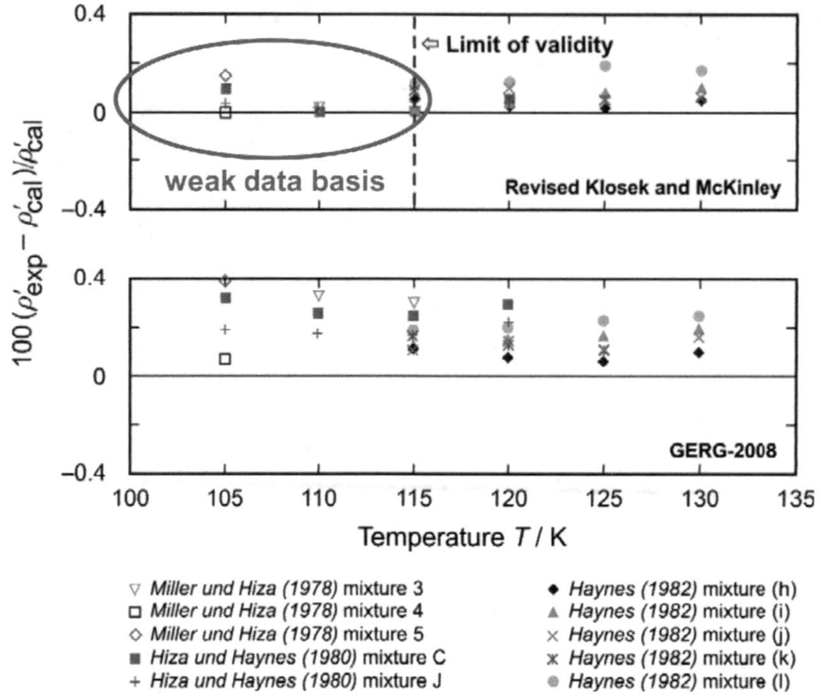

Cubic EOS | Simplified Analysis | Helmholtz Mixture Models | Challenges

Custody Transfer for LNG

14AIChE Spring Meeting, New Orleans, LA — RUB

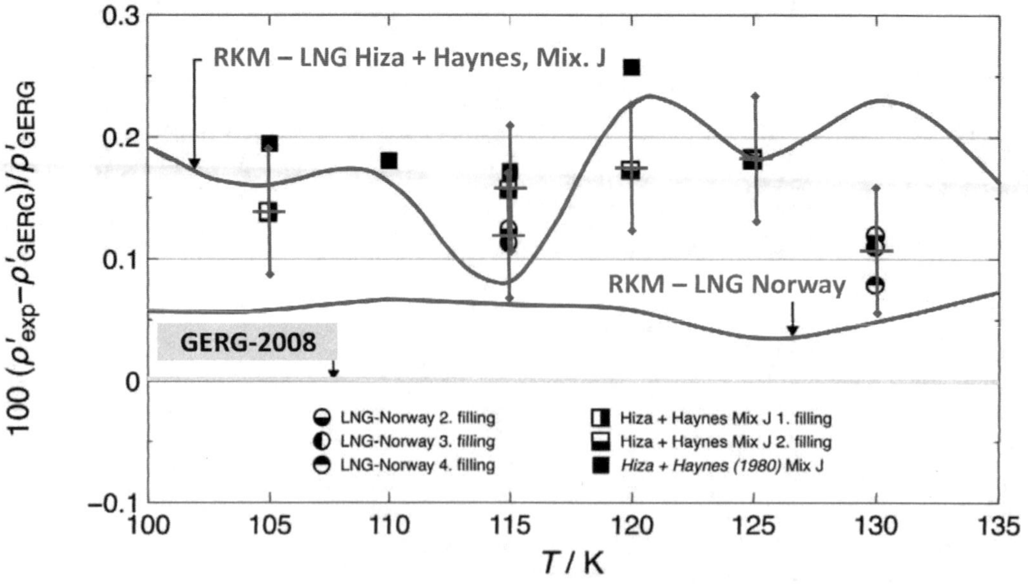

saturated liquid densities

Richter et al. (2013)

Cubic EOS | Simplified Analysis | Helmholtz Mixture Models | Challenges

Custody Transfer for LNG

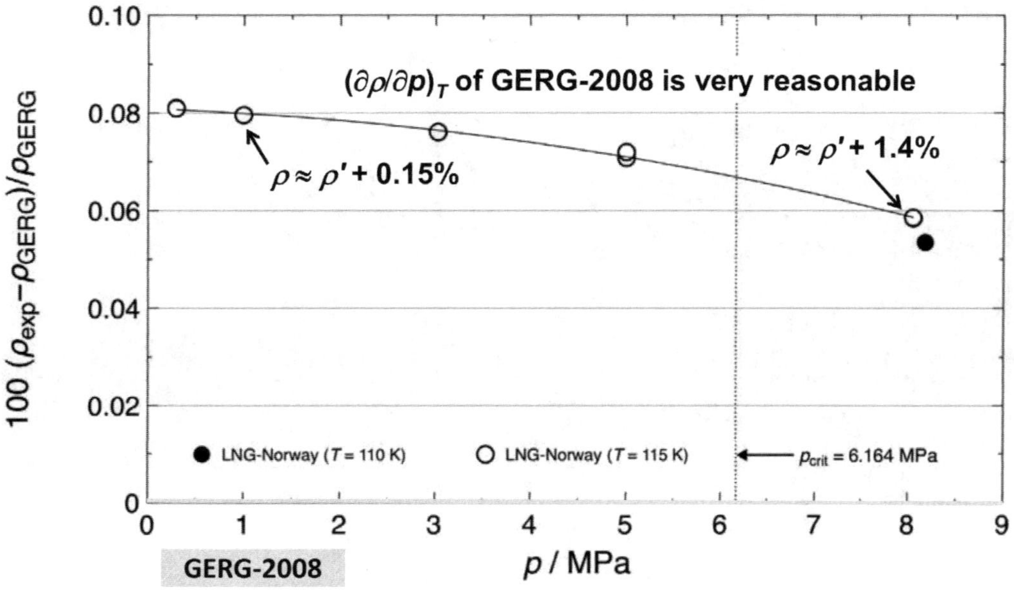

homogeneous liquid densities

Richter et al. (2013)

Improving GERG-2008 for Custody Transfer (?)

- Helmholtz mixture models are fitted exclusively to binary data
- Accurate data for LNG / multi-component mixtures required to validate the model
- Binary data (for several binary mixtures!) required to refit the model

$$\alpha(\delta,\tau,\bar{x}) = \sum_{i=1}^{N} x_i \left[\alpha_{oi}^{0}(\rho,T) + \ln x_i\right] + \sum_{i=1}^{N} x_i \alpha_{oi}^{r}(\delta_{\mathrm{m}},\tau_{\mathrm{m}}) + \sum_{i=1}^{N-1}\sum_{j=i+1}^{N} x_i x_j F_{ij} \alpha_{ij}^{r}(\delta_{\mathrm{m}},\tau_{\mathrm{m}})$$

Methane (CH_4)	n-Pentane (n-C_5H_{12})	Hydrogen (H_2)
Nitrogen (N_2)	Isopentan (i-C_5H_{12})	Carbon monoxide (CO)
Carbon dioxide (CO_2)	n-Hexane (n-C_6H_{14})	Hydrogen sulphide (H_2S)
Ethane (C_2H_6)	n-Heptane (n-C_7H_{16})	Water (H_2O)
Propane (C_3H_8)	n-Octane (n-C_8H_{18})	Oxygen (O_2)
n-Butane (n-C_4H_{10})	n-Nonane (n-C_9H_{20})	Argon (Ar)
Isobutane (i-C_4H_{10})	n-Decane (n-$C_{10}H_{22}$)	Helium (He)

Challenges

○ Limited suitability / accuracy; ◉ Accurate but with limitations; ● Accurate	NG			LNG			CCS			
	Processing	Transport Engin.	Custody Transfer	Liquefaction Engin.	Transport Engin.	Custody Transfer	Pre/Post Capture	Oxyfuel Capture	Transport	Storage (Geol.)
Engineering (SRK, PR, LKP)	○	○		○	○		○	○	○	○
Empirical, Simplified Analysis (S-GERG, AGA8-SC 92)			◉							
Empirical, Full Analysis (AGA8-DC 92)		●	●							
Empirical, Simplified Analysis (Revised Klosek McKinley)						◉				
Helmholtz Energy Mixture Model (GERG-2008, EOS-CG)	✗	●	●	●	●	◉	✗	●	●	✗

Avoiding Inconsistencies – Injection of CO_2

- Engineers working on pipelining of CO_2 use GERG-2008 / EOS-CG
- Reservoir engineers use cubic EOS (+ hydrate / electrolyte models)
- Severe inconsistencies, e.g., for injection of CO_2 (-rich mixtures)

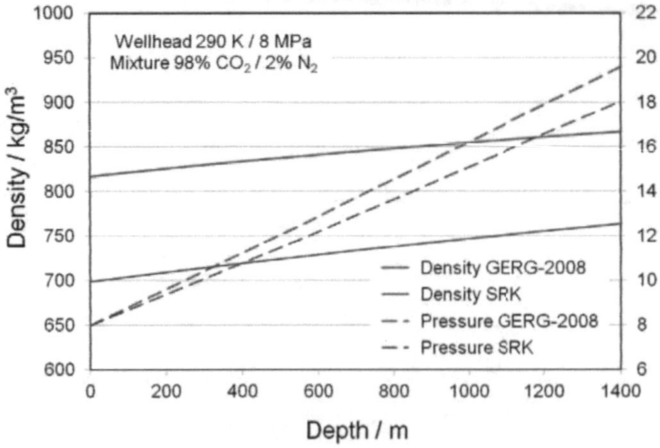

Pipeline engineer delivers
500 to/h ⇨ 577 m³/h at 19.6 MPa

Reservoir engineer receives
577 m³/h at 18.1 MPa ⇨ 440 to/h

Cubic EOS | Simplified Analysis | Helmholtz Mixture Models | Challenges

Roland Span | AIChE Spring Meeting | April 2014

Complex Phase Equilibria

- Helmholtz mixture models are able to predict three phase equilibria
- Numerically complex models with multiple solutions require highly advanced phase equilibrium algorithms
- Fully automatic calculation of three phase equilibria in user-friendly tools is still a challenge

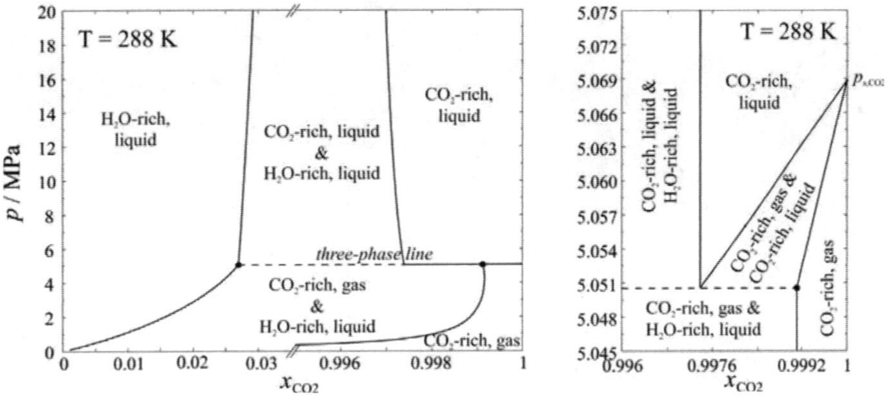

Solid Phases (Hydrates / Ice / Dry Ice)

14 AIChE Spring Meeting, New Orleans, LA RUB

Cubic EOS | Simplified Analysis | Helmholtz Mixture Models | Challenges

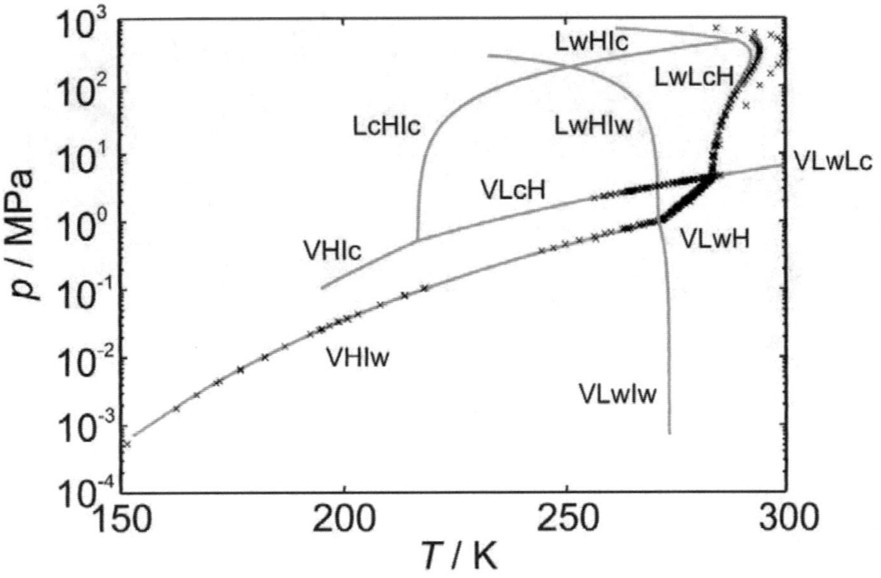

Carbon dioxide hydrates

Jäger et al. (2013)

Roland Span | AIChE Spring Meeting | April 2014

Solid Phases (Hydrates / Ice / Dry Ice)

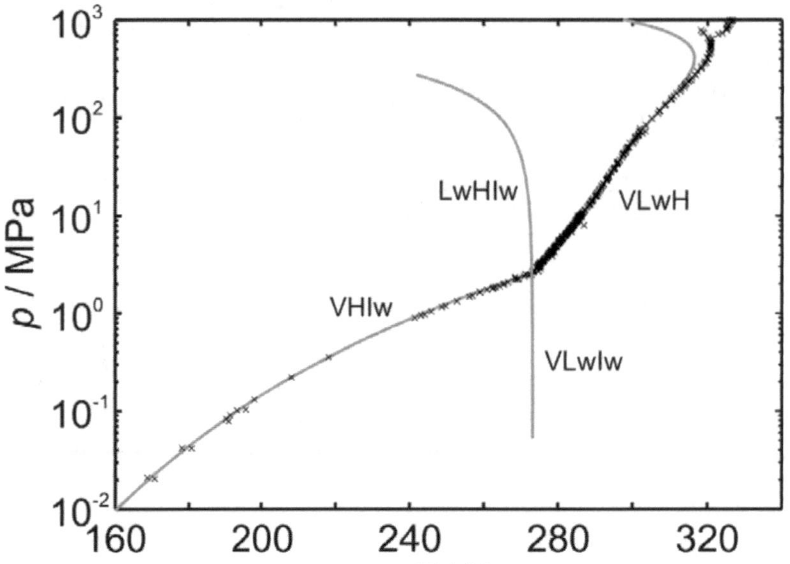

Methane hydrates

Jäger et al. (2013)

Demands Resulting from CCS

14 AIChE Spring Meeting, New Orleans, LA RUB

Matrix of Relevant Components

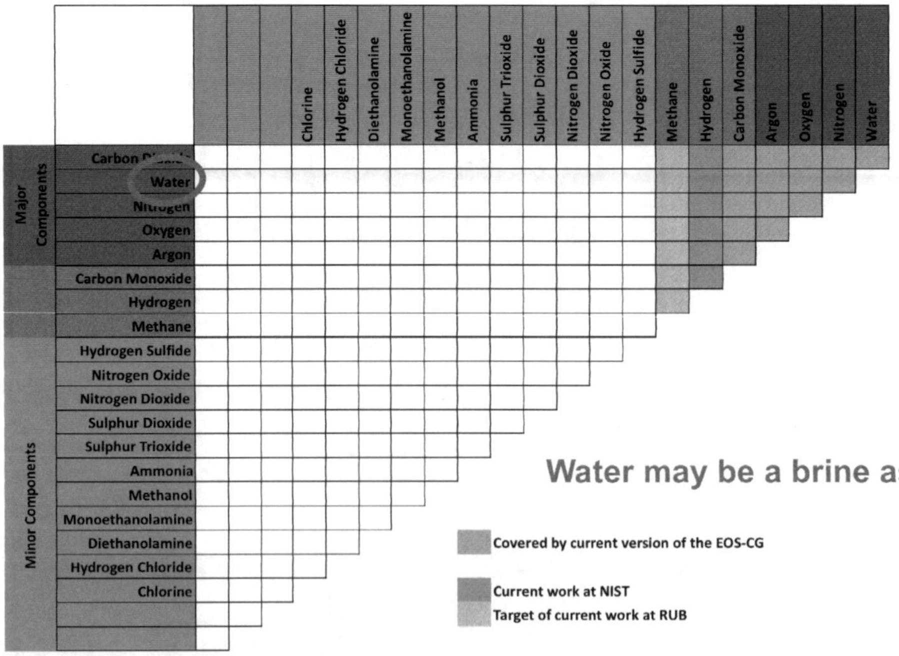

Water may be a brine as well!

Cubic EOS | Simplified Analysis | Helmholtz Mixture Models | Challenges

Roland Span | AIChE Spring Meeting | April 2014

Thank You for Your Attention!

14 AIChE Spring Meeting, New Orleans, LA | RUB

Roland Span | AIChE Spring Meeting | April 2014

Phase Equilibria with Solid Phases – Dry Ice

14 AIChE Spring Meeting, New Orleans, LA | RUB

- Fundamental equation for solid CO_2 (dry ice) developed
- Gibbs enthalpy as a function of pressure and temperature
- Fitted only to data for solid CO_2
- Approach by Tillner-Roth (1998) adapted

$$g(p,T) = h_0 - Ts_0 + \int_{T_0}^{T} c_p(T,p_0)\mathrm{d}T - T\int_{T_0}^{T} \frac{c_p(T,p_0)}{T}\mathrm{d}T + \int_{p_0}^{p} v(p,T)\mathrm{d}p$$

Roland Span | AIChE Spring Meeting | April 2014

Phase Equilibria with Solid Phases – Hydrates

- The model of Ballard und Sloan (2002) was chosen and slightly modified:

$$\mu_w^H(T,p,f_J) = g_w^\beta(T,p) - RT\sum_i v_i \ln\left[1+\sum_J C_{i,J}(T,p) f_J\right]$$

- Adjustable parameters of the model:

σ, ε

Potential-parameters

↳ fitted but almost unchanged

κ_1, κ_2

Pressure dependence of the molar volume

↳ **fitted** to $\mu(T,p)$-data generated from experimental data & fluid model

$g_w^{\beta,0}, h_w^{\beta,0}$

Reference state

↳ consistency to reference state of fluid model

Roland Span | AIChE Spring Meeting | April 2014

353345 Current Operational LNG Expanders

Abstract:

The liquefaction of Natural Gas requires a large amount of energy and many improvements were established since the first commercial natural gas liquefaction plant was built in 1917 in West Virginia. One of the main improvements for the efficiency is the increase in pressure. By compressing the incoming feed gas to a higher density the plant efficiency increases and the costs for the liquefaction reduces.

The Liquefied Natural Gas LNG is transported and stored at atmospheric pressure and it is necessary to expand the high pressure condensed LNG after the liquefaction to atmospheric pressure. This expansion is performed by cryogenic LNG expanders in current plants and by Joule-Thomson valves in older plants. The expansion across an LNG expander increases the output of the liquefied LNG and significantly reduces the costs for the liquefaction. The paper presents the design of the current operational LNG expanders.

351254 Use Off-Gases to Save Energy-Environment & Produce Chemicals

Abstract:

Abstract This paper will present the value addition by processing off-gases from various sources- to convert them into higher value chemicals and also the added benefit of reduced energy and environmental emissions. The presentation covers a variety of schemes, their relative merits, energy efficiency & economics to produce Ammonia from Hydrogen rich off-gases from various sources including Methanol Plants, CO plants, Ethylene plants. The presented schemes will include PSA option to Nitrogen wash, Secondary reforming with & without Gas Heated reforming & an Air Separation unit. The environmental benefit on Ethylene plant (mainly NOx reduction) will also be presented.

A FT-GTL technology for small-scale applications

Mark Peters, Diane Hildebrandt, David Glasser and Xinying Liu

Material and Process Synthesis (MaPS)
A Research Unit
@ UNISA

Learn without limits.

UNISA | college of science, engineering and technology

MaPS Engineering: a Research Unit @ UNISA

MaPS Engineering is a newly formed research unit based at the Science campus in Florida, with the vision of utilizing the resources in universities to serve industry, especially in the field of alternative energy.

MaPS has 20 permanent staff, among them over 10 holding a PhD's in Engineering or Science, and has over 20 postgraduate students.

Team members were all from the group winning the NST-BHP Billiton award for research for innovations through an NGO, NP, or CBO 2009-2010

XTL services provided

- XTL process design and optimization
  - Biomass-to-Liquids (BTL)
  - Coal-to-Liquids (CTL)
  - Gas-to-Liquids (GTL)
  - Waste-to-Liquids (WTL)
- Evaluation of the available XTL technologies
- XTL projects engineering
  - Opportunity Analysis and Business Proposals
  - Feasibility Study
  - Front End Engineering Design
  - Commissioning of the Fischer-Tropsch section

Conventional Fischer Tropsch Process

- Modern Fischer Tropsch process has been used only in MEGA plants
  - Secunda CTL: 160,000 bpd
  - PetroSA GTL: 36,000 bpd
  - Oryx GTL: 34,000 bpd
  - Pearl GTL: 140,000 bpd
  - Ningdong CTL: 90,000 bpd (under construction)

Mega Plants

- Mega plants perform at high efficiency within the
  - Flow sheet
  - Equipment used
  - Product they are designed to produce
- Rely on economy of scale to improve profitability

Is it too big?

- Huge investment involved
  - Sasol Two: R 2.5 B (8% of GDP of SA in 1976)
  - Sasol Three: R 3.3 B (7% of GDP of SA in 1979)
  - Pearl GTL: USD 18–19 B (11% of GDP of Qatar in 2010)
  - Long time for financing
  - Lots of politics involved
  - Long time to bring into market
- Flow sheet chosen normally conservative
  - To minimize the risk
  - Difficult to implement the latest research development
- Large amount of carbon resources needed
  - Billion tons of reserves needed to secure a long time life time
  - Large amount of CO_2 emitted from a single plant in the case of CTL

Is there a reason to think small-scale?

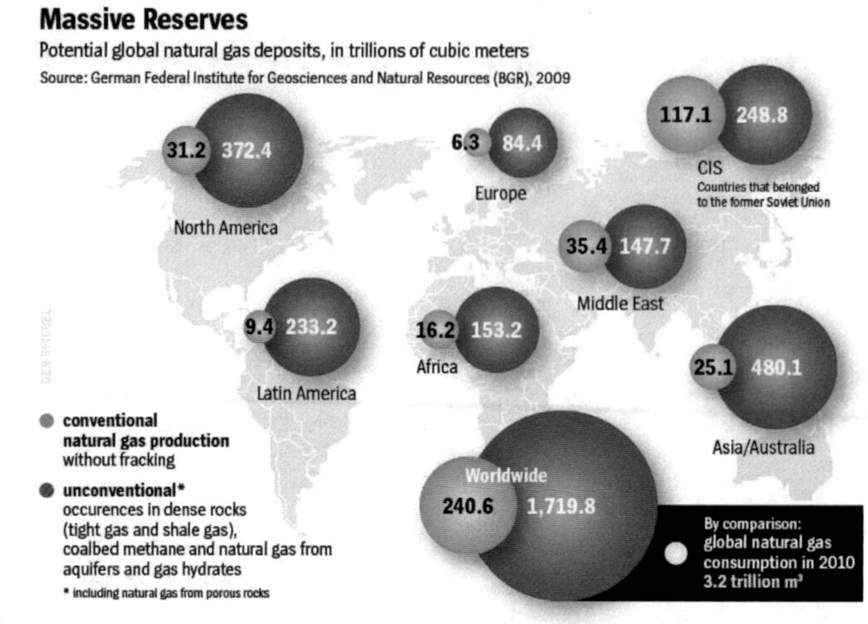

Why think small-scale?

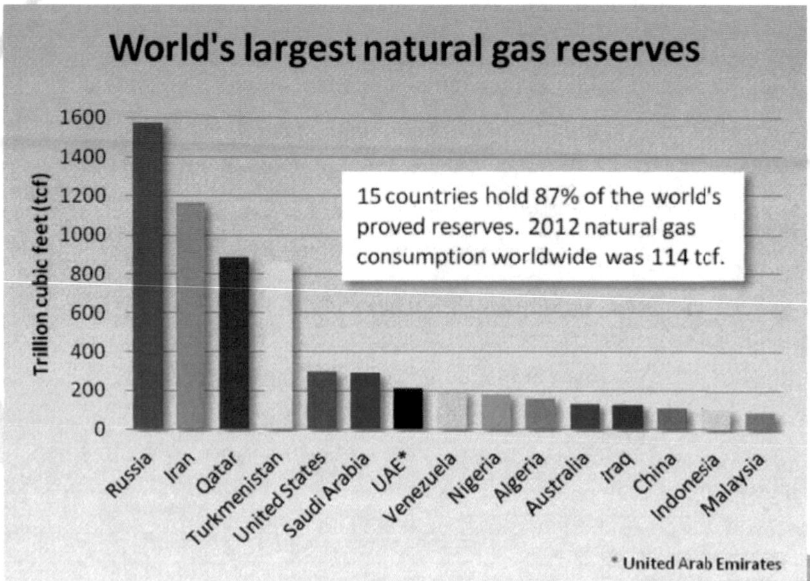

There are some reasons to think small

- There are many carbon resources that are not as big
  - Small carbon reserves
  - Coal bed methane
  - Coke oven gas
  - Underground coal gasification
  - Landfill gas
  - Flare gas
  - Soluble methane
  - Biomass
  - Municipal solid waste
  - etc.
- Smaller investors want to be involved
- Shortage of engineers and skilled-labor, especially in many developing countries
- Short project startup time gives a faster response to market
- Lower financial risk as investment is smaller
- To implement latest development from research

Modular Approach?

- A new approach is to build smaller modular plants.
- These have the advantages of being
  - less capital intensive
  - more flexible and
  - having a faster time-to-market
- Start giving returns on expenditure much sooner
- Later modules can incorporate newer ideas
- Good for managing risk
- Get higher efficiency by being able to incorporate new ideas
- Need a new flow sheet to make it work

Development of Conventional GTL Process: Reactions

- **Reforming**:
  - Methane / Natural Gas feed to produce Synthesis Gas ("Syngas"):
    - $CH_4 + H_2O \rightarrow CO + 3\ H_2$
    - $CH_4 + ½\ O_2 \rightarrow CO + 2\ H_2$
    - $CH_4 + 2\ O_2 \rightarrow CO_2 + 2\ H_2O$
- **Fischer Tropsch**:
  - Syngas converted to hydrocarbons and water:
    - $CO + 2\ H_2 \rightarrow -CH_2- + H_2O$

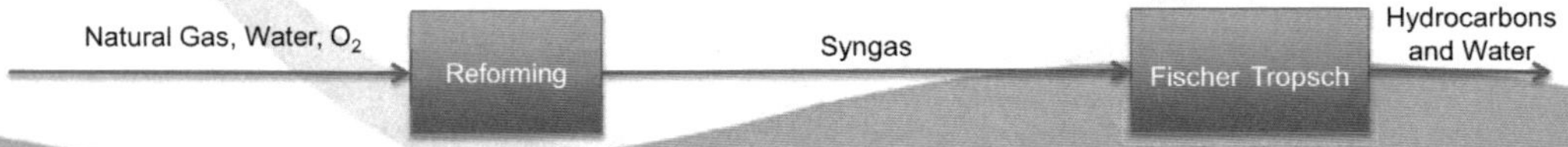

Development of Conventional GTL Process: Produce the Right Syngas

- **Pure O_2 is required**
  - Need an Oxygen Plant
- **H_2/CO is wrong**
  - Corrected by water gas shift reaction
  - $CO + H_2O \rightarrow CO_2 + H_2$
- **High moisture content in syngas**
  - Water is to be knocked out
- **Feed stock is always dirty**
  - Syngas is also dirty, clean up is needed to protect downstream catalyst and for easy operation

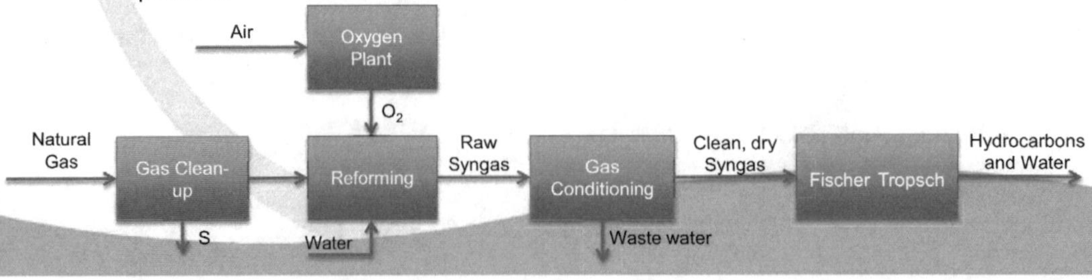

Development of Conventional GTL Process: Produce the Right Products

- **The Fischer Tropsch reaction unable achieve 100% conversion and produces a wide range of hydrocarbons from C_1 to C_{many}**
  - Normally Only C_{5+} are needed for fuel
    - Naphtha, Kerosene, Diesel, Fuel oil, Wax
  - Light hydrocarbons are useful chemicals
    - Ethylene, Propene, Butene, etc.
  - Unreacted syngas and light hydrocarbons can be recycled through reforming process
    - CO_2 need to be knocked out during gas conditioning
- **Need for Separation and Product Workup**

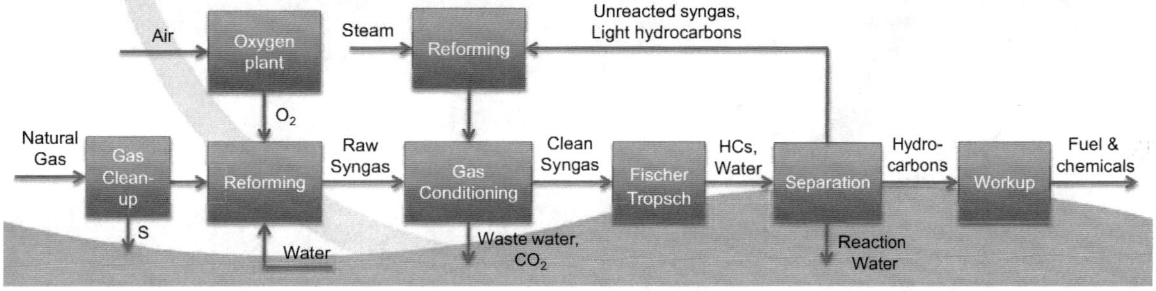

Development of Conventional GTL Process: Consider the Start-up and Activation

- **Inert gas needed for start-up**
  - Normally N_2 is used, which is also produced from the oxygen plant
- **Hydrogen needed for catalyst reduction/activation, and product workup**
  - H_2 can be produced from syngas via water gas shift reaction

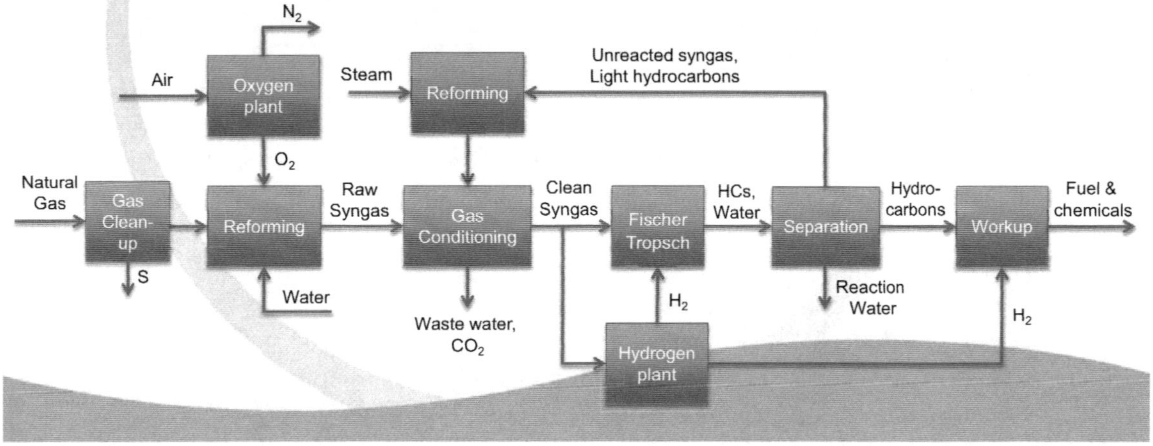

Development of Conventional GTL Process: And Utilities

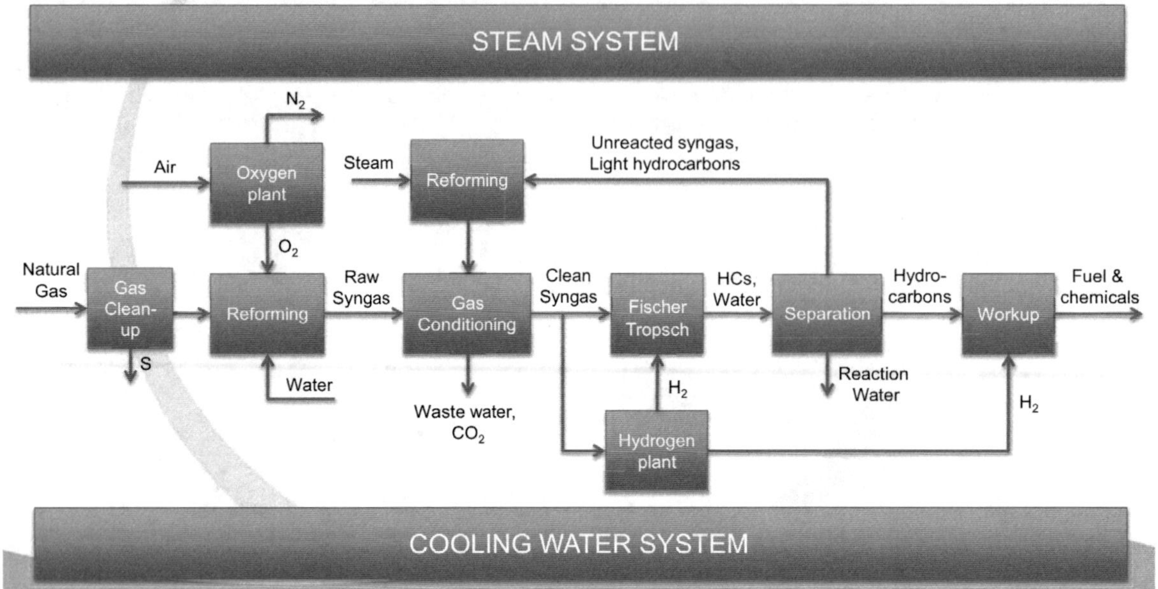

Development of Conventional GTL Process: A Real Case

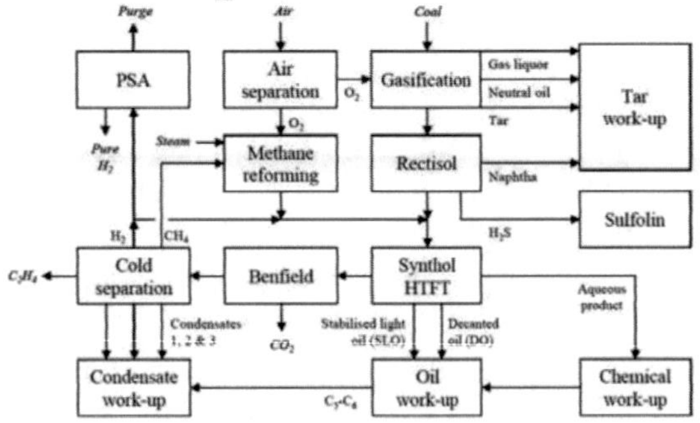

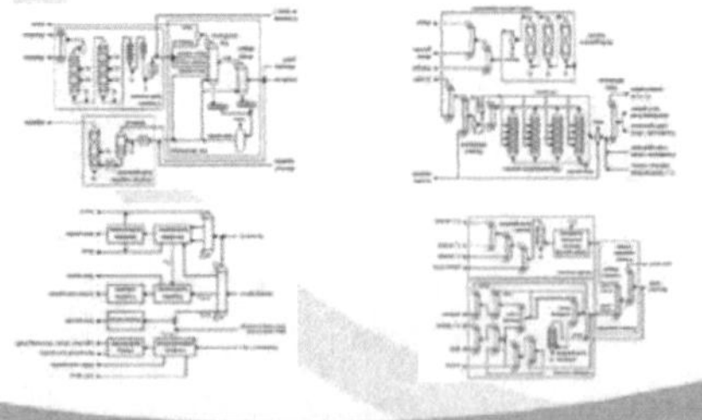

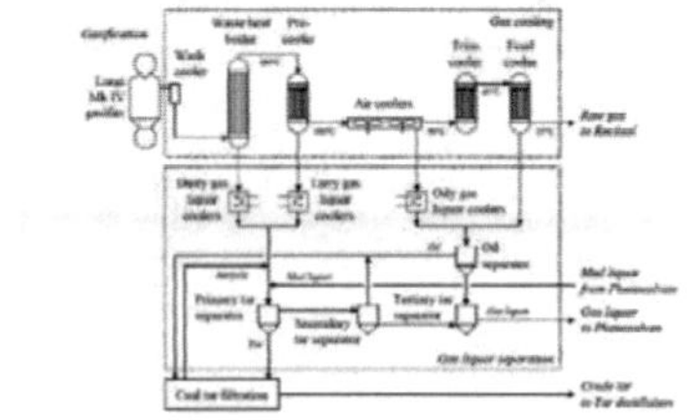

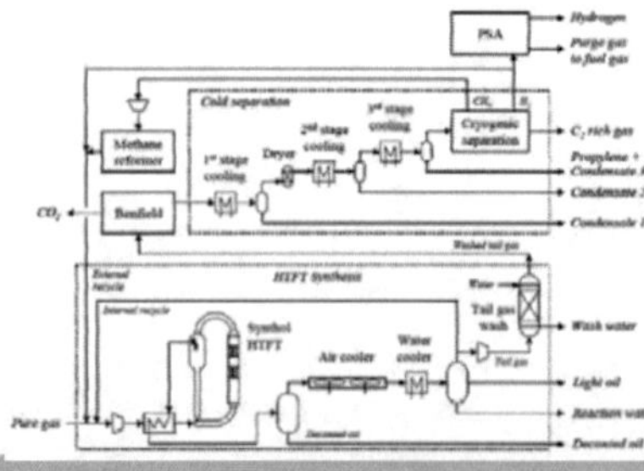

Small-scale approach: Can we simplify it?

- **Do we need a recycle?**
  - By pushing conversion & heavy hydrocarbon selectivity higher, recycle can be removed
  - Unconverted syngas and light hydrocarbons can be used to provide energy (heat or electricity)

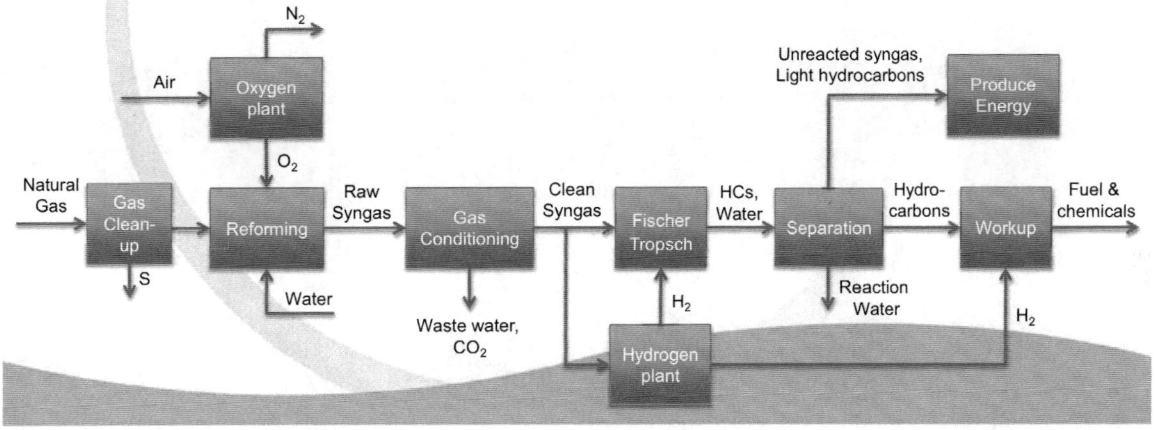

Small-scale approach: Can we simplify it?

- **As there is no recycle, we don't have to use pure oxygen for reforming**
  - Air plant can be smaller or be removed
- **We can tolerate more CO_2 in the FT section**
  - Removing CO_2 in gas conditioning is not a must

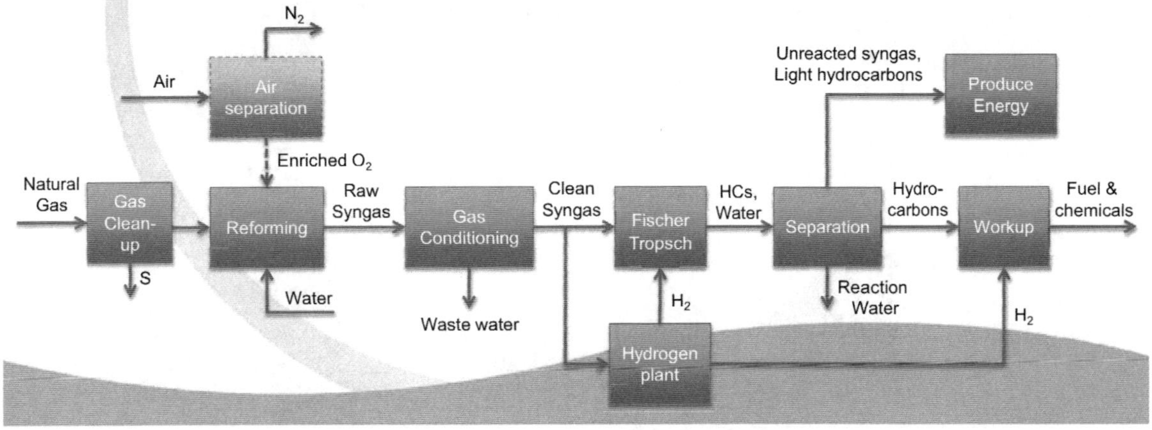

Small-scale approach: Can we simplify it?

- **Is producing syncrude sufficient?**
  - Produce syncrude and sell to refineries

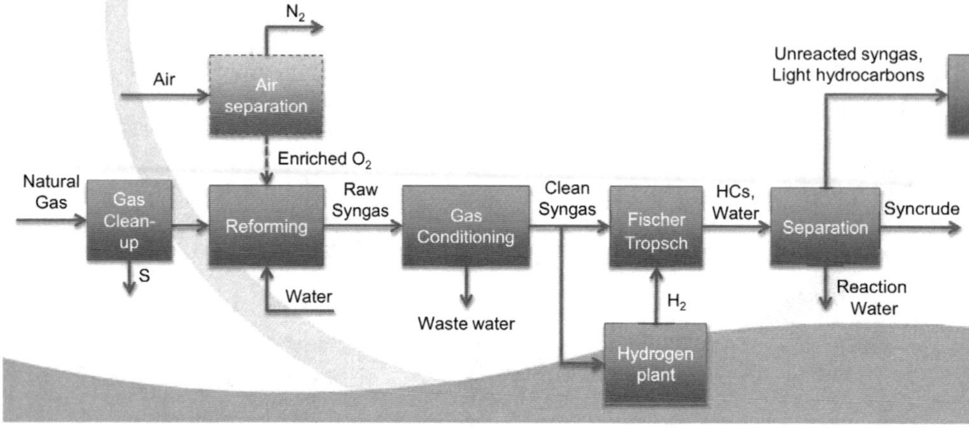

Small-scale approach: Can we simplify it?

- **Is Hydrogen a must?**
  - By developing new catalyst, catalyst can be reduced by syngas, thus no need for Hydrogen

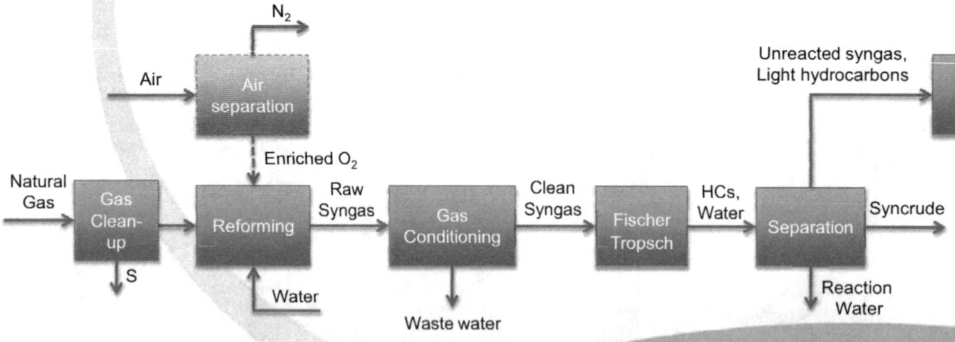

Small-scale approach: Can we simplify it?

- **Is Air Separation necessary?**
  - By developing new catalyst, it can be dried by syngas
  - Reactor can be flushed with clean natural gas in shutdowns

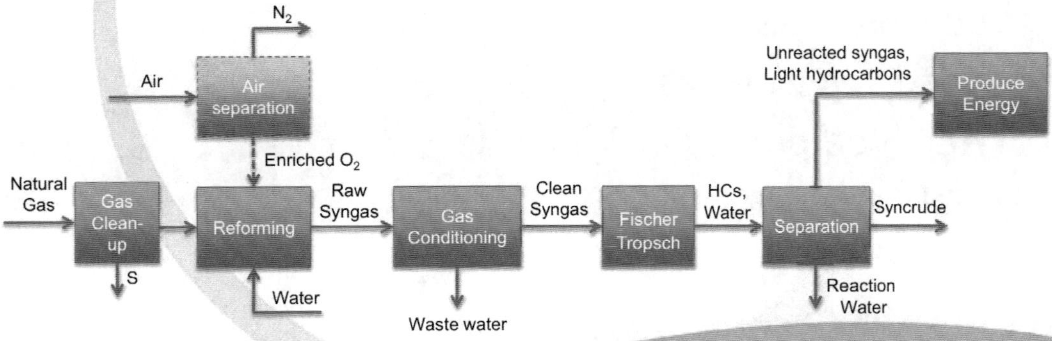

Now we have a simplified process

- **More Robust, modular, once-through**
- **Less Capital intensive**
- **Request for catalyst:**
  - Active at low syngas partial pressure
  - High heavy hydrocarbon selectivity
  - Tolerant to CO_2
  - Reduction using syngas
  - Easy to regenerate

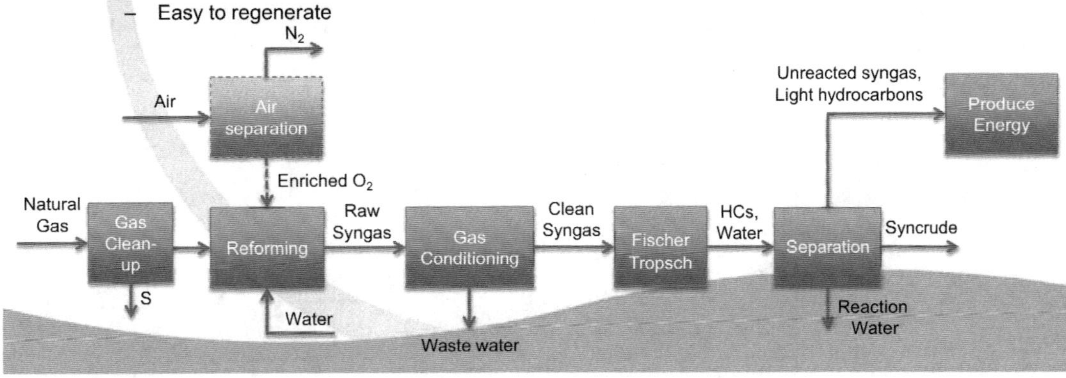

500bpd GTL: Feasibility Study

- Request from client:
  - Conduct a Feasibility Study of a 500bpd GTL demo unit
  - Location: USA
  - Feed: pipeline gas
  - Required products: Syncrude & Wax
  - MaPS responsibility: FT Section (using modular approach)
  - Rest of plant to be done by client-chosen vendors

500bpd GTL: Feasibility Study

- Feasibility Study included:
  - Concept design
    - Operating conditions (T, P, syngas ratio, etc.)
    - Catalyst selection (Fe vs Co)
    - Environmental considerations
  - Equipment listing and sizing
  - Cost estimates and quotes
  - Syncrude and Wax Product offtake

Flow Sheet

- **Flow sheet of entire plant:**
  - Choose ATR conditions
  - Enriched O_2 VS Air
  - Minimize utilities required
  - No heat/electricity needed (by client): thus flare tail-gas

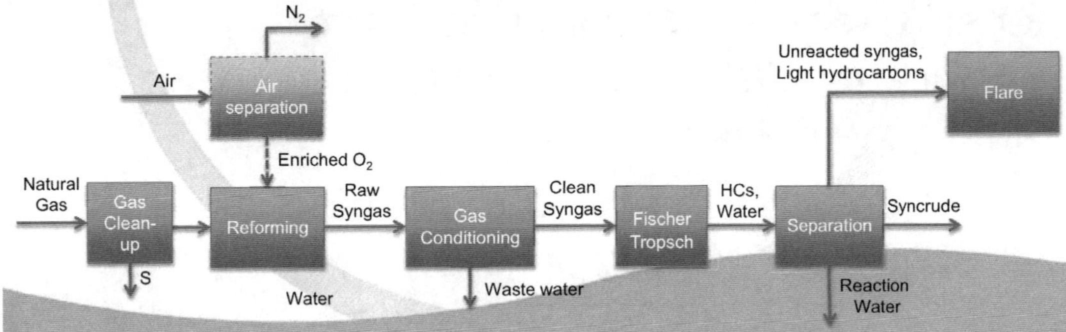

Economics

- CAPEX comparison with a recent large-scale (96,000bpd) GTL case for USA (feasibility study, 2012)

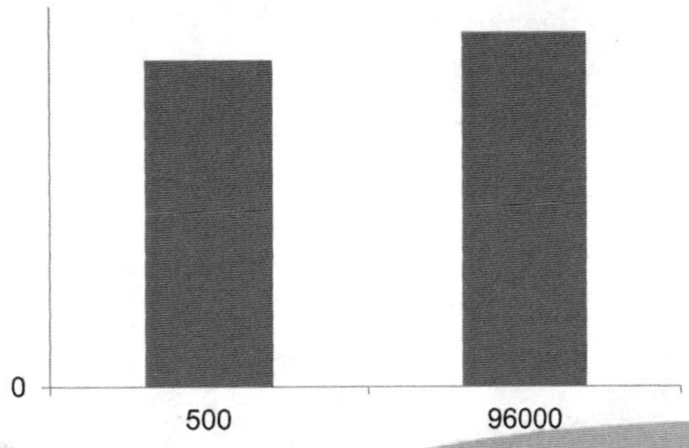

Problems encountered

- Products – what to do with them?
  - Refineries?
  - Other end users?
- Process Engineering aspect has progressed for small-scale modular plants. Other engineering disciplines need to also. This is vital to the economics
- The large-scale design and manufacturing approach can not be applied

Biomass to Liquid Pilot Plant

- Funded by SA NRF and China MoST
  - Capacity: up to 1 L syncrude production a day
  - Collaboration with Hebei Academy of Agricultural and Forestry Sciences
    - In Shijiazhuang, Hebei, China
- Full demonstration of biomass to syncrude process
  - Air blow biomass gasification
  - Syngas conditioning
  - Fischer Tropsch Synthesis
- Successful commissioned in 2012

Gasification

Down draft air blow ambient pressure biomass gasification unit
Load: 50 kg of compressed wood
Capacity: 10 Nm^3/hr

Gasification → Water scrubber → Corn cob drier → Buffer tank → Raw gas clean up → Compressor

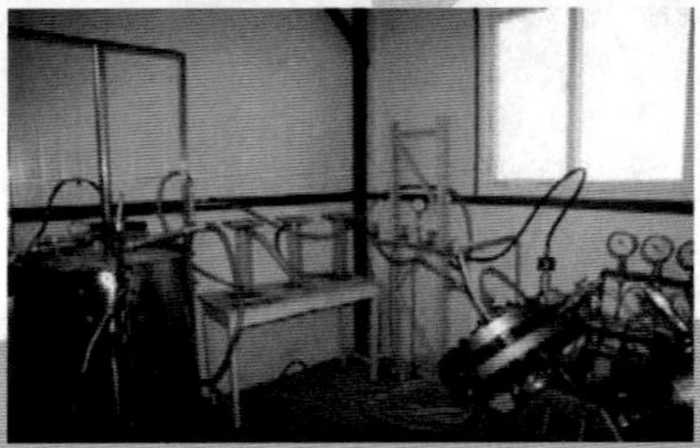

29

Typical gasification results (v%)

	N_2	CH_4	CO	H_2	O_2	H_2/CO
1	62.3	6.0	13.2	20.0	1.7	1.5
2	59.9	2.3	21.0	13.6	0.9	0.65
3	41.5	5.7	11.0	18.7	1.2	1.7
4	46.6	4.8	13.3	16.0	0.7	1.2
5	46.9	4.8	13.5	15.6	0.6	1.2

30

Guard Bed

Remove H_2S → Hydrolysis → Remove H_2S → Remove O_2 → Drier

- Total S level is lowered
  - From ~ 10 ppmv to < 10 ppbv
- O_2 < 0.1 v%
- Over 2000+ hrs operation experience achieved
  - No deactivation observed

Fischer Tropsch Synthesis

- Single tube fixed bed reactor
  - i.d. 50 mm
  - Length 900 mm
  - Max catalyst bed 600 mm
    - 1.2 L
- Heated by
  - heating oil (< 260 C)
  - Nitrogen (250 – 400 C)
- Reaction pressure up to 4 MPa
- Two steps products separation
  - Step one: 150-190 C
  - Step two: 4 C

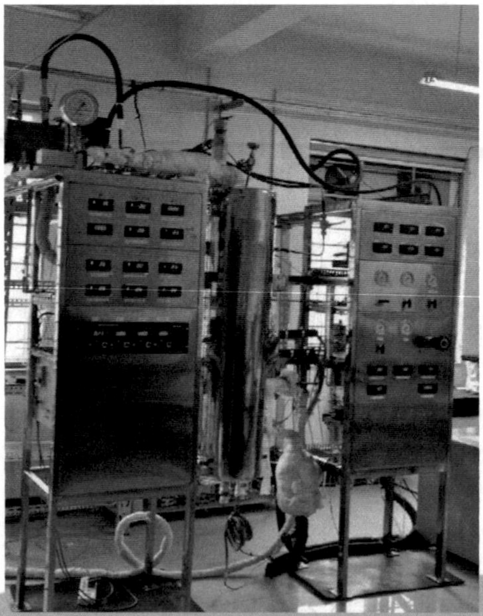

32

Liquid products

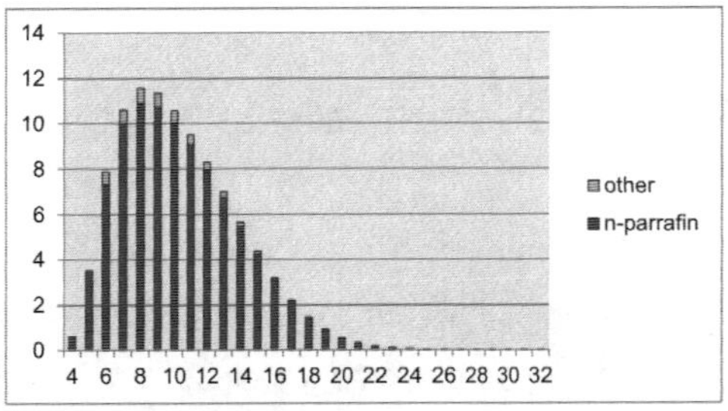

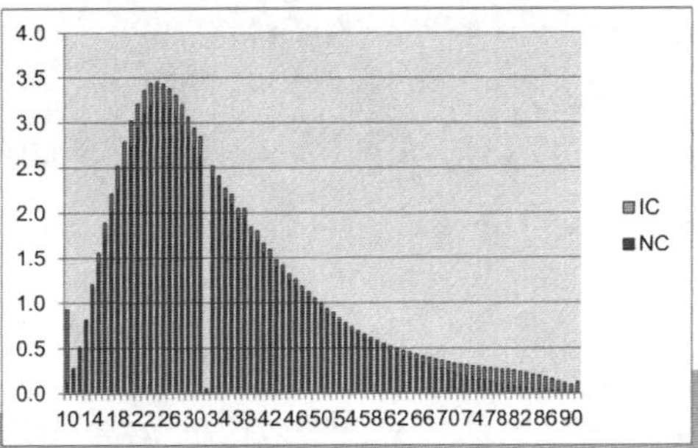

33

Simulation results based on pilot data

- At the right operation condition, 1 barrel of syncrude (including wax) can be produced using 1 ton of biomass (LHV ~ 20 MJ/kg), without energy input from outside.
- More fuel can be produced if electricity is provided from grid

Current and future work

- Container scale waste to liquid demonstration plant in design phase
  - 1 barrel syncrude production a day from municipal solid waste
  - Collaboration with Nuclear Energy Cooperation South Africa (NECSA)
  - To be constructed later 2014
  - Targeting
    - Fully independent plant, or
    - Maximum fuel production
- Further simplify the process by looking at the utility requirement

Thank you

Learn without limits.

UNISA college of science, engineering and technology

Study of the active site over Rh substituted pyrochlores ($La_2Zr_2O_7$) for dry (CO_2) reforming of CH_4

Devendra Pakhare[1], Viviane Schwartz[2], Victor Abdelsayed[3], Daniel Haynes[3], Dushyant Shekhawat[3], James Spivey[1]

1 – Department of Chemical Engineering, Louisiana State University, Baton Rouge, LA 70803, USA

2 – Center for Nanophase Materials Science, Oak Ridge National Laboratory, Oak Ridge, TN 37831, USA

3 – National Energy Technology Laboratory, Collins Ferry Rd., Morgantown, WV 26507, USA

Outline

- Dry reforming
- Pyrochlores as catalysts
- CO_2 and CH_4 activation: FTIR
- Conclusion

Dry (CO_2) Reforming of Methane

U.S. DEPARTMENT OF ENERGY
Center for Atomic-Level Catalyst Design
an Office of Basic Energy Sciences
Energy Frontier Research Center

Dry Reforming of Methane (DRM)

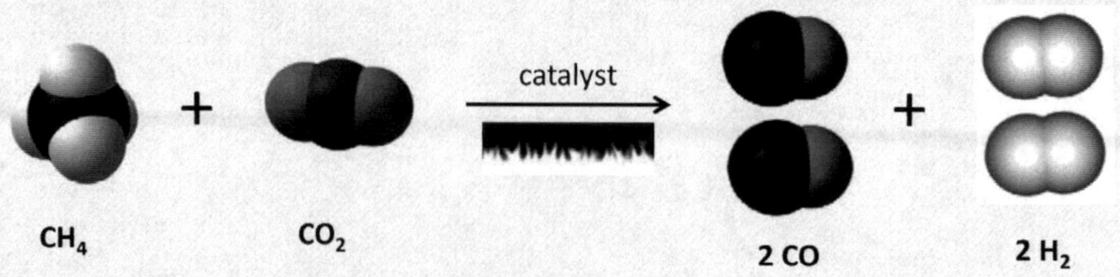

- Involves two extremely stable reactant molecules.
- Highly endothermic reaction (ΔH_{298K} = +59.1 kcal mol^{-1})
- Thus operates at high temperatures.

Introduction

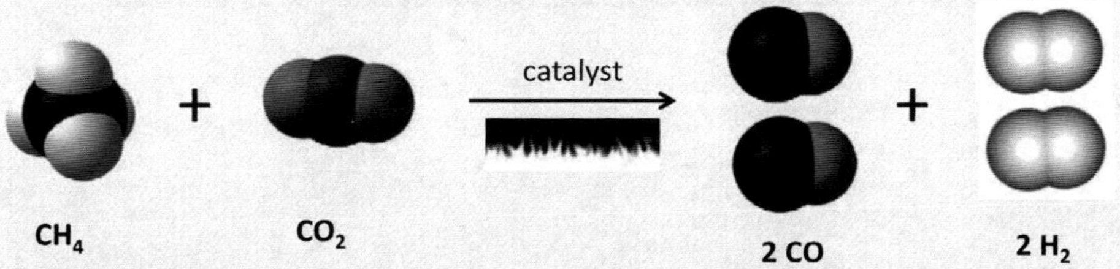

- Produces syngas with H_2/CO ratio = 1.
- Usually RWGS occurs simultaneously.

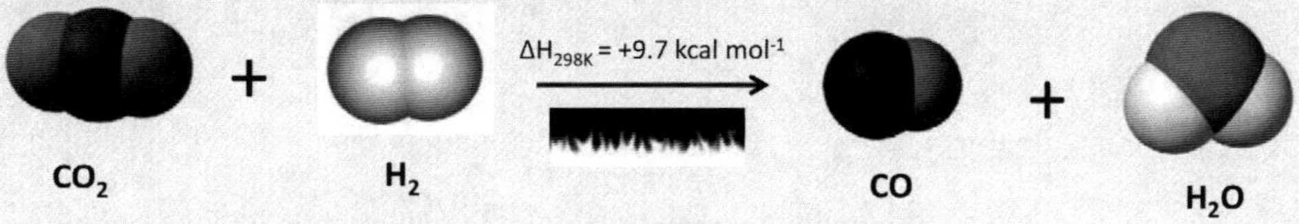

- Thus DRM results in H_2/CO ratio < 1.

Carbon deposition

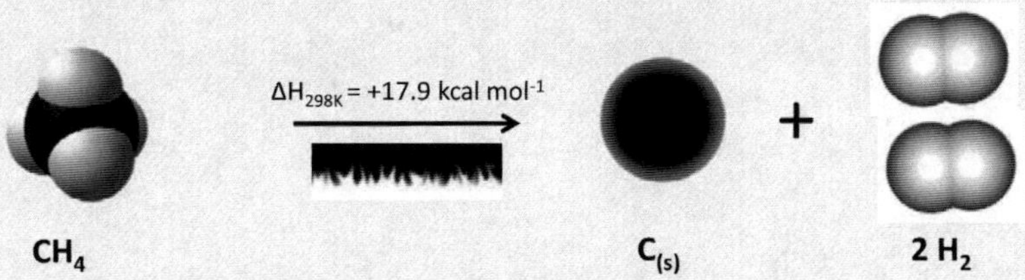

methane decomposition

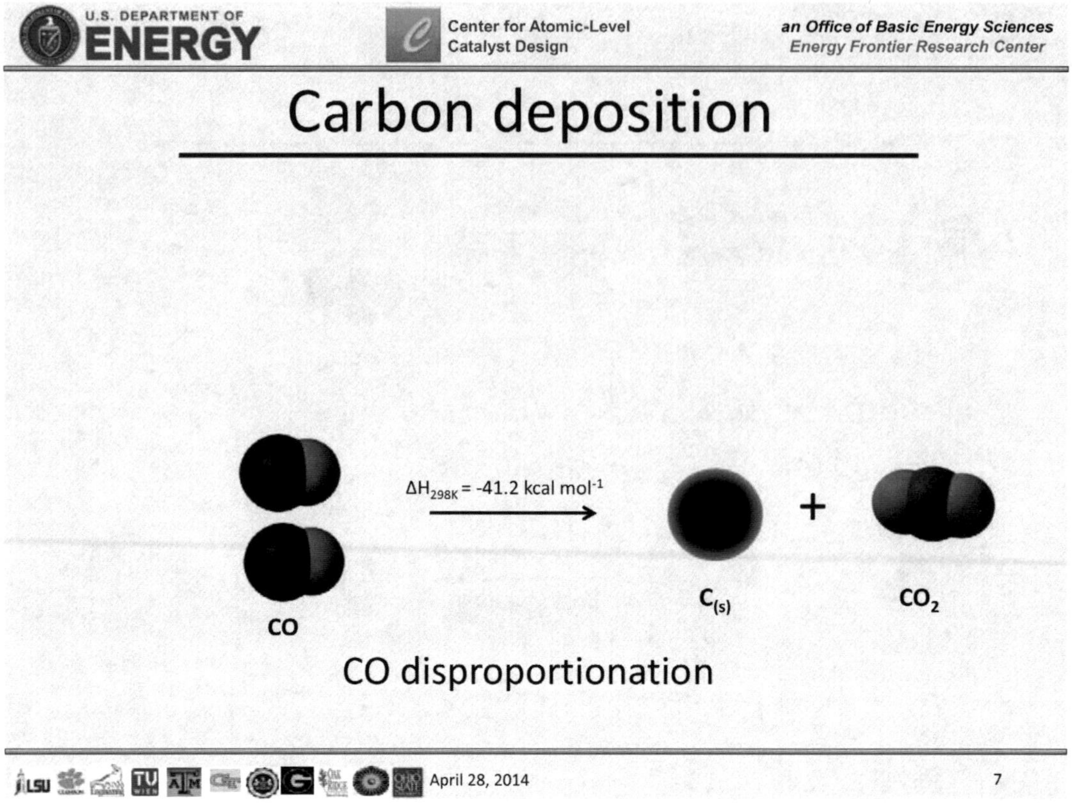
U.S. DEPARTMENT OF ENERGY
Center for Atomic-Level Catalyst Design
an Office of Basic Energy Sciences
Energy Frontier Research Center
Carbon deposition
ΔH_{298K} = -41.2 kcal mol^{-1}
CO
$C_{(s)}$
+
CO_2
CO disproportionation
April 28, 2014
7

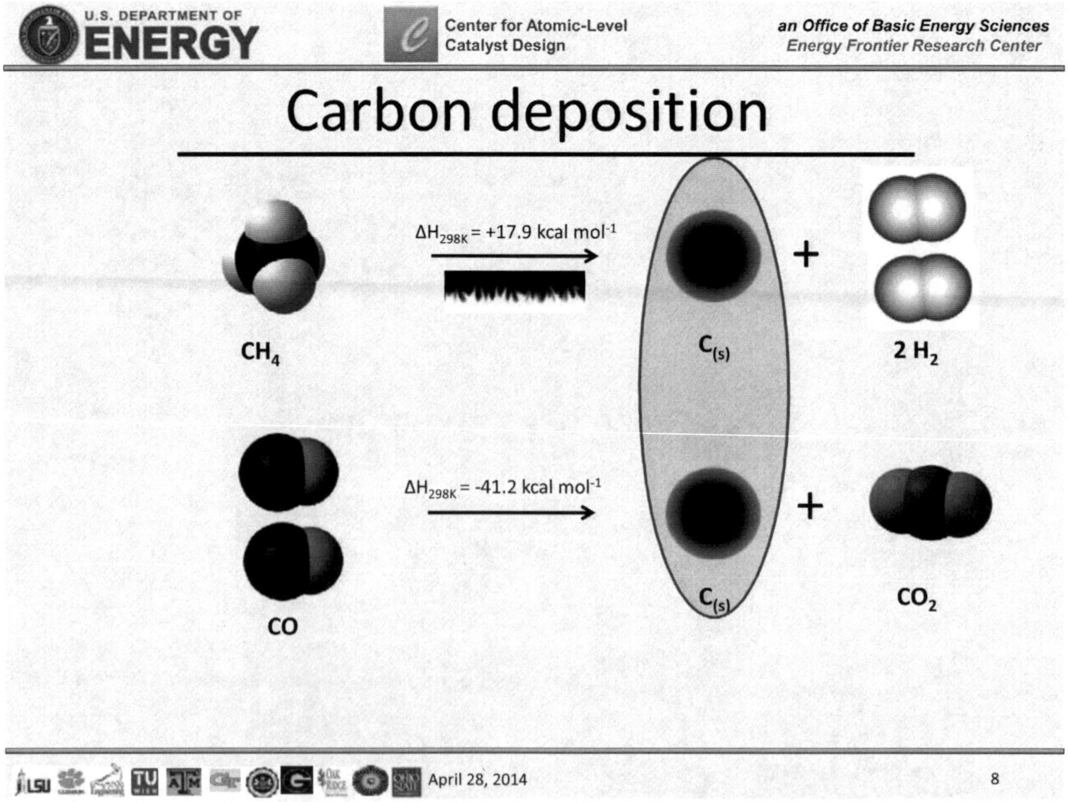
U.S. DEPARTMENT OF ENERGY
Center for Atomic-Level Catalyst Design
an Office of Basic Energy Sciences
Energy Frontier Research Center
Carbon deposition
ΔH_{298K} = +17.9 kcal mol^{-1}
CH_4
$C_{(s)}$
+
2 H_2
ΔH_{298K} = -41.2 kcal mol^{-1}
CO
$C_{(s)}$
+
CO_2
April 28, 2014
8

Pyrochlores as catalysts

Catalyst used in the study

- Pyrochlores : metal oxide crystals.
- General formula: $A_2B_2O_7$
- We use $La_2Zr_2O_7$ with Rh on (Zr) B-site

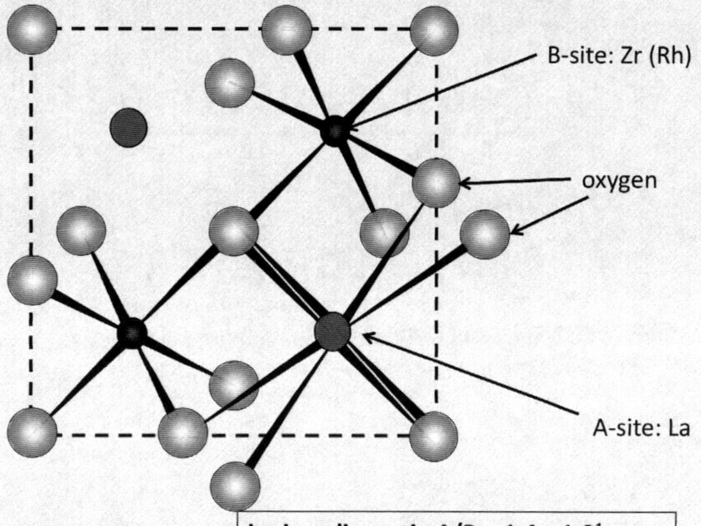

ionic radius ratio A/B = 1.4 – 1.8[1]

[1] D. Sedmidubsky et al., *Journal of Chemical Thermodynamics* 37 (2005) 1098.
D.J. Haynes et al., *Catal. Today*, 136 (2008) 206-213.

Why Pyrochlores??

- Thermally stable.
- Oxygen conductivity inhibits carbon formation.
- One paper on pyrochlores for DRM (Ashcroft et al., 1993).
  - A-site composed of Nd/ Sm/ Eu/ Gd
  - B-site composed of Ir or Ru.
  - e.g., $Eu_2Ir_2O_7$
  - Pyrochlores decomposed above 340°C.
- We partially substitute B-site with Rh.

U.S. DEPARTMENT OF ENERGY
Center for Atomic-Level Catalyst Design
an Office of Basic Energy Sciences Energy Frontier Research Center

Catalyst used in this study

- 0 wt% Rh → **LZ**
- 2 wt% Rh → **LRhZ (2%)**
- 5 wt% Rh → **LRhZ (5%)**

CO_2 activation by La: FTIR

1. Theory of La-oxycarbonates

U.S. DEPARTMENT OF ENERGY
Center for Atomic-Level Catalyst Design
an Office of Basic Energy Sciences Energy Frontier Research Center

Theory of La-oxycarbonates

- Main reaction

$$La_2O_3 + CO_2 \rightarrow (La_2O_2^{2+})CO_3^{2-}$$

- La-carbonates differ in structural orientation (type I, Ia, II).
  - Type I has square layers and is tetragonal.
  - Type Ia is monoclinic distortion of type I.
  - Type II has a hexagonal unit cell.

R.P. Turcotte, J.O. Sawyer, L. Eyring, Inorg. Chem. 8 (1969) 238.
S. Irusta, L.M. Cornaglia, E.A. Lombardo, Mater. Chem. Phys. 86 (2004) 440-447.

Theory of La-oxycarbonates

- Main reaction

$$La_2O_3 + CO_2 \rightarrow (La_2O_2^{2+})CO_3^{2-}$$

- La-carbonates differ in structural orientation (type I, Ia, II).
  - Type I has square layers and is tetragonal.
  - Type Ia is monoclinic distortion of type I.
  - Type II has a hexagonal unit cell.

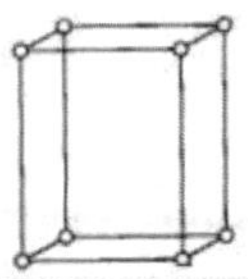

R.P. Turcotte, J.O. Sawyer, L. Eyring, Inorg. Chem. 8 (1969) 238.
S. Irusta, L.M. Cornaglia, E.A. Lombardo, Mater. Chem. Phys. 86 (2004) 440-447.

U.S. DEPARTMENT OF ENERGY
Center for Atomic-Level Catalyst Design
an Office of Basic Energy Sciences
Energy Frontier Research Center

Theory of La-oxycarbonates

- Main reaction

$$La_2O_3 + CO_2 \rightarrow (La_2O_2^{2+})CO_3^{2-}$$

- La-carbonates differ in structural orientation (type I, Ia, II).
  - Type I has square layers and is tetragonal.
  - Type Ia is monoclinic distortion of type I.
  - Type II has a hexagonal unit cell.

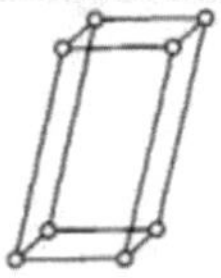

R.P. Turcotte, J.O. Sawyer, L. Eyring, Inorg. Chem. 8 (1969) 238.
S. Irusta, L.M. Cornaglia, E.A. Lombardo, Mater. Chem. Phys. 86 (2004) 440-447.

U.S. DEPARTMENT OF ENERGY | Center for Atomic-Level Catalyst Design | an Office of Basic Energy Sciences Energy Frontier Research Center

Theory of La-oxycarbonates

- Main reaction

$$La_2O_3 + CO_2 \rightarrow (La_2O_2^{2+})CO_3^{2-}$$

- La-carbonates differ in structural orientation (type I, Ia, II).
  - Type I has square layers and is tetragonal.
  - Type Ia is monoclinic distortion of type I.
  - Type II has a hexagonal unit cell.

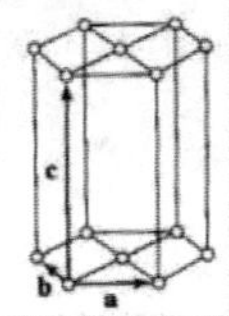

R.P. Turcotte, J.O. Sawyer, L. Eyring, Inorg. Chem. 8 (1969) 238.
S. Irusta, L.M. Cornaglia, E.A. Lombardo, Mater. Chem. Phys. 86 (2004) 440-447.

U.S. DEPARTMENT OF ENERGY | Center for Atomic-Level Catalyst Design | an Office of Basic Energy Sciences Energy Frontier Research Center

CO_2 activation by La: FTIR

2. Catalyst pre-treatment

FTIR experimental procedure for CO_2 adsorption

- ✓ Ramped to 550°C in flowing 4% H_2/He.
- ✓ Reduced at 550°C for 30 min in flowing 4% H_2/He.
- ✓ Flush with He for 15 min and collect background spectra.
- ✓ Flow 5% CO_2/He for 30 min and collect spectra.
- ✓ Flow He for 30 min and collect spectra.

CO_2 activation by La: FTIR

3. Series spectra

U.S. DEPARTMENT OF ENERGY | Center for Atomic-Level Catalyst Design | *an Office of Basic Energy Sciences Energy Frontier Research Center*

Series spectra for **LRhZ (2%)**

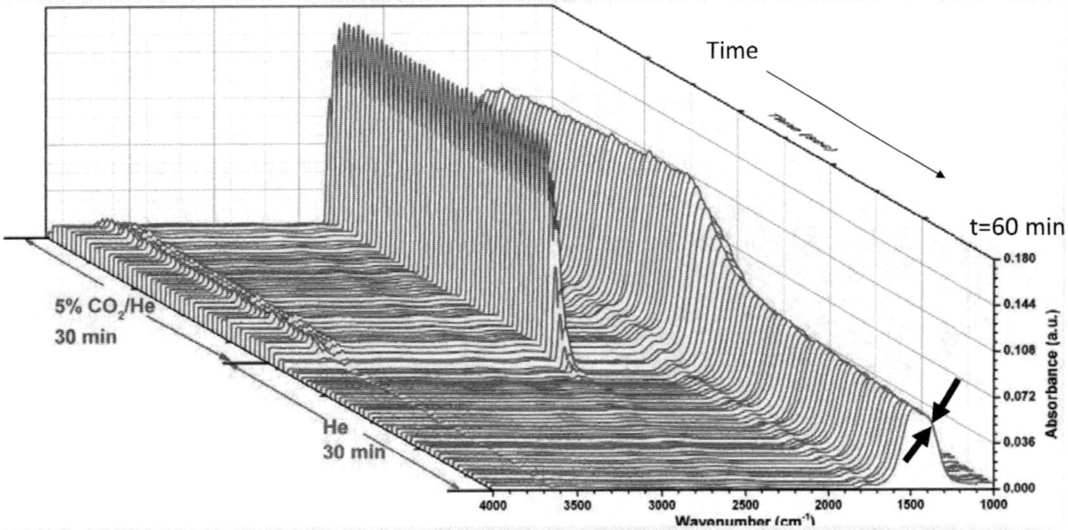

U.S. DEPARTMENT OF ENERGY | Center for Atomic-Level Catalyst Design | *an Office of Basic Energy Sciences Energy Frontier Research Center*

CO_2 activation by La: FTIR

4. Adsorbed species

U.S. DEPARTMENT OF ENERGY | Center for Atomic-Level Catalyst Design | *an Office of Basic Energy Sciences Energy Frontier Research Center*

Type-Ia

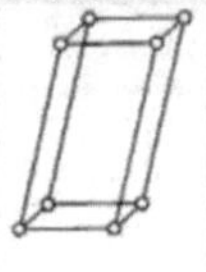

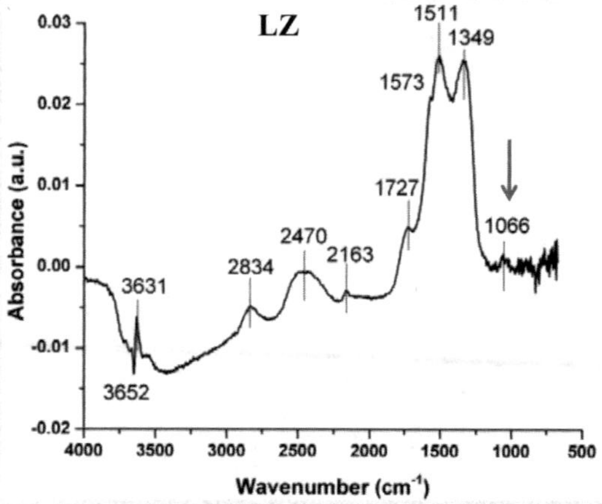

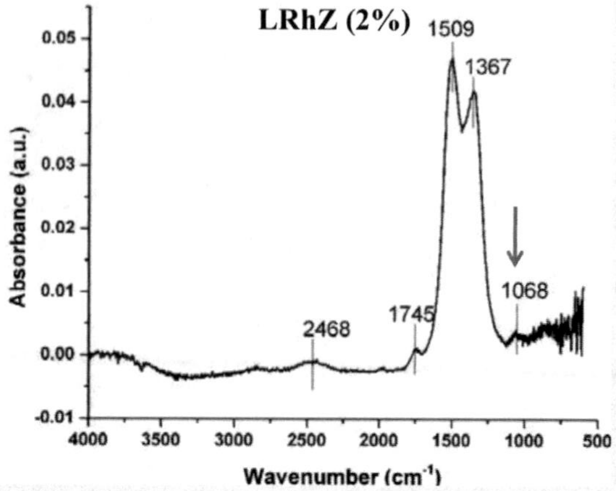

April 28, 2014 23

U.S. DEPARTMENT OF ENERGY | Center for Atomic-Level Catalyst Design | *an Office of Basic Energy Sciences Energy Frontier Research Center*

Type-Ia

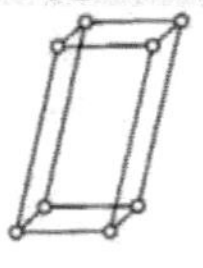

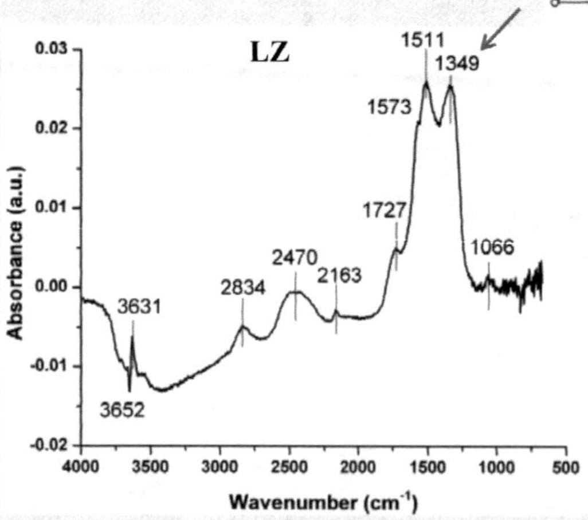

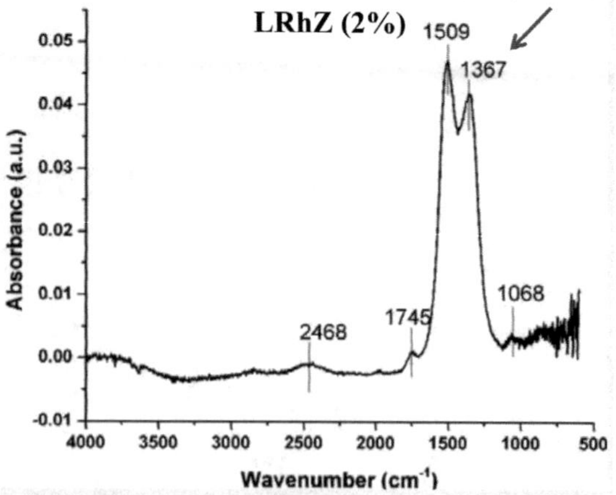

April 28, 2014 24

U.S. DEPARTMENT OF ENERGY

Center for Atomic-Level Catalyst Design

an Office of Basic Energy Sciences
Energy Frontier Research Center

Type-II

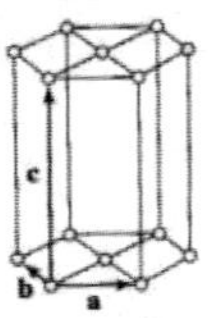

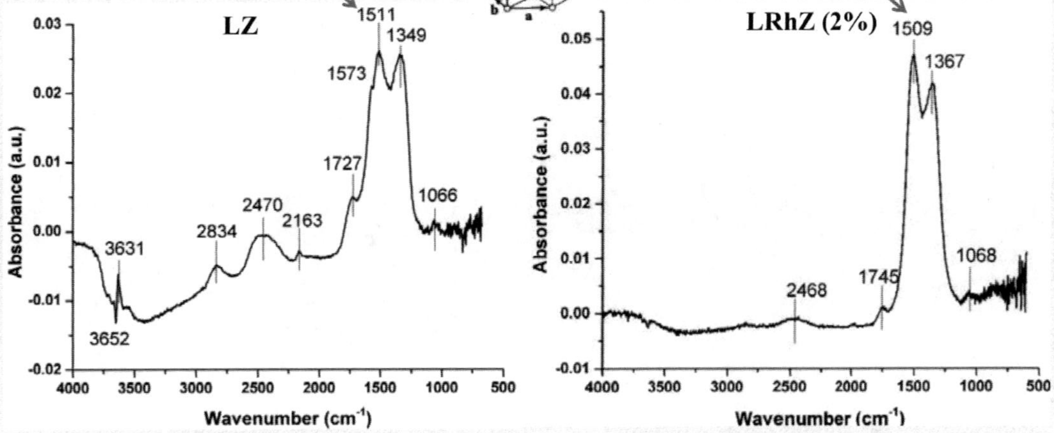

U.S. DEPARTMENT OF ENERGY

Center for Atomic-Level Catalyst Design

an Office of Basic Energy Sciences
Energy Frontier Research Center

Carbonyl group

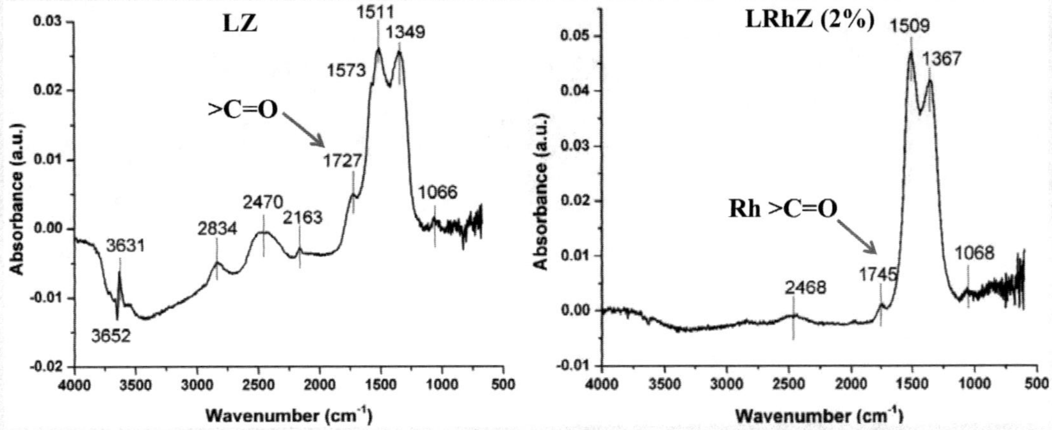

Hydroxyls reduce CO_2 to CO and C-H stretching

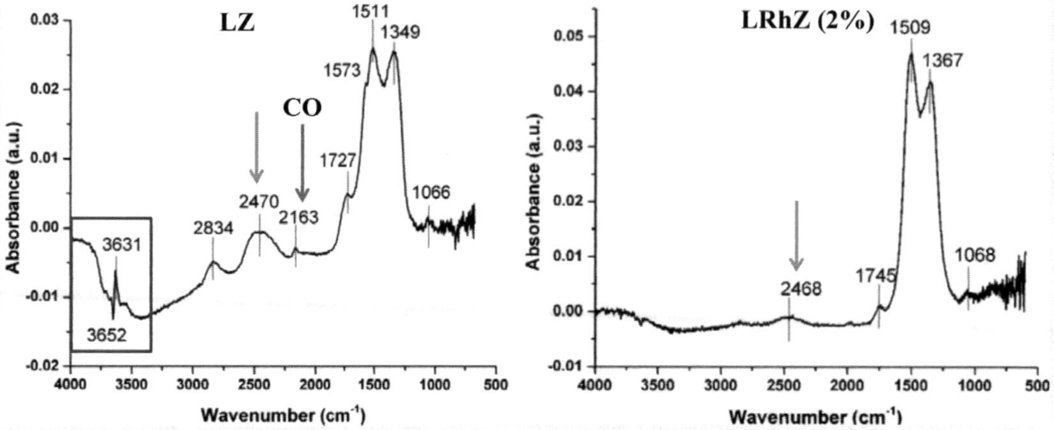

U.S. DEPARTMENT OF ENERGY
Center for Atomic-Level Catalyst Design
an Office of Basic Energy Sciences
Energy Frontier Research Center

CO_2 activation by La: FTIR

5. Pulsing CH_4 to study the reactive carbonates

U.S. DEPARTMENT OF ENERGY | Center for Atomic-Level Catalyst Design | *an Office of Basic Energy Sciences Energy Frontier Research Center*

Pulsed CH_4 reacting with pre-adsorbed CO_2 over **L2RhZ**

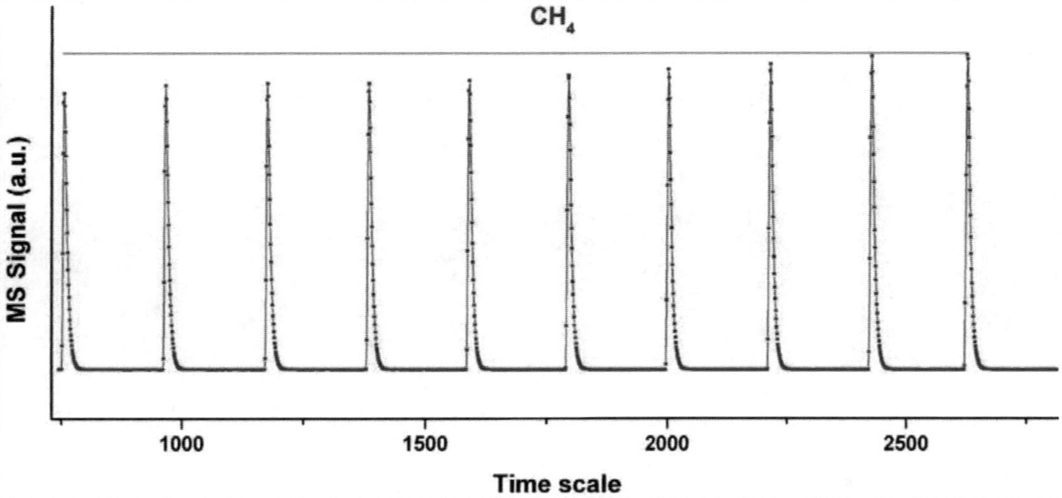

April 28, 2014 29

U.S. DEPARTMENT OF ENERGY | Center for Atomic-Level Catalyst Design | *an Office of Basic Energy Sciences Energy Frontier Research Center*

IR spectra after pulsing CH_4 for **LZ**

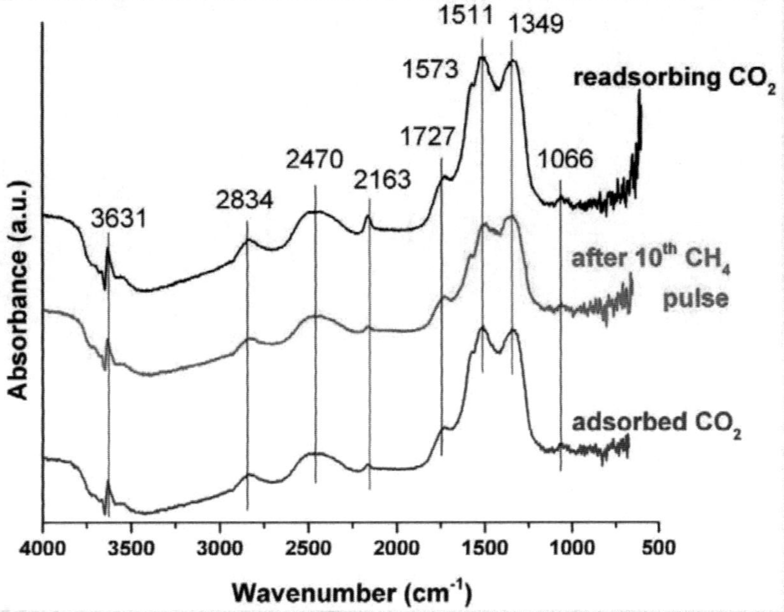

April 28, 2014 30

IR spectra after pulsing CH_4 for **L2RhZ**

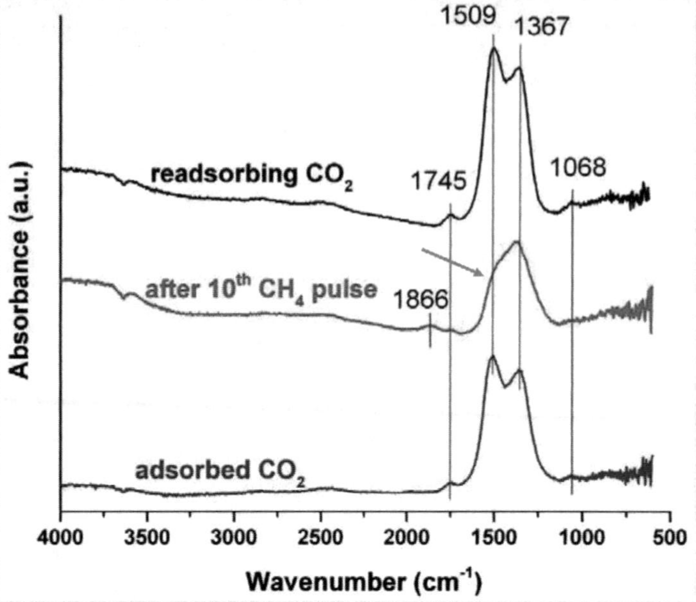

U.S. DEPARTMENT OF ENERGY
Center for Atomic-Level Catalyst Design
an Office of Basic Energy Sciences
Energy Frontier Research Center

Conclusions

Conclusions

- Basic La sites on pyrochlore surface activate CO_2 to form carbonates.

Conclusions

- Basic La sites on pyrochlore surface activate CO_2 to form carbonates.
- Carbonates differ in the structural orientation.

U.S. DEPARTMENT OF ENERGY | Center for Atomic-Level Catalyst Design | *an Office of Basic Energy Sciences Energy Frontier Research Center*

Conclusions

- Basic La sites on pyrochlore surface activate CO_2 to form carbonates.
- Carbonates differ in the structural orientation.

April 28, 2014 35

U.S. DEPARTMENT OF ENERGY | Center for Atomic-Level Catalyst Design | *an Office of Basic Energy Sciences Energy Frontier Research Center*

Conclusions

- Basic La sites on pyrochlore surface activate CO_2 to form carbonates.
- Carbonates differ in the structural orientation.
- The type-II carbonates are the dynamic surface species.

April 28, 2014 36

U.S. DEPARTMENT OF ENERGY | Center for Atomic-Level Catalyst Design | *an Office of Basic Energy Sciences Energy Frontier Research Center*

Conclusions

- Basic La sites on pyrochlore surface activate CO_2 to form carbonates.
- Carbonates differ in the structural orientation.
- The type-II carbonates are the dynamic surface species.
- **Rh-La interface is the active site for oxidation of carbon and reduction of carbonates.**

April 28, 2014 37

U.S. DEPARTMENT OF ENERGY | Center for Atomic-Level Catalyst Design | *an Office of Basic Energy Sciences Energy Frontier Research Center*

Acknowledgement

- This material is based upon work supported as part of the Center for Atomic Level Catalyst Design, an Energy Frontier Research Center funded by the U.S. Department of Energy, Office of Science, Office of Basic Energy Sciences under Award Number DE-SC0001058.

Center for Atomic Level Catalyst Design (CALCD-EFRC)

U.S Department of Energy

April 28, 2014 38

Acknowledgement

- A portion of this research was conducted at the Center for Nanophase Materials Sciences, which is sponsored at Oak Ridge National Laboratory by the Scientific User Facilities Division, Office of Basic Energy Sceinces, U.S. Department of Energy.

Oak Ridge National Laboratory

U.S Department of Energy

U.S. DEPARTMENT OF ENERGY
Center for Atomic-Level Catalyst Design
an Office of Basic Energy Sciences Energy Frontier Research Center

Thank you

CO_2 activation by La: In-situ XPS

Carbonates oxidize La

In-situ XPS experimental procedure for CO_2 adsorption

- ✓ Catalyst loaded and spectra was collected at 550°C.
- ✓ Catalyst at 550°C in vacuum and dosed with 10% H_2/He.
- ✓ Reduced at 550°C for 30 min in 10% H_2/He.
- ✓ Vacuumed at 550°C.
- ✓ Dosed with 10% CO_2/He for 30 min and vacuumed again.
- ✓ Collect spectra.

In-situ XPS: La 3d spectra for fresh **L2RhZ**

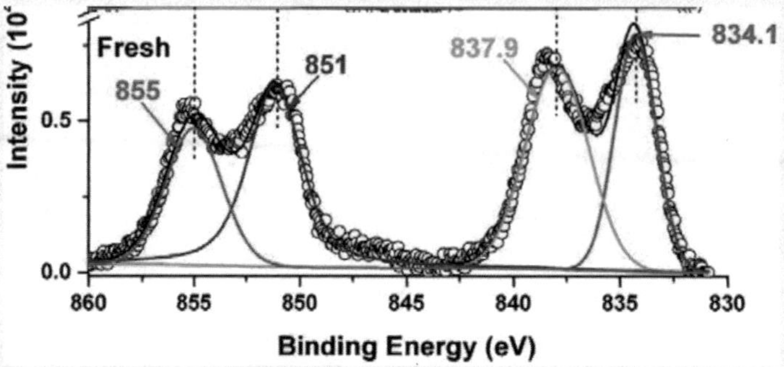

U.S. DEPARTMENT OF ENERGY
Center for Atomic-Level Catalyst Design
an Office of Basic Energy Sciences
Energy Frontier Research Center

In-situ XPS: La 3d for CO_2 treated **L2RhZ**

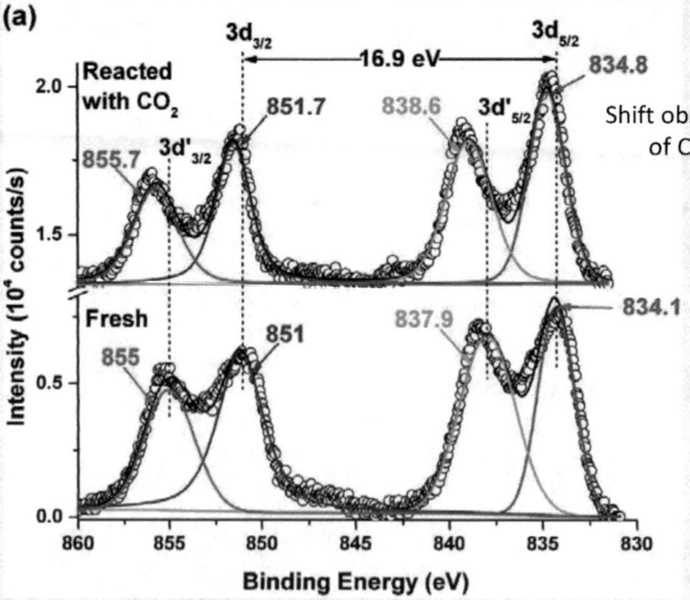

Shift observed as a result of CO_2 treatment

U.S. DEPARTMENT OF ENERGY | Center for Atomic-Level Catalyst Design | *an Office of Basic Energy Sciences Energy Frontier Research Center*

Catalyst characterization

X-Ray Diffraction (XRD)

45

U.S. DEPARTMENT OF ENERGY | Center for Atomic-Level Catalyst Design | *an Office of Basic Energy Sciences Energy Frontier Research Center*

Perovskite peak for LRhZ (5%)

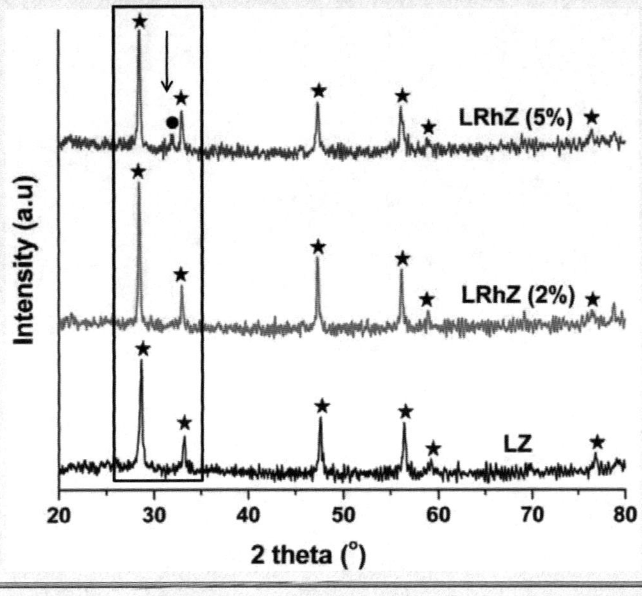

46

U.S. DEPARTMENT OF ENERGY | Center for Atomic-Level Catalyst Design | an Office of Basic Energy Sciences Energy Frontier Research Center

Possible perovskite peak for LRhZ (2%)

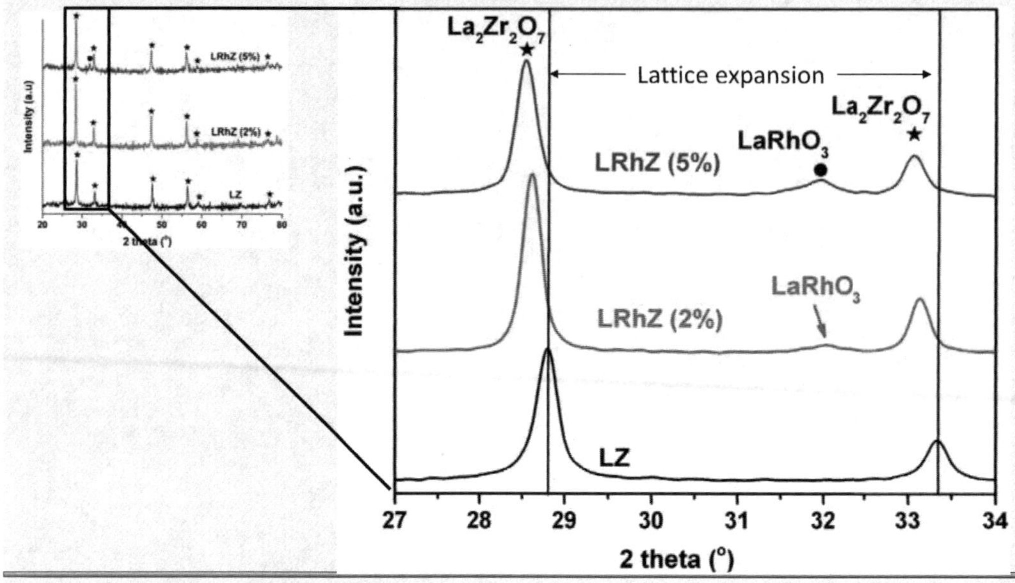

U.S. DEPARTMENT OF ENERGY | Center for Atomic-Level Catalyst Design | an Office of Basic Energy Sciences Energy Frontier Research Center

IR spectra after pulsing CH_4 for **L5RhZ**

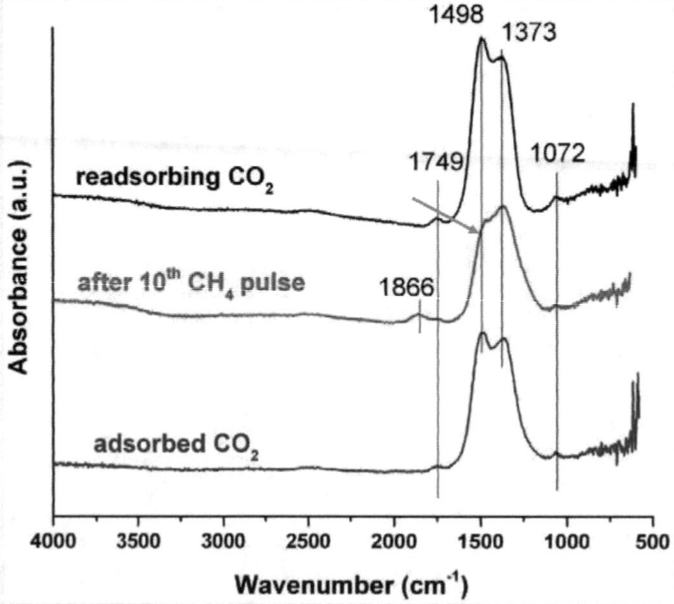

U.S. DEPARTMENT OF ENERGY

Center for Atomic-Level Catalyst Design

an Office of Basic Energy Sciences
Energy Frontier Research Center

Pulsed CH_4 reacting with pre-adsorbed CO_2 over **L2RhZ**

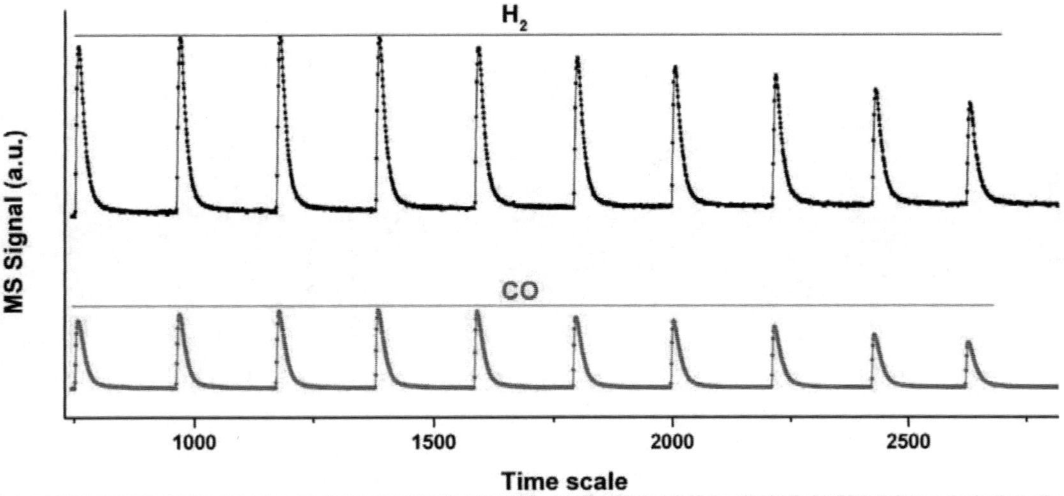

U.S. DEPARTMENT OF ENERGY

Center for Atomic-Level Catalyst Design

an Office of Basic Energy Sciences
Energy Frontier Research Center

Transient pulse with Ar tracer (L2RhZ)

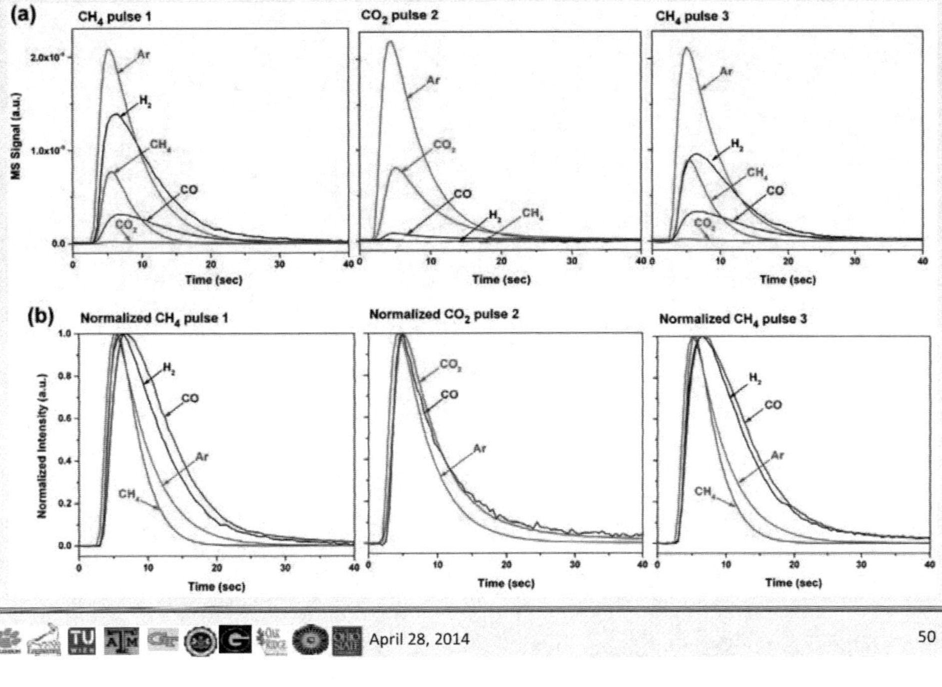

Transient pulse with Ar tracer (L5RhZ)

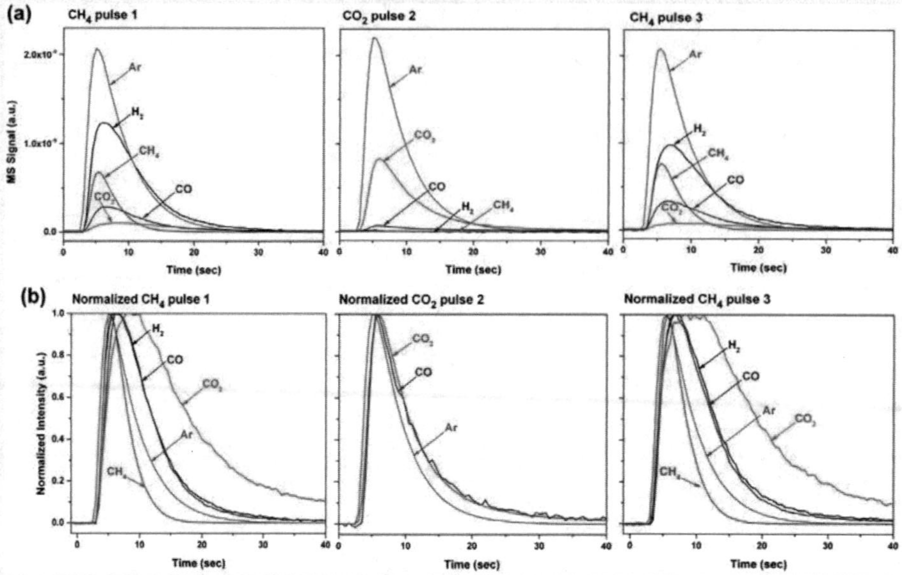

U.S. DEPARTMENT OF ENERGY
Center for Atomic-Level Catalyst Design
an Office of Basic Energy Sciences
Energy Frontier Research Center

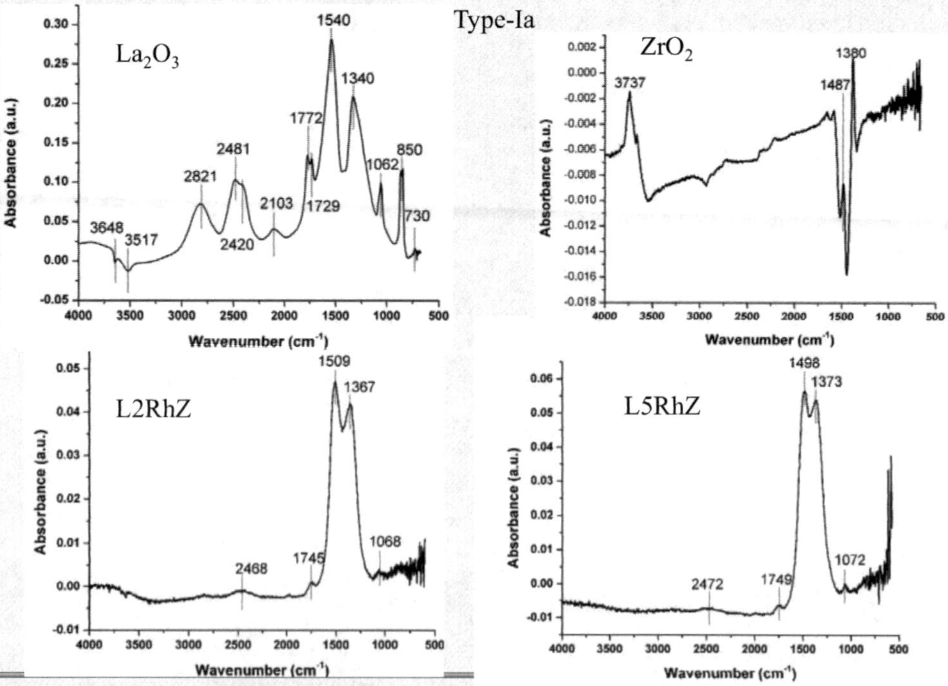

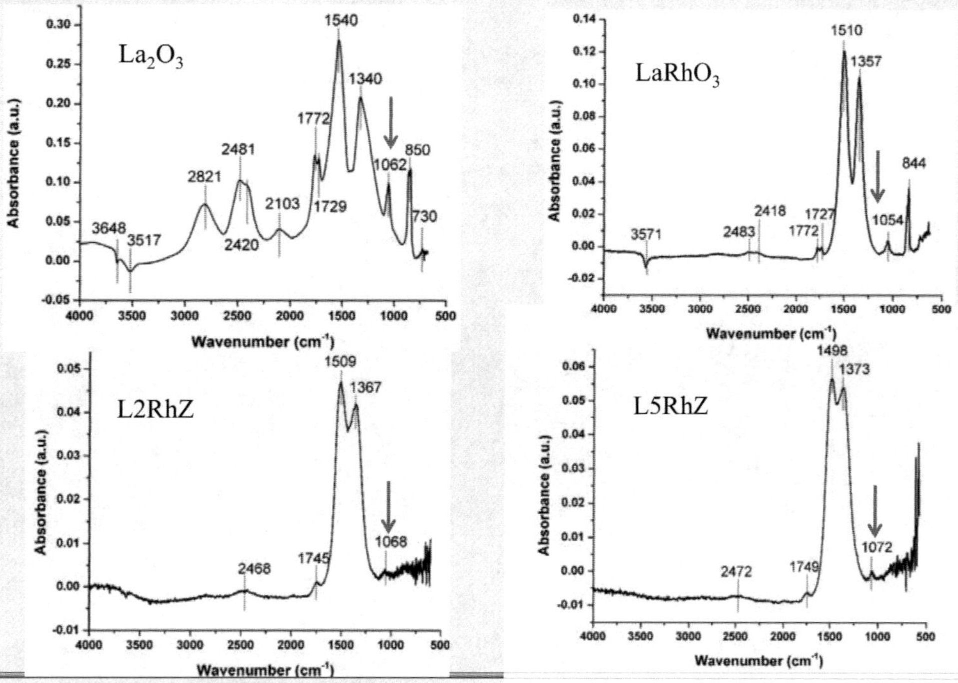

Equilibrium → high temperatures required

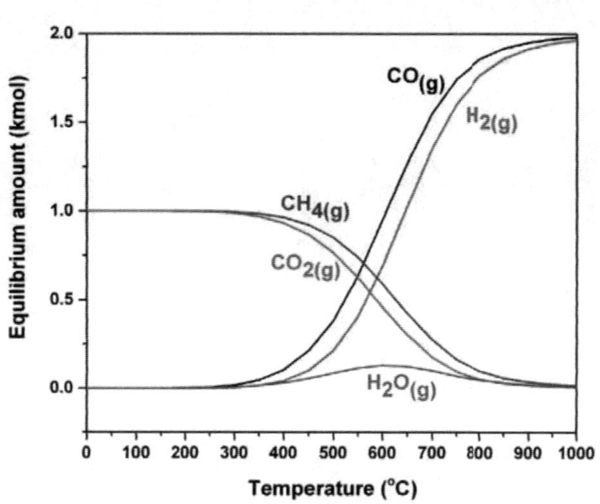

Center for Atomic-Level Catalyst Design

an Office of Basic Energy Sciences
Energy Frontier Research Center

Equilibrium → high temperatures required

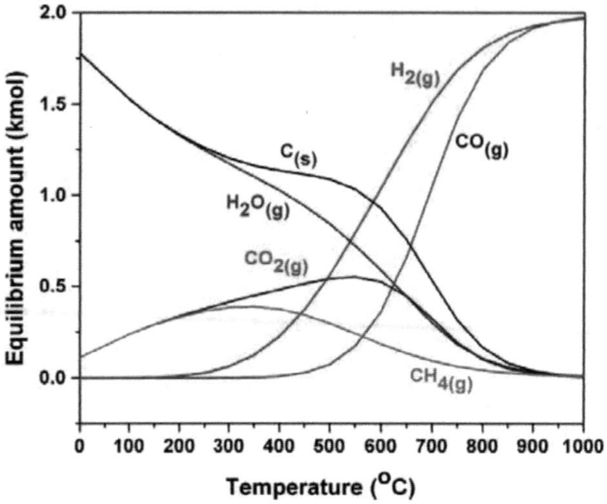

55

Center for Atomic-Level Catalyst Design

an Office of Basic Energy Sciences
Energy Frontier Research Center

Equilibrium → high temperatures required

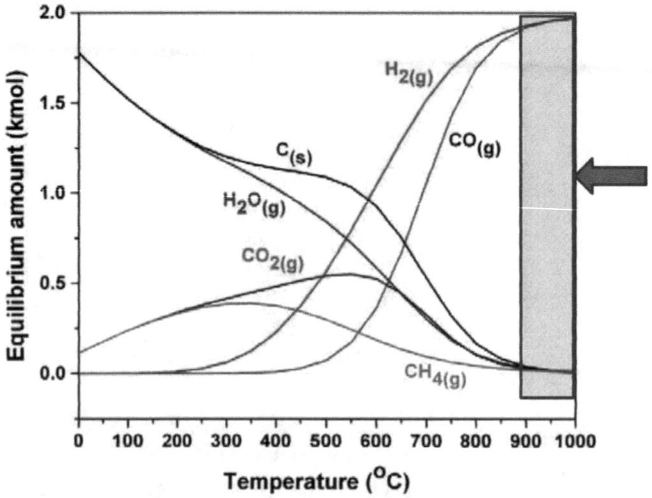

56

TEXAS A&M UNIVERSITY at QATAR

Design of Synthetic Fuels and Value-Added Chemicals Derived from Natural Gas via Combined Experimental and Statistical Methodology

Rehan Hussain, Elfatih Elmalik, Haider Ramadhan, Nimir Elbashir

Tuesday April 1th 2014
2014 AIChE Spring Meeting
New Orleans, LA, USA

AIChE The Global Home of Chemical Engineers

CHEMICAL ENGINEERING PROGRAM

TEXAS A&M UNIVERSITY at QATAR

Outline

- **Introduction**
- Background
- Approach & Methodology
- Experimental and modeling results
- Ongoing & Future Work

CHEMICAL ENGINEERING PROGRAM 2 TEXAS A&M UNIVERSITY at QATAR

State of Qatar

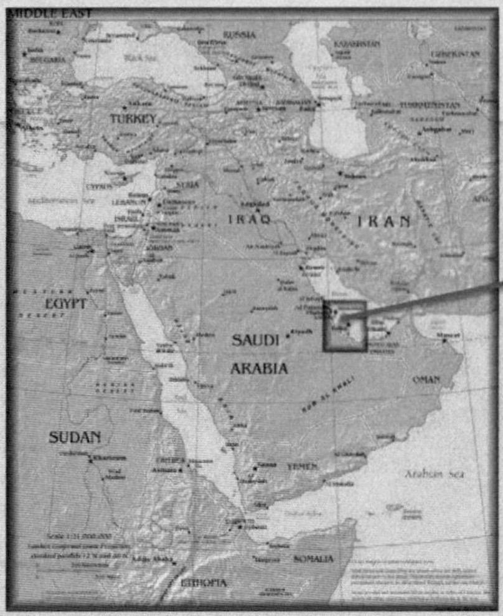

Area: 4,416 sq mi
Population: ~2 million

- Qatar has the world's **third largest** proven reserves of natural gas, approximately **896 trillion ft³**
- Qatar had the world's **highest GDP per capita**, while the economy grew by **18.8%** in 2011 - the fastest in the world

CHEMICAL ENGINEERING PROGRAM 3 TEXAS A&M UNIVERSITY at QATAR

Qatar's Role in the Energy Market

World's Primary Energy Supply

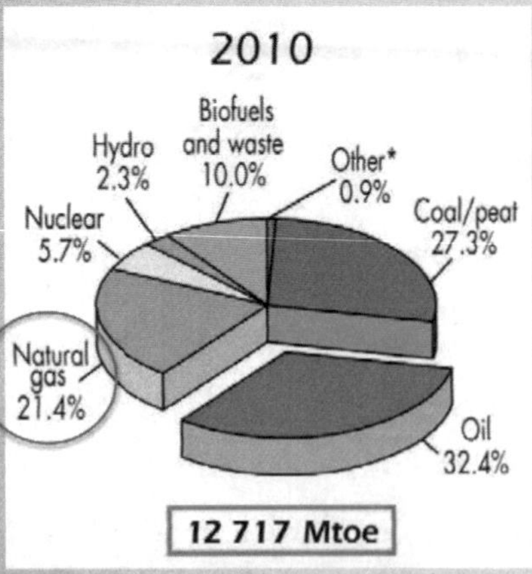

Source: International Energy Agency
www.iea.org

Natural gas

Producers	bcm	% of world total
Russian Federation	677	20.0
United States	651	19.2
Canada	160	4.7
Qatar	151	4.5
Islamic Rep. of Iran	149	4.4
Norway	106	3.1
People's Rep. of China	103	3.0
Saudi Arabia	92	2.7
Indonesia	92	2.7
Netherlands	81	2.4
Rest of the world	1 126	33.3
World	3 388	100.0

2011 data

Net exporters	bcm
Russian Federation	196
Qatar	119
Norway	99
Canada	63
Algeria	49
Indonesia	46
Netherlands	33
Turkmenistan	29
Nigeria	26
Malaysia	22
Others	152
Total	834

2011 data

CHEMICAL ENGINEERING PROGRAM 4 TEXAS A&M UNIVERSITY at QATAR

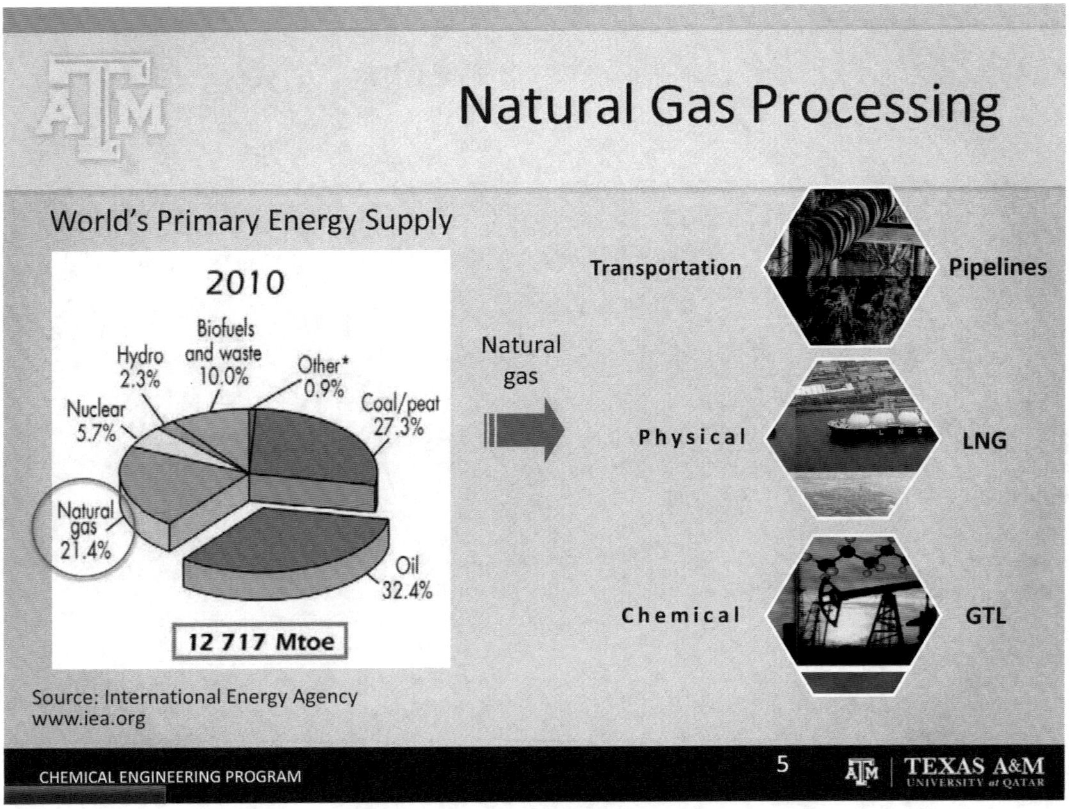

Natural Gas Processing
World's Primary Energy Supply
2010
Hydro 2.3%
Biofuels and waste 10.0%
Other* 0.9%
Nuclear 5.7%
Coal/peat 27.3%
Natural gas 21.4%
Oil 32.4%
12 717 Mtoe
Source: International Energy Agency www.iea.org
Natural gas
Transportation
Pipelines
Physical
LNG
Chemical
GTL
CHEMICAL ENGINEERING PROGRAM
5
TEXAS A&M UNIVERSITY at QATAR

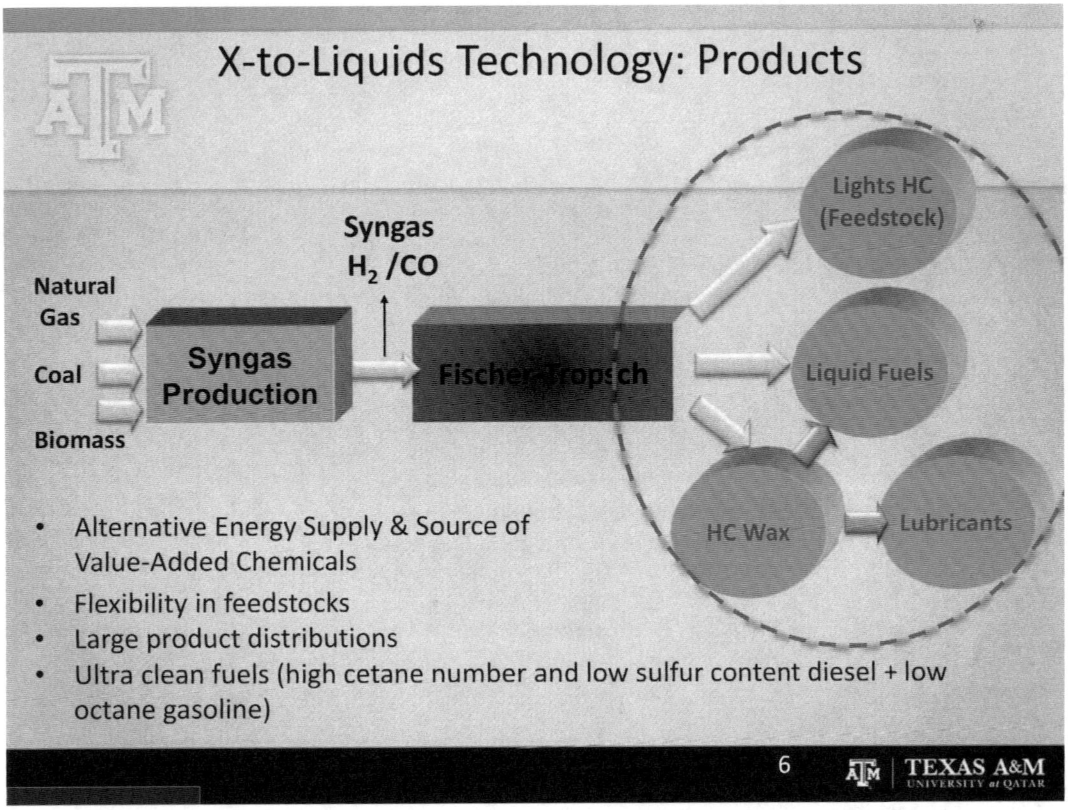

X-to-Liquids Technology: Products
Natural Gas
Coal
Biomass
Syngas Production
Syngas
H2 /CO
Fischer-Tropsch
Lights HC (Feedstock)
Liquid Fuels
HC Wax
Lubricants
• Alternative Energy Supply & Source of Value-Added Chemicals
• Flexibility in feedstocks
• Large product distributions
• Ultra clean fuels (high cetane number and low sulfur content diesel + low octane gasoline)
6
TEXAS A&M UNIVERSITY at QATAR

GTL Technology in Qatar

"The GTL Capital of the World"!

ORYX GTL

قطر للبترول
Qatar Petroleum

sasol
reaching new frontiers

135,000 bpd
kerosene
+ base oil

24,000 bpd diesel
+ 9,000 bpd naphtha
+1,000 bpd LPG

CHEMICAL ENGINEERING PROGRAM 7 TEXAS A&M UNIVERSITY at QATAR

Outline

- Introduction
- **Background**
- Approach & Methodology
- Experimental and modeling results
- Ongoing & Future Work

CHEMICAL ENGINEERING PROGRAM 8 TEXAS A&M UNIVERSITY at QATAR

Hydrocarbon Groups

Group	Structure
normal-Paraffins	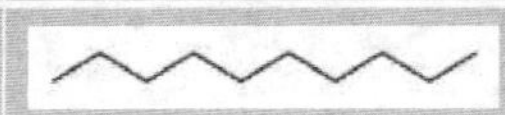
iso-Paraffins	
Naphthenes (cyclo-Paraffins)	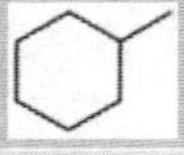
mono-Aromatics	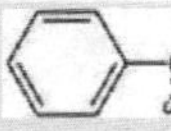
di-Aromatics	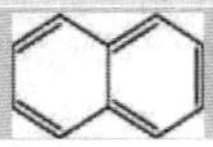
Naphthenic-mono-Aromatics	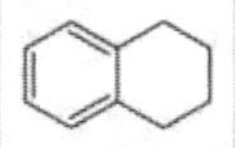

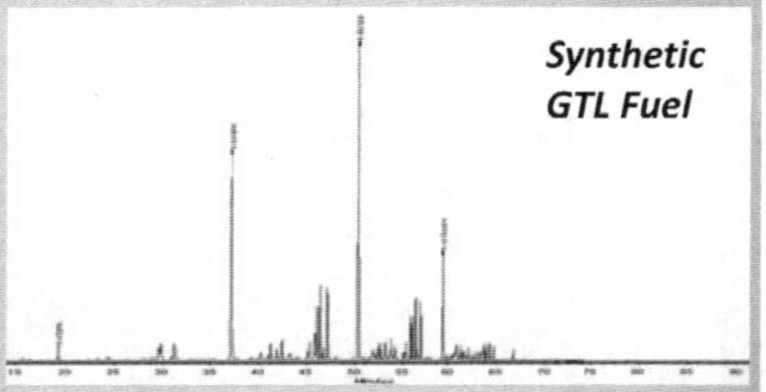

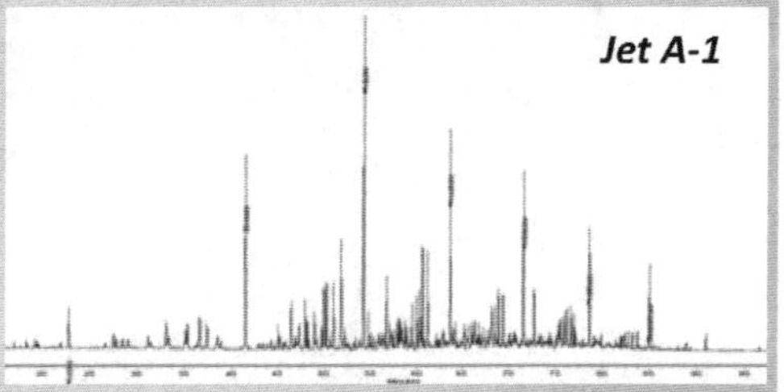

GC data courtesy of Shell

Hydrocarbon Groups

Species & carbon number distribution in an oil-derived jet fuel (Jet A-1) versus a GTL Synthetic Paraffinic Kerosene (SPK) .

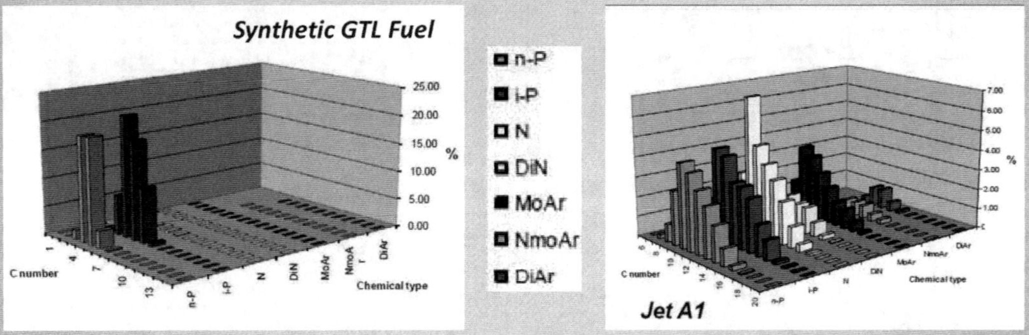

GCxGC data courtesy of Shell

Fuel Properties

- The different compositions of oil-derived and synthetic jet fuels leads to a difference in physical and chemical properties, which may be crucial in determining performance
- ASTM standards such as D1655 and D7566 specify property limits (ranges) which jet fuels must obey to withstand flight conditions (e.g. fuel tank size, low temperatures). They do not specify composition!

Property-Performance Interlinks

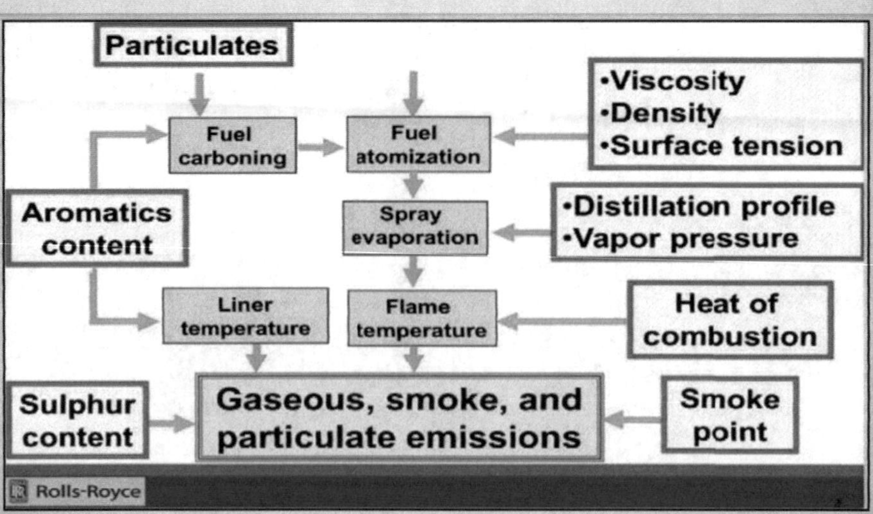

Picture courtesy of Dr. John Moran from Rolls-Royce

QATAR SCIENCE & TECHNOLOGY PARK
Member of Qatar Foundation

Rolls-Royce

Consortium

Funding Agencies

TEXAS A&M UNIVERSITY at QATAR

QATAR AIRWAYS القطرية

- A unique collaboration between industry and academia partners.
- Each partner worked on specific topics and collaborated towards the overall objective.
- The testing was split up as follows:

Technical Guidance

AIRBUS

DLR

The University Of Sheffield.

Properties Testing | **Combustion Testing** | **Performance Review**

CHEMICAL ENGINEERING PROGRAM 13 TEXAS A&M UNIVERSITY at QATAR

Texas A&M Qatar Fuel Characterization Lab

- Built a world class research lab to support the development of the fuel technology capabilities of Qatar for GTL processes.

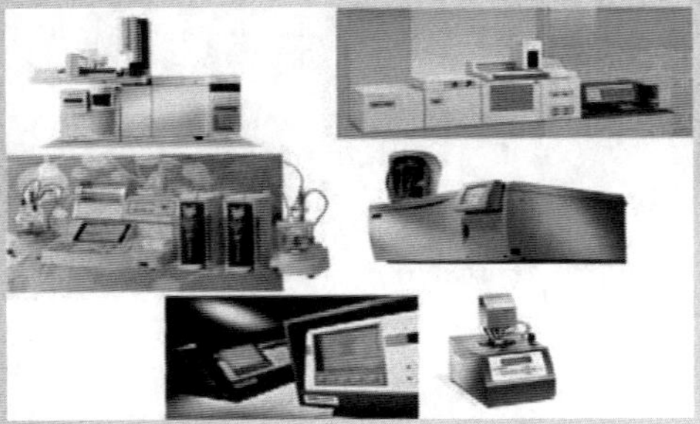

- Over 20 types of analytical equipment including advanced Gas Chromatographs (GCs)
- Trained over 25 graduate and undergraduate researchers since 2009

CHEMICAL ENGINEERING PROGRAM 14 TEXAS A&M UNIVERSITY at QATAR

Outline

- ➢ Introduction
- ➢ Background
- ➢ **Approach & Methodology**
- ➢ Experimental and modeling results
- ➢ Ongoing & Future Work

CHEMICAL ENGINEERING PROGRAM 15 TEXAS A&M UNIVERSITY at QATAR

Approach - Design of GTL Synthetic Fuels

The overall scope of our work is as follows:

- Identify the chemical components for making synthetic fuel blends that satisfy specific product needs (e.g. gasoline, diesel, jet fuel)
- Develop experimental protocols for characterization of the physical and chemical properties of the blends
- Model the relationships between fuel properties and blend composition to predict the properties of a given blend

CHEMICAL ENGINEERING PROGRAM 16 TEXAS A&M UNIVERSITY at QATAR

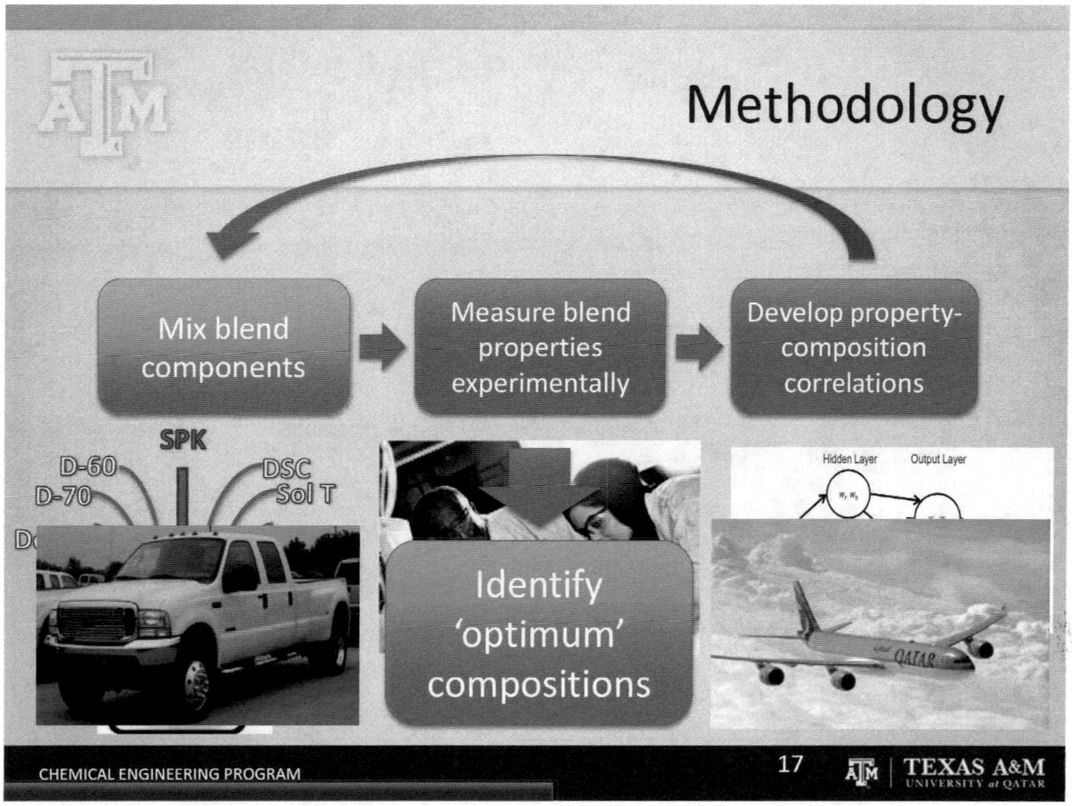

Outline

- ➢ Introduction
- ➢ Background
- ➢ Approach & Methodology
- ➢ **Experimental and modeling results**
- ➢ Ongoing & Future Work

CHEMICAL ENGINEERING PROGRAM 18 TEXAS A&M UNIVERSITY at QATAR

Case Study: Role of hydrocarbon building blocks

- 'Fuel blends' were made using 3 solvents, representing different hydrocarbon structural groups:

  ***normal*-paraffin:** decane
  ***iso*-paraffin:** Sol T
  ***cyclo*-paraffin:** decalin

- 32 blends were formulated, with compositions spread out across the ternary diagram

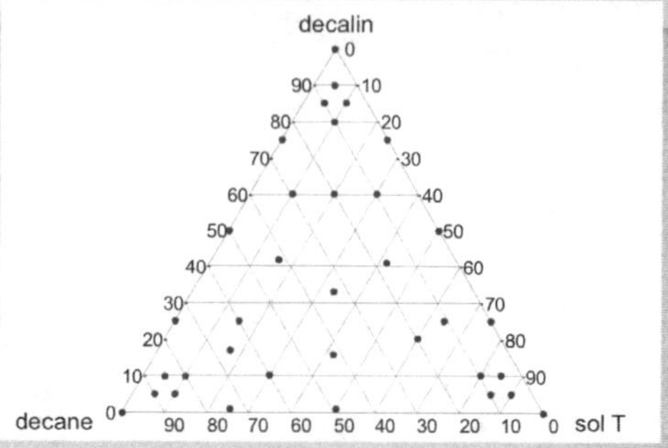

- 5 properties were chosen:
  - density
  - viscosity
  - heat content
  - flash point
  - freezing point

Elmalik et al., *Ind. Eng. Chem. Res.* 2014, 53, 1856-1865

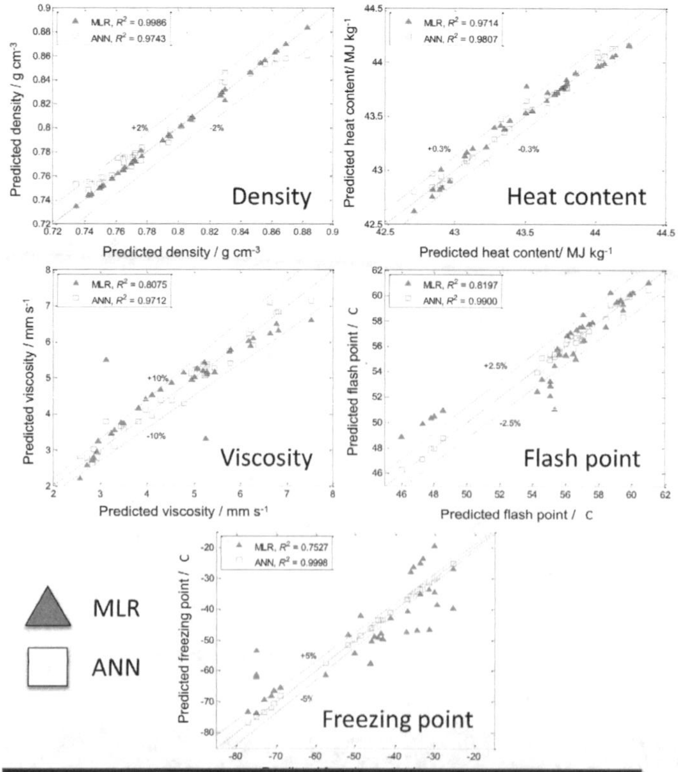

Results – MLR vs ANN

- Density, heat content, and viscosity vary linearly with composition
- Flash and freezing point show significant deviation from the MLR model
- ANN can predict non-linear properties well

Results – Focus on Freezing Point

Ternary system – decane, decalin, SolT

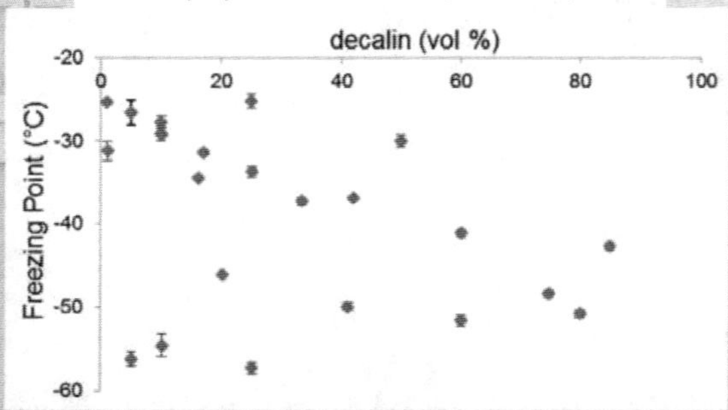

Binary system – decane, decalin

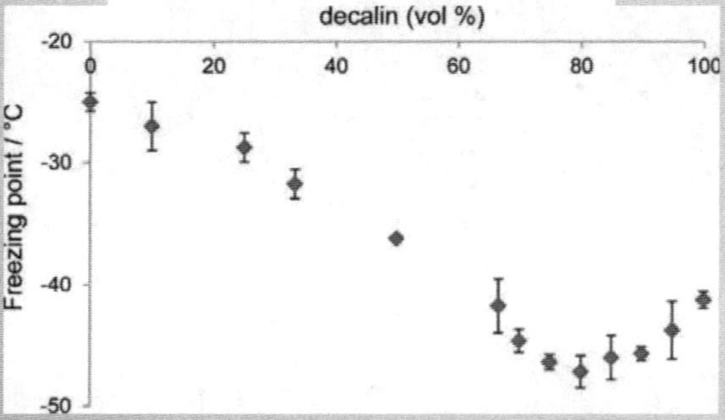

- Freezing point appears to be the most non-linear of the properties measured
- We confirmed this by measuring the freezing points of a binary decane-decalin system
- The binary system shows classic freezing point depression behavior

CHEMICAL ENGINEERING PROGRAM 21 TEXAS A&M UNIVERSITY at QATAR

Crystallization Images

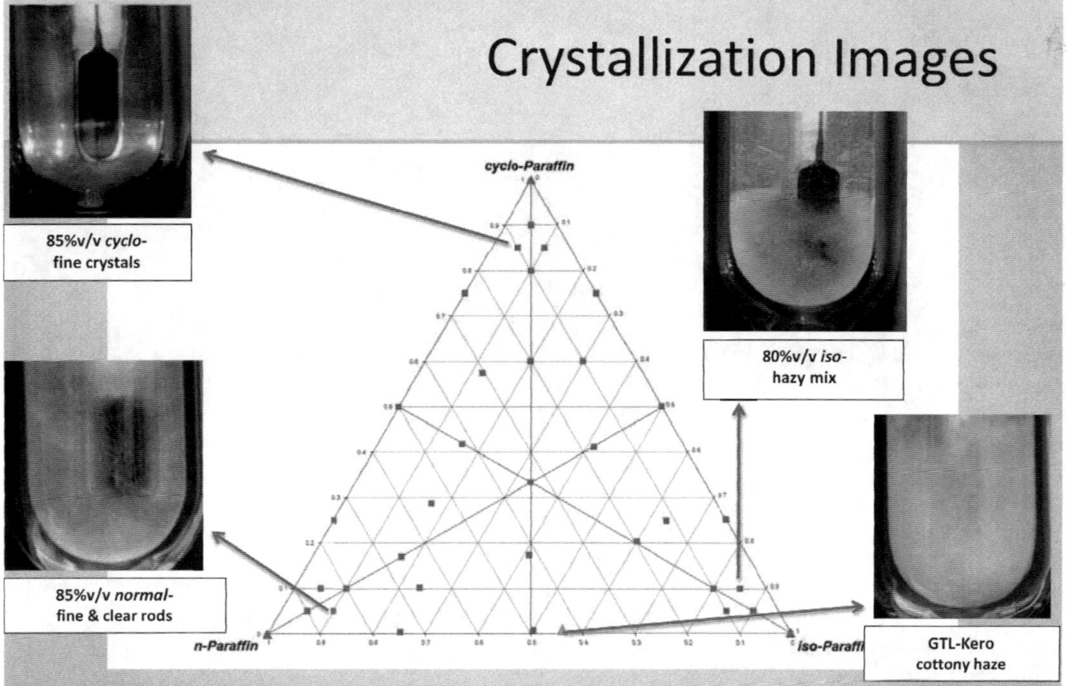

85%v/v *cyclo-* fine crystals

85%v/v *normal-* fine & clear rods

80%v/v *iso-* hazy mix

GTL-Kero cottony haze

CHEMICAL ENGINEERING PROGRAM 22 TEXAS A&M UNIVERSITY at QATAR

Region of Optimum Properties

- Density
- Heat Content
- Viscosity
- Flash Point
- Freezing Point

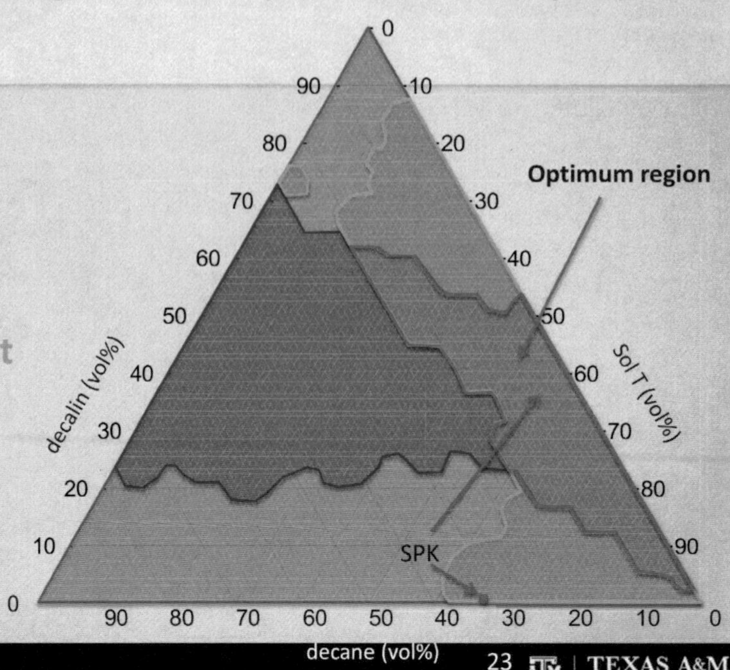

CHEMICAL ENGINEERING PROGRAM 23 TEXAS A&M UNIVERSITY at QATAR

Adding Aromatics to Synthetic GTL Fuel

Advantages:

- Helps fuel meet density and lubricity limits
- Elastomer compatibility for sealing o-rings

Disadvantages:

- Negative environmental impact

Investigated the effect of adding an aromatic component to the previously studied ternary blends

CHEMICAL ENGINEERING PROGRAM 24 TEXAS A&M UNIVERSITY at QATAR

Adding Aromatics to Synthetic GTL Fuel

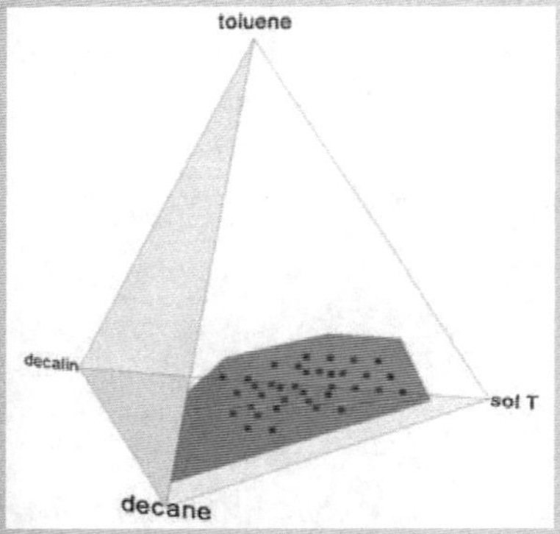

- Toluene was used to represent aromatic hydrocarbons
- 36 blends were prepared, with the toluene concentration limited to a maximum of 20 vol % and the decalin concentration varied between 0-15 vol % to realistically reflect the aromatic and cyclo-paraffin content of jet fuel
- 3 properties were chosen: density, viscosity, and freezing point

CHEMICAL ENGINEERING PROGRAM 25 TEXAS A&M UNIVERSITY at QATAR

Results - aromatics

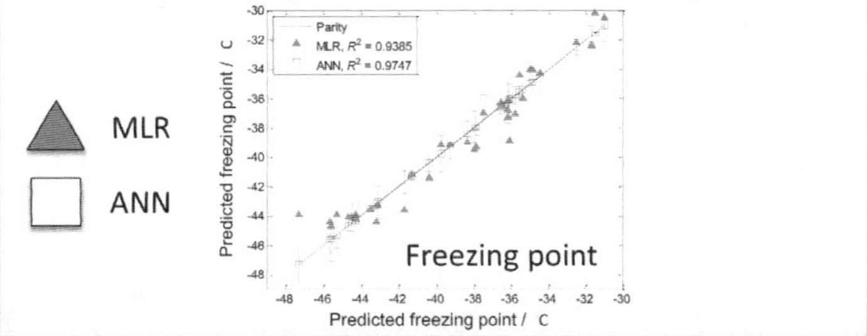

- Both the MLR and ANN models fit the three properties quite well
- The low decalin content compared to the ternary blends indicates that this component shows the biggest deviations from linear freezing point behaviour

CHEMICAL ENGINEERING PROGRAM 26 TEXAS A&M UNIVERSITY at QATAR

Results - aromatics

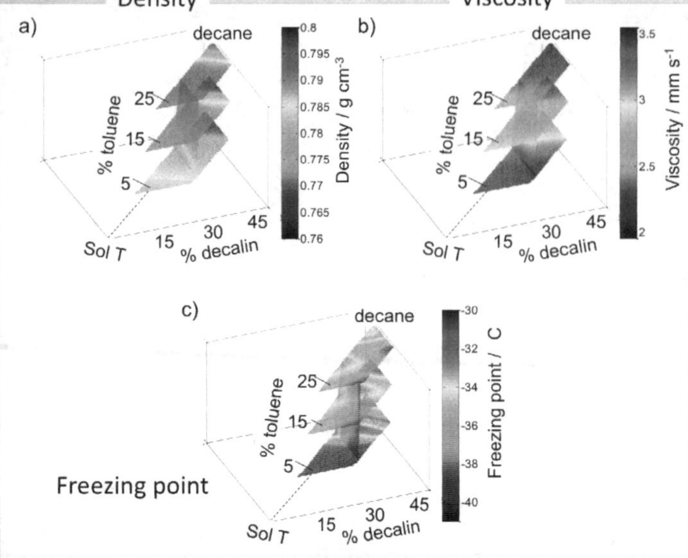

- All freezing points were above the maximum allowable value of -47 C in the ASTM
- For blends with higher Sol T content, increasing the amount of toluene appears to elevate the freezing point

CHEMICAL ENGINEERING PROGRAM 27 TEXAS A&M UNIVERSITY at QATAR

Outline

- ➢Introduction
- ➢Background
- ➢Approach & Methodology
- ➢Experimental and modeling results
- ➢**Ongoing & Future Work**

CHEMICAL ENGINEERING PROGRAM 28 TEXAS A&M UNIVERSITY at QATAR

Current Projects

Understanding the Role of Aromatics in Synthetic Jet Fuel-Experimentation, Statistical Analysis and Visualization. Funded by Qatar National Research Fund (QNRF) under the Undergraduate Research Experience Program (UREP) in collaboration with Shell. LPI: Dr. Nimir Elbashir; PI: Mansoor Al-Shamri (Shell) to support three undergraduate research students.

Project Focus: This study focuses on the role of different types of mono- and bi-aromatic additives in enhancing the properties of synthetic jet fuels obtained from natural gas via GTL. Expected finish: May 2014

CHEMICAL ENGINEERING PROGRAM 29 TEXAS A&M UNIVERSITY at QATAR

Current Projects

Design of Synthetic Fuels and Value-added Chemicals Derived from Natural Gas via Combined Experimental and Process Integration Methodology. Funded by QNRF under National Priorities Research Program (NPRP). LPI: Dr. Nimir Elbashir; PIs: Dr. Mahmoud El-Halwagi (TAMU) and Dr. Rafiqul Gani (DTU).

Project Focus: This research work will facilitate a better link between experimental data and model development to enable the design of the next generation of synthetic fuel blends. Expected finish: June 2016

CHEMICAL ENGINEERING PROGRAM 30 TEXAS A&M UNIVERSITY at QATAR

Ongoing and Future Work

- Extend our developed methodology to include fuel cuts such as diesel and gasoline, as well as conventional fuels
- Build a database of fuel data for use in the development of accurate composition-property models for a wide range of fuels and additives
- Conduct a systematic assessment of the relationship between carbon number and fuel properties of different fuels, using fractional distillation to obtain different carbon number cuts
- Develop new methods for visualization of large multi-dimensional data sets to allow easy identification of compositions of interest for optimizing fuel properties with a view to industry certification

CHEMICAL ENGINEERING PROGRAM 31 TEXAS A&M UNIVERSITY at QATAR

Acknowledgements

- Prof Rafiq Gani (DTU)
- Prof Mahmoud El-Halwagi (TAMU – College Station)
- Dr Joanna Bauldreay (Shell Global Solutions/Aviation Fuels)
- Haider Ramadhan
- Graduate and undergraduate students at Texas A&M Qatar
- Funding: Qatar National Research Fund, Qatar Science & Technology Park, Shell

الصندوق القطري لرعاية البحث العلمي Qatar National Research Fund Member of Qatar Foundation

QATAR SCIENCE & TECHNOLOGY PARK Member of Qatar Foundation

DTU

TEXAS A&M UNIVERSITY

TEXAS A&M UNIVERSITY at QATAR

CHEMICAL ENGINEERING PROGRAM 32 TEXAS A&M UNIVERSITY at QATAR

Thank you for listening!

CHEMICAL ENGINEERING PROGRAM

336F Texas A&M Engineering Building
Education City
PO Box 23874
Doha, Qatar

Tel. +974.4423.0017
Fax +974.4423.0065

chen@qatar.tamu.edu

http://chen.qatar.tamu.edu

352260 Monetising Gas Hydrates: New Mode for Natural Gas Storage and Transportation Against LNG Storage and Transportation

Abstract:

Gas Hydrates are considered in the petroleum industry e.g. natural gas transportation through pipelines, during natural gas production from the crude oil wells. But this paper focuses on converting the nuisance of gas hydrates to an attractive means of transportation. New Technology development has been focusing on using gas hydrates as a way of converting gas to solids.

The Natural Gas Hydrate (NGH) is a clatherate compound in icy state and stabilised around -20 degree Celsius under atmosphere. This paper summarises about the cost effectiveness of the NGH value chain against the LNG (Liquefied Natural Gas) value chain and also in this paper entire processes and operations to create NGH pellets are designed and discussed. The calculations show that this process has a high initial plant cost but on a long term this process of NGH production is cost effective compared to LNG liquefaction process. Moreover, it is easy to handle the NGH at -20 to -30 degree Celsius compared to LNG at around -160 degree Celsius.

Thus, this unique method of transporting the natural gas in bulk solid form leads to reduction of the total investment cost for the entire value chain. Moreover, NGH is expected to be a plausible solution in monetising small to medium gas fields which are considered to be difficult to be developed economically using LNG.

352647 Towards the Development of a Control-Relevant Model of the Hydraulic Fracturing Process to Investigate Various Control Strategies

Abstract:

Hydraulic fracturing or 'fracking' is an increasingly important technique for developing energy resources across both the U.S. and the world, in which high pressure fluid is injected into the reservoir to crack the rock formation and provide highly conductive channels for oil and gas transport back to the wellbore. During this process, the underground environment is characterized by complex fluid dynamics, geo-mechanics, and multiphase flow phenomena. Surface equipment exerts control over the process from between hundreds of feet to several miles from the moving front of the fracture(s) below ground. The fluid pumped underground is a slurry consisting of water, proppant (often a special type of sand) to keep the created fractures open after the process finishes, and chemical additives such as cross-linkers, acids, and surfactants to maintain fluid viscosity, reduce corrosion, and perform other duties. During the fracturing process, pumping rate and fluid properties (viscosity and the concentrations of proppant and other additives) can be varied at the surface. The main objectives are to keep the fracture growth within a predefined region of the reservoir open by distributing the proppant evenly throughout the fracture so that oil and gas can flow after the fracturing fluids have leaked into the formation. Complications can include excessive growth of the fracture external to the target region that results in the proppant being spread so thinly that it is ineffective to keep the fracture open. Conversely, if a high concentration of proppant is added to the fracturing fluid mix too early in the process the condition called 'screen-out' can occur wherein the proppant enters the advancing fracture front and bridges the tip preventing further fracture growth and causing rapid pressure buildup in the system. A related issue is proppant settling, that is a large sand bank or dune forms near the entrance of the fracture at the wellbore resulting in only a small open region around the wellbore after the fluid leaks off. These three potential proppant-related challenges may have a large negative impact on oil well production.

Applying process control techniques to regulate the hydraulic fracturing process, via setting optimal pumping schedules, is a challenging problem not only due to the remoteness of the control action, but also due to the limited feedback information available because of the sensing system associated with the bottom-hole situation. To better understand these challenges, a mathematical model of the physics of the fracturing system is developed to determine the effect of different pumping schedules on the final fracture conductivity, the most widely used calculated performance metric. The physics-based model incorporates equations governing fracture propagation, fluid dynamics, proppant transport, and bank formation. The resulting system of nonlinear, partial differential equations is solved numerically using a moving grid method to track the fracture growth and particle tracking to predict the proppant movement and settling. Additionally, the 'shut-in' process, during which pumping is stopped and the rock formation walls are allowed to close as the fluid leaks off is a part of this model development. The final geometry of the channel available to oil and gas flow is predicted. Using this model, improved pumping schedules can be developed and tested to achieve near-optimal fracture conductivity.

352284 Shale Gas as a Potential Source of Unconventional Reserve for Crude Oil

Abstract:

With a decrease in the conventional reserves available for the recovery of oil, the alternative fossil fuel extraction industry is on a growth. As the development of unconventional oil and gas resources has been progressed, the heterogeneity and complexity of shale gas and tight oil reservoirs have become apparent. In the U.S. primarily shale gas and tight oil is expected to continue growing dramatically in coming decades. According to the U.S. Energy Information Administration, an assessment of 137 shale formations are identified in 41 countries other than those identified in U.S. At the local level, though the environmental impacts of shale gas and tight oil development may turn out to be significant.

Shale gas and tight oil are unconventional sources of petroleum that are found within the reservoirs with a very low permeability. The oil and gas contained within these reservoir rocks typically will not flow to the wellbore at the economic rates without assistance from technologically advanced drilling and completion processes. Usually, horizontal drilling coupled with multi stage fracturing is used to access these difficult to produce reservoirs. Unconventional natural gas and oil deposits are difficult to characterize overall, but in general are often lower in resource concentration, more dispersed over large area, and require well stimulation or some other extraction or conversion technologies.

353319 Dual Layer Spinning with Lumen Layer Containing PAI/Silica/PEI Hollow Fiber Sorbent for CO2 Separation by Rapid Temperature Swing Adsorption

Abstract:

Amine-loaded cellulose acetate (CA) polymer/silica hollow fiber sorbents for post-combustion CO_2 recovery were fabricated by utilizing a post-spinning amine infusion technique. Building on prior work focusing on amine-loaded CA and poly(amide-imide) (PAI)/silica hollow fiber sorbents, direct dual layer spinning with PAI lumen layer PAI/silica/ poly(ethyleneimine) (PEI) hollow fiber sorbent are reported here for the first time. Post-spinning amine infusion of PEI is used to incorporate amine sites during the solvent exchange steps after dual layer fiber spinning. After infusing the 50 wt% silica hollow fiber sorbents with 10% PEI solutions in methanol, several amine-loaded fibers were assembled into shell-and-tube modules and exposed on the shell side at 1 atm and 65 °C to simulated flue gas with an inert tracer (14 mol% CO_2, 72 mol% N_2 and 14 mol% He at 100% R.H.). The fibers are found to have a humid breakthrough CO_2 capacity of 1.0 mmol/g-fiber sorbent. The humid CO_2 pseudo-equilibrium uptake of the module is 3.0 mmol/g-fiber sorbent. This approach provides a versatile platform that may facilitate the ultimate scale-up of practical amine sorbents for flue gas CO_2 recovery applications.

ExxonMobil

Application of Controlled Freeze Zone™ Technology for Commercialization of Sour Gas Resources

Ananda Krishna Nagavarapu
Jaime A. Valencia
Scott D. Kelman

ExxonMobil Upstream Research Company

Presentation Outline

- **Background**
- **Technology**
- **Commercial Demonstration Plant**
- **Results and Discussion**
  - ➤ Steady State Operation
  - ➤ Influence of Operating Pressure
  - ➤ Influence of Feed Composition
  - ➤ Robustness and Operability
- **Summary**

Global Perspectives:
Natural Gas Demand & Impact

- **Natural gas is the cleanest burning hydrocarbon fuel**
- **Strong demand growth for electrical power generation**
  - ➢ Emits 60% less CO_2 than coal
- **Anticipated to become the 2nd largest global fuel by 2040**
- **Meeting that demand will require the development of increasingly sour global gas reserves**
- **Geosequestration or use in enhanced oil recovery are the preferred options in managing GHG emissions**
- **Saturated sulfur markets will need an alternative → acid gas injection**

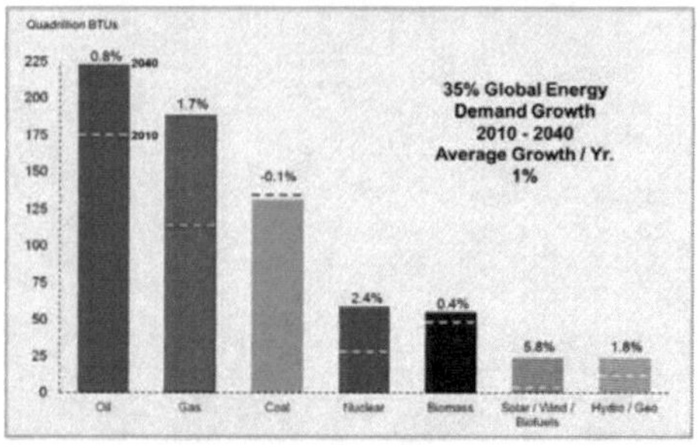

- **Controlled Freeze Zone™ (CFZ™) technology can remove the contaminants and facilitate their reinjection**

Controlled Freeze Zone™ Technology

A single step cryogenic distillation process for the removal of CO_2 and H_2S from natural gas

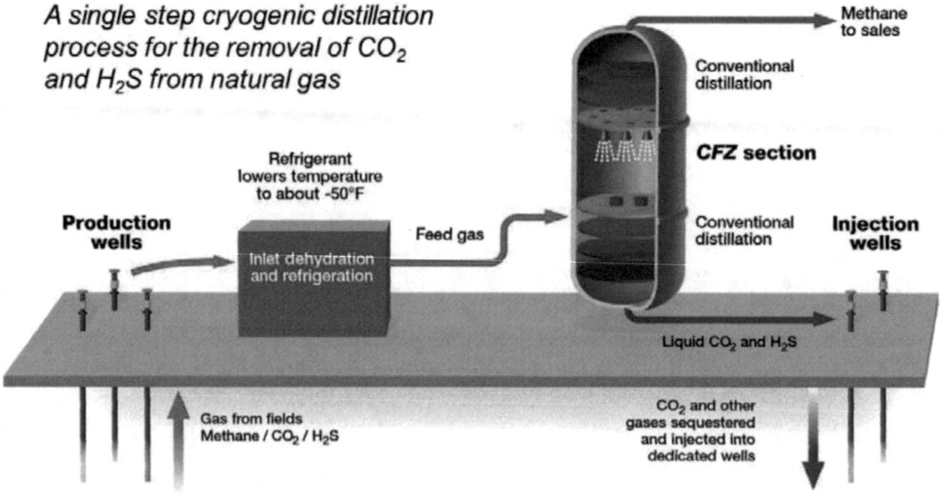

Thermodynamic Considerations

Rather than avoiding solidification of CO_2,
***control* it and *confine* it to specially designed section in distillation column**

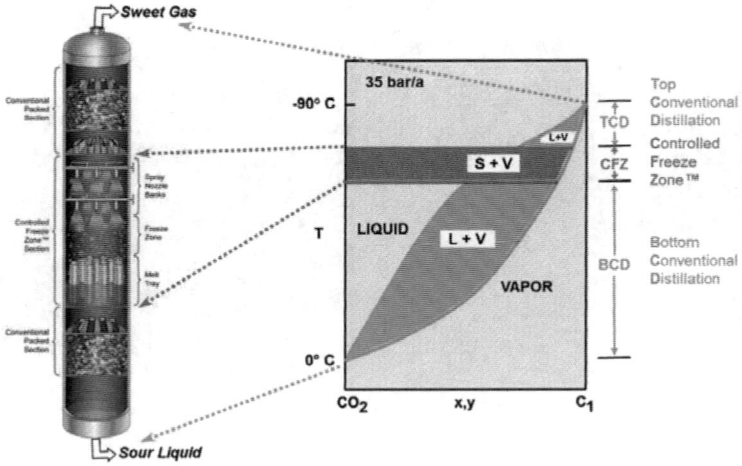

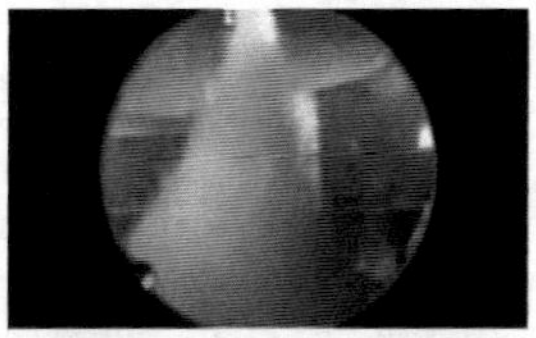

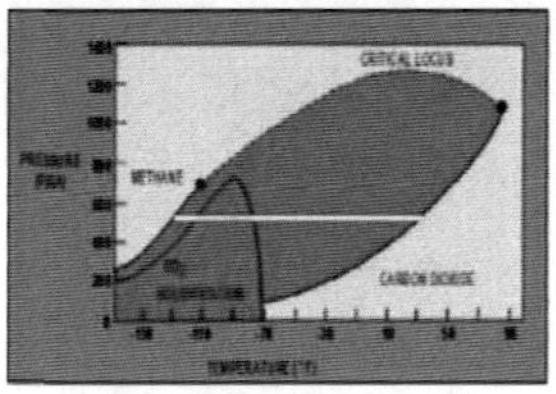

ExxonMobil

Economic Benefits

CFZ™ /AGI is the lowest cost route for handling sour gas with a minimal environmental footprint

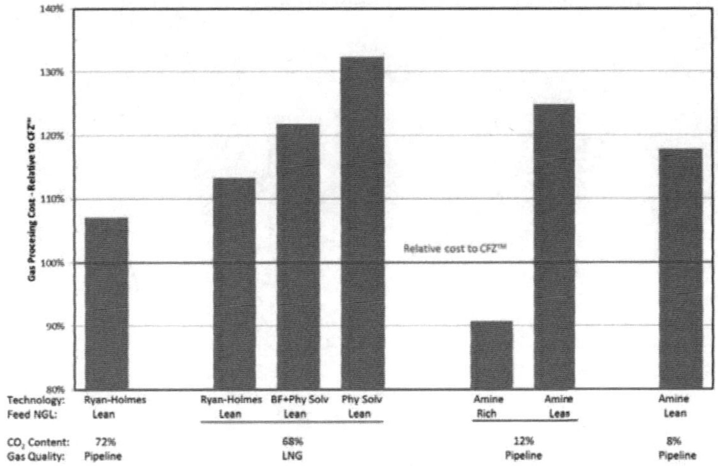

- **Lower capital investment costs**
- **Lower operating expenses**
- **Higher efficiency**
- **More compact**
- **Environmental benefits**

ExxonMobil

Commercial Demonstration Plant (CDP)

- 13.5 MMSCFD (400 kNm3/d) commercial demonstration plant built in Wyoming, USA
- Designed to process feeds with 15 to 70% CO_2 and up to 35% H_2S
- Integration with acid gas injection
- Extensive test program to gather all necessary data to confidently scale-up, design and operate facilities up to 1 BSCFD (28 MNm3/d)
- Test program started in early 2012 and was completed in 2013

Simplified Process Schematic

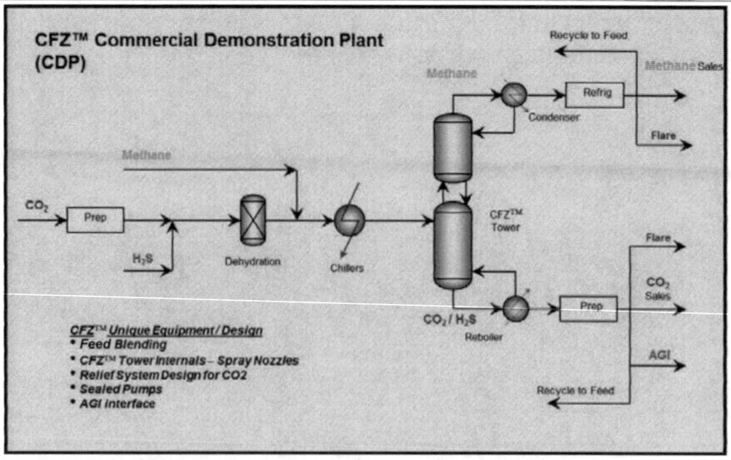

Three gas streams are used to blend the desired feed:

- Methane sales gas (with a small amount of nitrogen)
- CO_2 sales gas
- Acid gas (65 mol% H_2S and 35 mol% CO_2)

Composition and Temperature Profiles

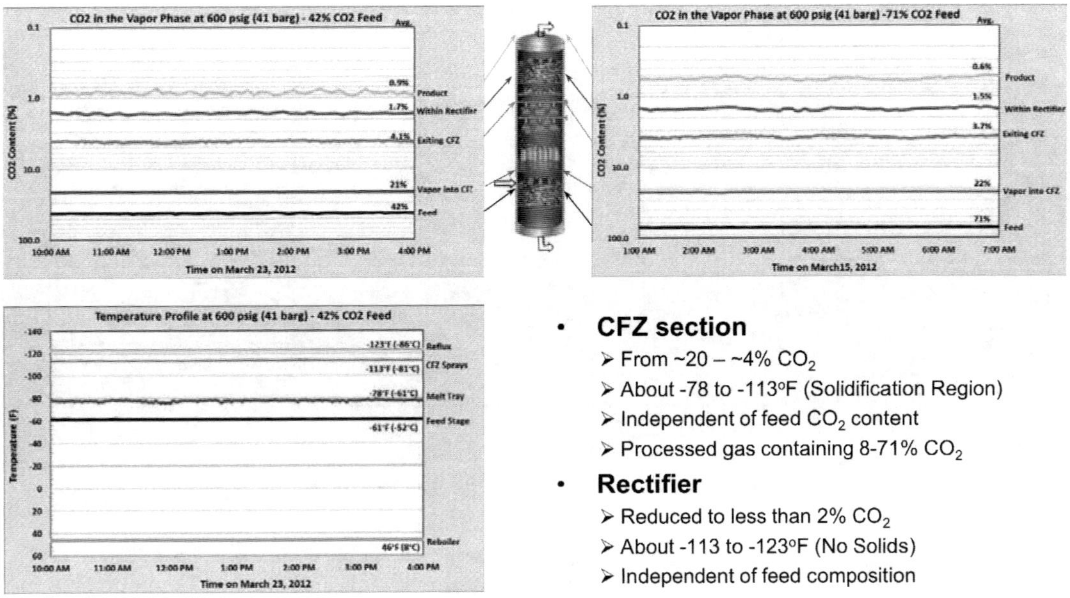

- **CFZ section**
  - From ~20 – ~4% CO_2
  - About -78 to -113°F (Solidification Region)
  - Independent of feed CO_2 content
  - Processed gas containing 8-71% CO_2
- **Rectifier**
  - Reduced to less than 2% CO_2
  - About -113 to -123°F (No Solids)
  - Independent of feed composition

Influence of Operating Pressure

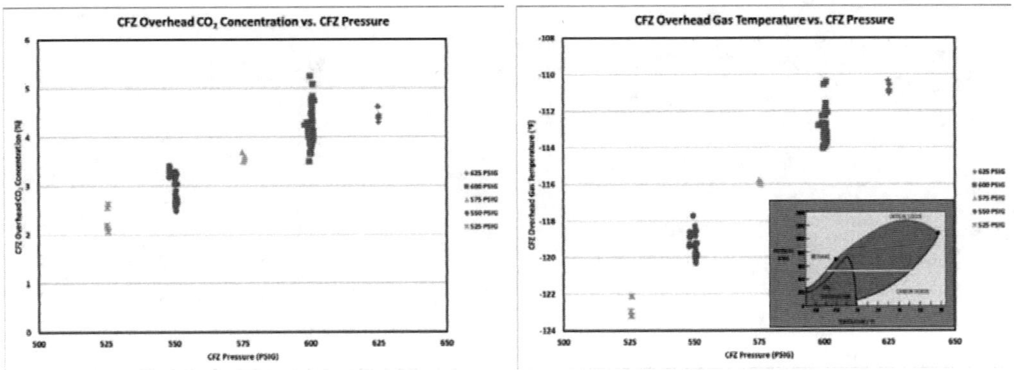

- **CFZ overhead CO_2 concentration**
  - Increases with increasing pressure
  - Typical distillation characteristics
- **CFZ overhead gas temperature**
  - Increase with increasing pressure
  - Freezing envelope shrinks with increasing pressure

Operating at lower pressures:
Colder temperatures, freezing envelope expands, but purities improve

Influence of Operating Pressure

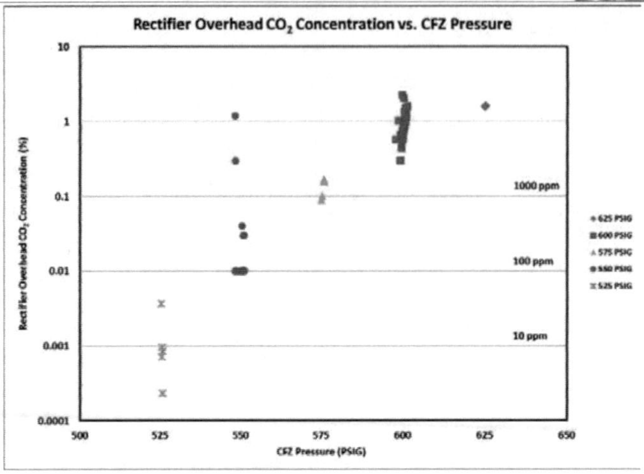

- **CO_2 concentration of product gas**
  - Increases with increasing pressure
  - Pipeline gas specification continuously achieved
  - LNG specifications met at lower pressures

Presence of H_2S in Feed

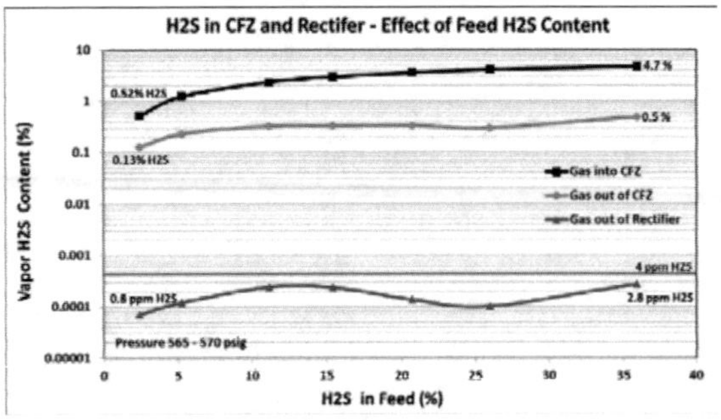

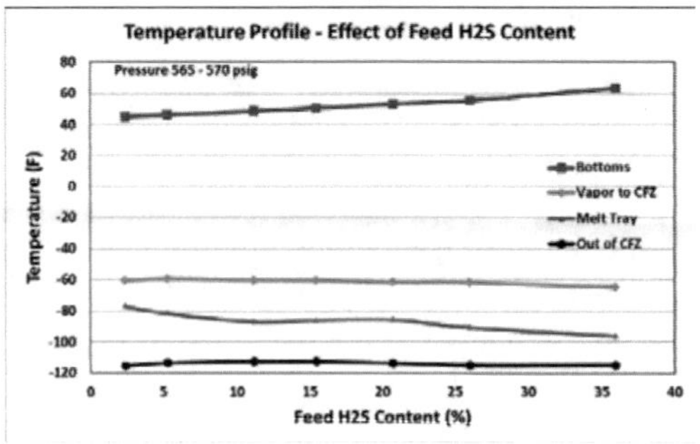

- **Achieved effective separation**
  - Less than 4 ppm H_2S in product gas
  - Met specification
- **As H_2S in feed increases**
  - H_2S content in CFZ section increases
    - Different from CO_2 trends seen previously
  - Melt tray temperature decreases
    - Presence of third component
  - Freezing envelope shrinks

Presence of Moisture in Feed

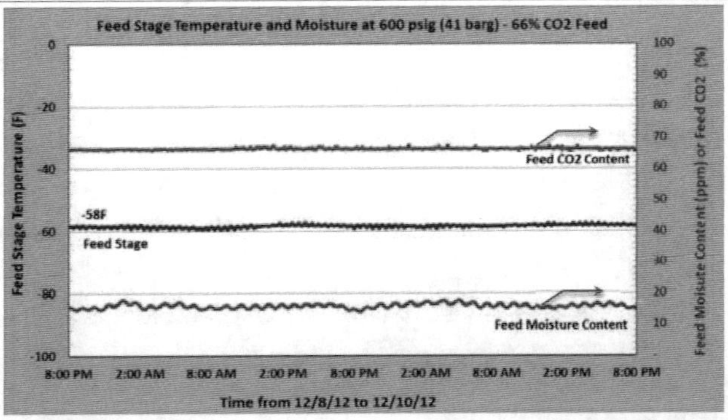

- **Liquid CO_2 can act as a dehydrating agent**
  - Has a significant water carrying capacity which increases with temperature
- **Feed gas with 2/3 CO_2**
  - 10-20 ppm feed moisture for a month
  - No adverse effect
- **Similar results for feed gas containing 1/3+% CO_2**
- **Potential for glycol-based dehydration units**

Process Robustness and Operability

- **The CDP was subjected to a number of upset conditions including:**
  - Sudden increase in feed temperature
  - Sudden decrease in feed rate
  - Loss of melt tray level
  - Loss of CFZ sprays or rectifier reflux
  - Loss of heating medium flow to the reboiler
- **Recovery procedures implemented after simulating these conditions**
- **Unit recovered well and returned to normal conditions**
- **Product specifications were maintained throughout upsets**
- **Corrective actions were developed and tested and will serve as operational guidelines for commercial applications**

Effect of Spray Loss

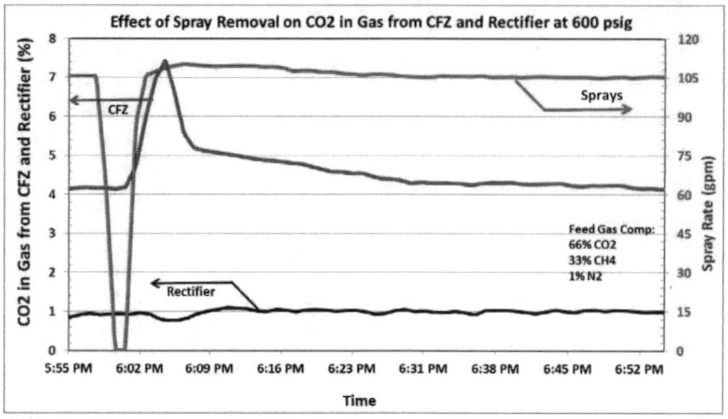

- **Sprays cut for approximately 4 minutes**
  - Representative of time required to bring redundant pump online
- **CFZ overhead CO_2 concentration increased temporarily above 7%**
- **Normal operating conditions within 30 mins of pump restart**
- **Rectifier product within specification**

Summary

- **Demonstrated effective separations for feeds ranging from 8 to 71% CO_2 and up to 36% H_2S**
  - < 2% CO_2 pipeline quality easily achieved at all conditions, < 4 ppm H_2S when processing H_2S
  - < 50 ppm CO_2 LNG feed quality achieved at some conditions
- **Demonstrated integration of CFZ™ with acid gas injection**
- **Demonstrated process robustness to variety of upsets**
- **Demonstrated viability of glycol-based dehydration systems for feed containing 1/3+ CO_2**
- **Economic evaluations show CFZ™ technology is the most cost effective option for developing lean sour gas resources where acid gas injection is to be used to minimize GHG emissions**

CFZ

352313 Modeling and Simulation of Removal of CO2 and N2 from Natural Gas Using Aspen Plus Platform

Abstract:

Due to increasing environmental awareness and stringent governmental regulations over carbon emissions, more emphasis is being laid on using clean fuels as energy sources. Due to its less carbon emissions when compared to other fuel sources, natural gas is considered as a major source of fuel in meeting the world's energy demands.

Methane is the major component of natural gas along with other components such as ethane, propane, C_4+, CO_2, H_2S, nitrogen. Before put into commercial use or transportation, to meet the required specifications, natural gas must be treated for removal of these unwanted components. On industrial scale, acid gas removal system involves counter-current absorption with amine solution, nitrogen removal involves either cryogenic separation or through warm gas rejection and heavier hydrocarbons are separated through cryogenic expansion process. This paper mainly deals with acid gas removal system in natural gas plants. In industries usage of amine solutions causes corrosion and requires higher amounts of energy supply. Alternative methods like vacuum–PSA, membrane technology, ionic liquids for this CO_2 removal are still under research. The present work involves simulation of CO_2 and nitrogen removal process from natural gas using Aspen Plus and comparison study of CO_2 removal efficiencies of amine solutions at different concentrations, tray geometries and thermodynamic models.

Model Based Analysis of Plasticization Behaviour of Polymeric Membranes used for High Pressure CO_2 Separation in Natural Gas Processing

J. K. Adewole [a, b], *A. L. Ahmad* [a*], *S. Ismail* [a], *C. P. Leo* [a], *A. S. Sultan* [b]

[a]*School of Chemical Engineering, Engineering Campus, Universiti Sains Malaysia, 14300 Nibong Tebal, Seberang Prai Selatan, Pulau Pinang, Malaysia*

[b]*Center for Petroleum & Minerals, King Fahd University of Petroleum & Minerals, Dhahran 31261, Saudi Arabia*

Abstract

Plasticization resistant property is very important in measuring the performance of polymer membranes for high pressure CO_2 separation from natural gas or separation in aggressive environment. Many polymeric membranes have been synthesized and tested for high pressure CO_2 separation. Thus, a quick qualitative comparison will be necessary in the early phases of a membrane gas separator design to facilitate material selection. In this paper, we developed a procedure that integrates experimental and mathematical models to determine the plasticization pressure from permeability measurement data. This procedure permits the use of a graphical method for membrane materials selection for high pressure operations. It was demonstrated that the plasticization behaviour of the current generation of plasticization resistant polymer membranes can be approximately described by a quadratic (2nd order) empirical equation. A theoretical justification was proposed to support this idea. The model was tested on more than ninety polymer membrane samples and results correlate well with experimental results that have been reported in the literature.

Three model parameters (a_1, a_2 and a_3) with values ranging from 0.0001 to 2.9175, -62.083 to -0.0055 and 0.7862 to 786.62, respectively were obtained. The values of these parameters were indications of the degrees of resistance to penetrant-induced plasticization and performance stability in aggressive feed streams. For example lower values of a_1 revealed a relatively higher plasticization pressure and consequently better performance stability during the separation process.

Samples application of the model is considered and recommendations for their use to design polymer membranes and evaluation of their performance under a penetrant induced palsticization environment are provided. The results of the model interpretation provide consistent set of parameters that can be included in simulation software for modeling and optimization of membrane gas separation involving aggressive feed streams.

---

* Corresponding author: chlatif@eng.usm.my; Tel.: +6045941012, fax +6045941013

Surface Area Measurements for Spray Absorption of CO_2 in Amine Solvents

Yash Tamhankar[1], Rob Whiteley[1], Mike Resetarits[2],
Tony Cai[2], & Clint P. Aichele[1]

[1]School of Chemical Engineering, Oklahoma State University
[2]Fractionation Research Incorporated (FRI)

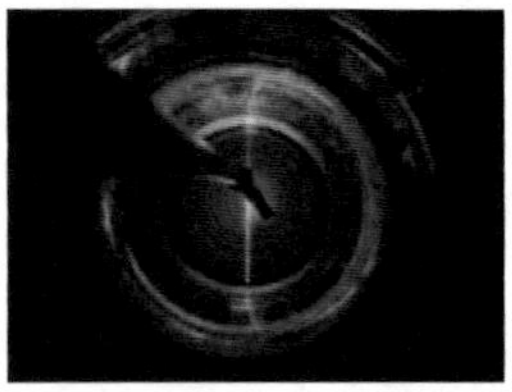

Outline

- Research Motivation
- Set Up
- Methods & Matrix
- Results
- Conclusions & Future Work

AIChE Meeting - Spring 2014

Aqueous Amine Absorption

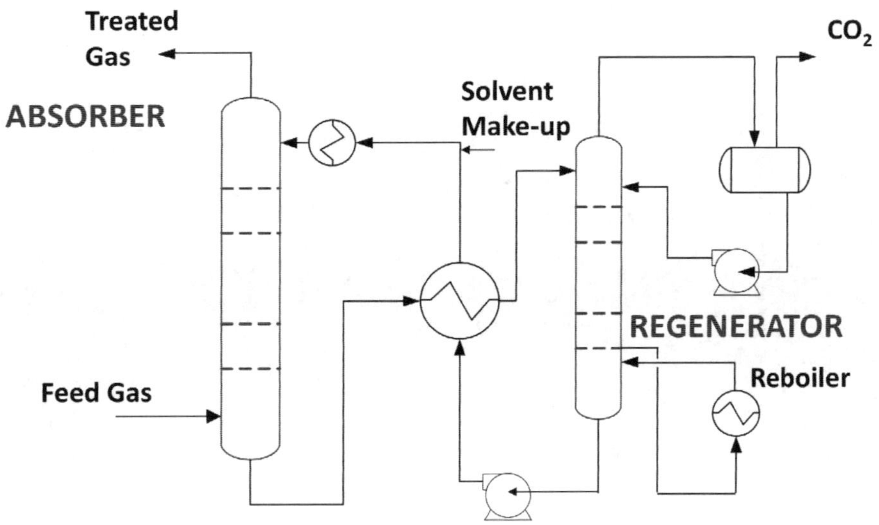

Need: Reduce CAPEX + OPEX → **Spray Column - Absorber**

Motivation for Spray Columns

	Mellapak M250Y	Spray Column
Gas rate (kmol/s)	25	25
$\Delta P/Z$ (in H_2O/ft)	0.208	0.028
Column height (m)	11.5	11.5
Column diameter (m)	20.3	20.3
*Installed blower cost (MM$)	0.526	0.193
*Operating blower cost (MM$/year)	0.734	0.397

*Calculations based on the models of: Tsai, R. E. (2010). Mass Transfer Area of Structured Packing. Chemical Engineering. Austin, TX, University of Texas at Austin. Ph.D. Dissertation.

Effect of Solvent Concentration

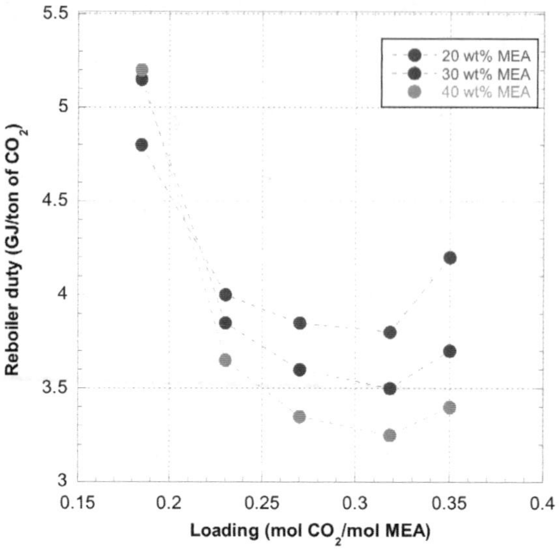

Ref: Abu-Zahra, M. R. M., L. H. J. Schneiders, et al. (2007). "CO2 Capture from Power Plants: Part I. A Parametric Study of the Technical Performance based on Monoethanolamine." International Journal of Greenhouse Gas Control 1(1): 37-46.

AIChE Meeting - Spring 2014

Research Goals & Objectives

Goals: How much CO_2 capture is possible with sprays?

- Efficiency - ascertain
  - Dropsize distribution
  - Surface area
  - Interfacial area
- Effect of solvent concentration on efficiency
  - How does dropsize distribution & surface area change with MEA concentration?

Objectives

- Measure drop size distribution for various MEA concentrations at multiple locations inside the spray plume
- Calculate surface area availability based on measured drop size distributions for varying MEA concentrations

AIChE Meeting – Spring 2014

Outline

- Research Motivation
- Set Up
- Methods & Matrix
- Results
- Conclusions & Future Work

AIChE Meeting - Spring 2014

Column

Ref: Tamhankar, Y.S., et al., Atomization and Sprays, 2014. **24**(2): p. 115-128.

AIChE Meeting – Spring 2014

Overall Set Up

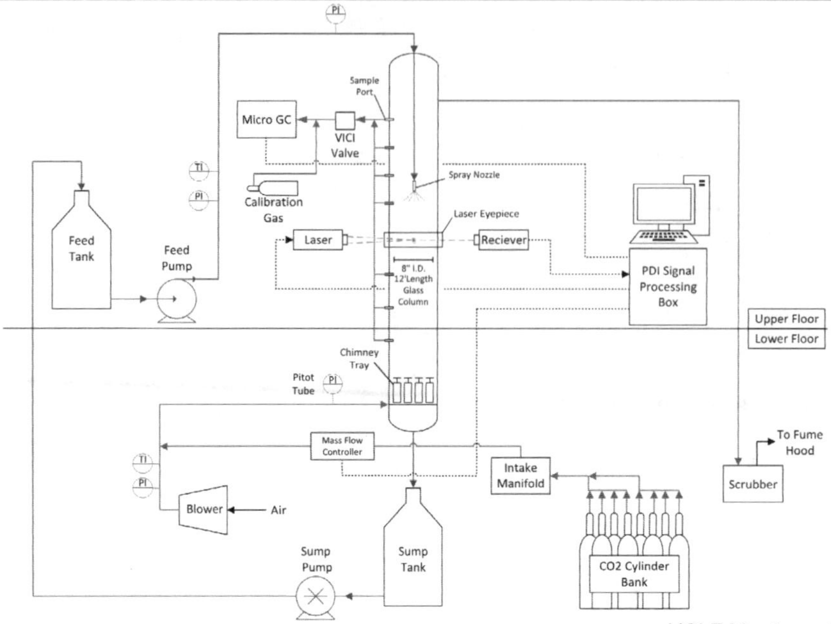

AIChE Meeting – Spring 2014

Outline

- Research Motivation
- Set Up
- Methods & Matrix
- Results
- Conclusions & Future Work

AIChE Meeting - Spring 2014

Method I – $6/D_{32}$

Sp. Surface Area – horizontal plane (m^2 of cum. drop surface /m^3 of cum. drop volume)

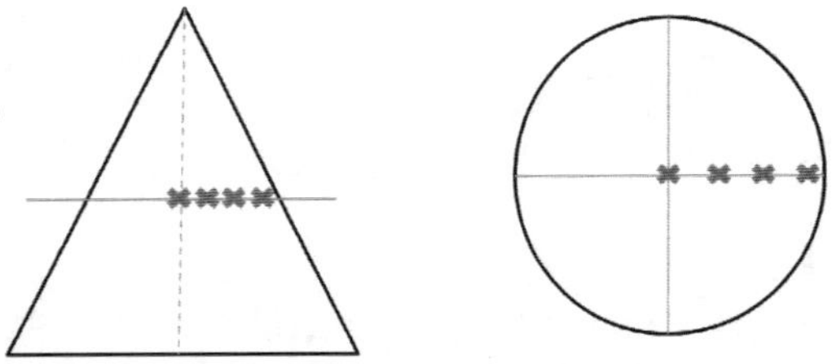

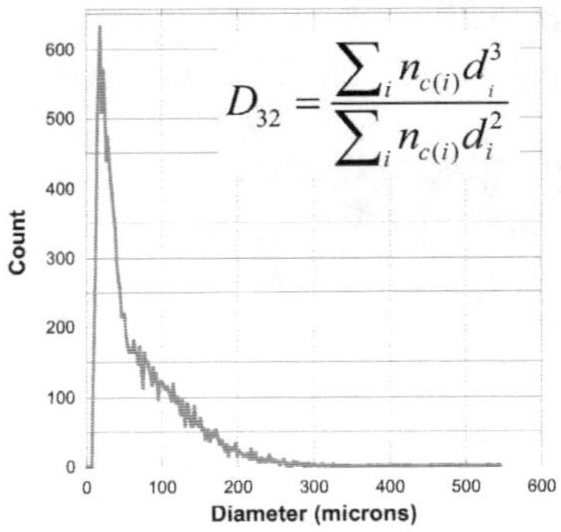

$$D_{32} = \frac{\sum_i n_{c(i)} d_i^3}{\sum_i n_{c(i)} d_i^2}$$

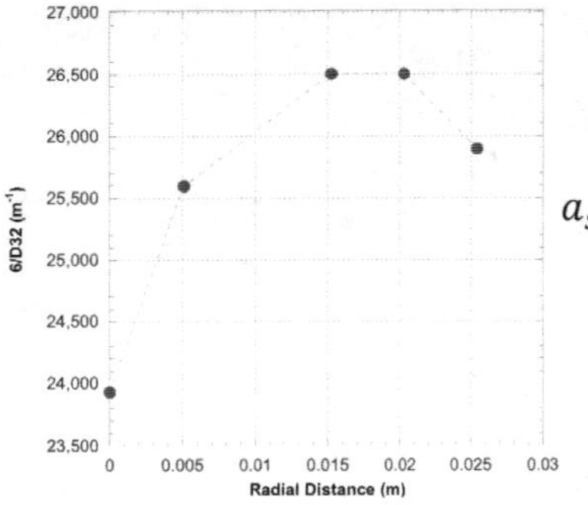

$$a_S = 2\,\pi \int_{r=0}^{r=r'} \frac{6}{d_{32}}(r)dr$$

AIChE Meeting – Spring 2014

Method II – T_S & T_V

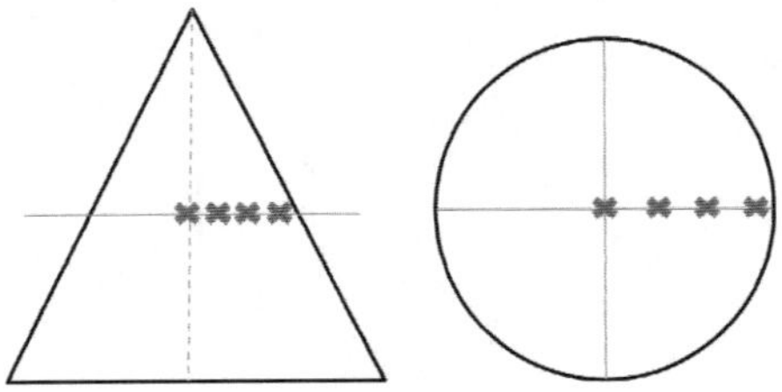

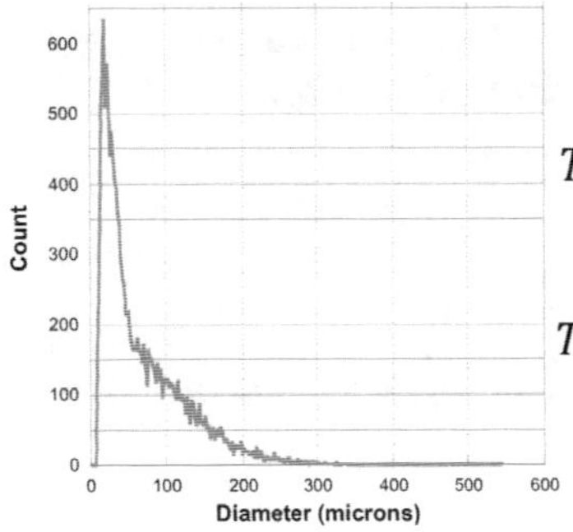

$$T_S = \pi \sum_i n_i\, d_i^{\,2}$$

$$T_V = \frac{\pi}{6} \sum_i n_i\, d_i^{\,3}$$

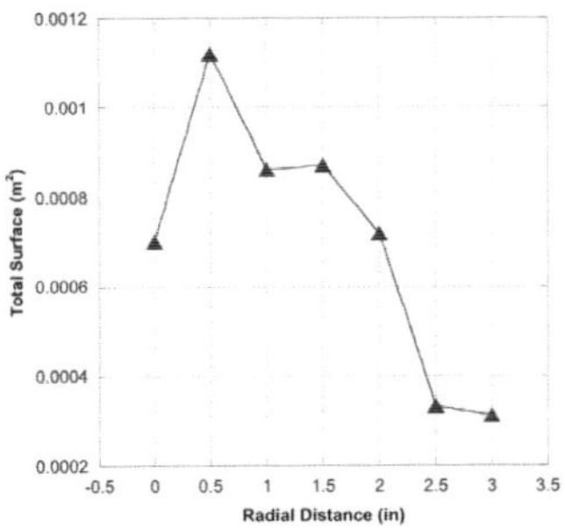

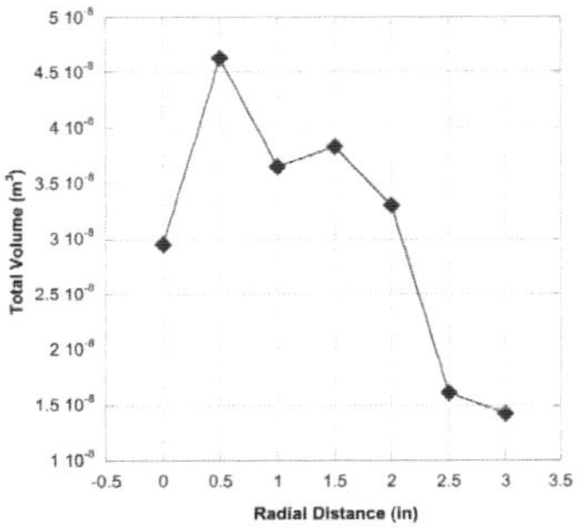

$$a_S = \frac{2\pi \int_{r=0}^{r=r'} T_S(r)\,dr}{2\pi \int_{r=0}^{r=r'} T_V\,(r)dr}$$

AIChE Meeting – Spring 2014

PDI Measurements - Matrix

Nozzle:
Full cone: BETE
1/8 MPL 0.30N

PDI Measurements:
Axial – 0.05 m (2")
Radial – 0 to 0.07 m (3") in increments of 0.002 m (0.5"; till outside the spray plume)

Liquid Rate: 7.57 x 10^{-4} m^3/min (0.2 gpm) at 28 psig
Liquid Temperature: 30 °C (86 °F)

Gas Rate: -

Wt% MEA: 20, 30, & 40

BETE MPL 0.30 N
Full Cone, 60°
Water:
0.12 – 0.35 gpm
10 – 80 psig
200 – 120 μm

Outline

- Research Motivation
- Set Up
- Methods & Matrix
- Results
- Conclusions & Future Work

Radial Variation - SMD

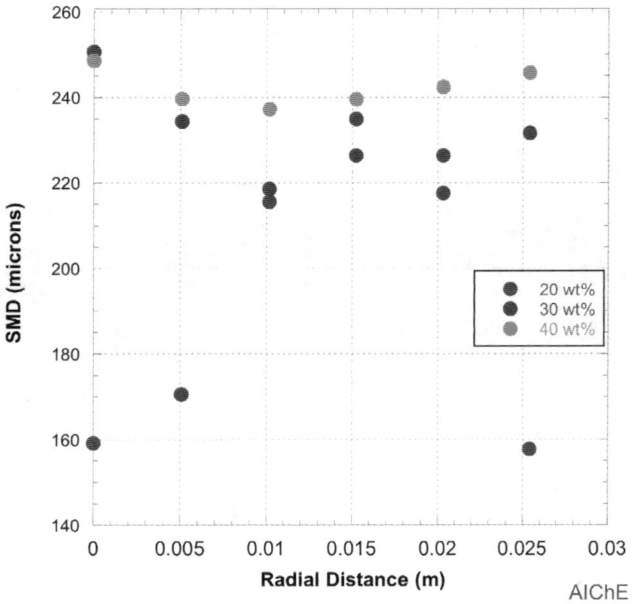

AIChE Meeting – Spring 2014

Radial Variation - SMD

Correlation vs. Vendor vs. Measured

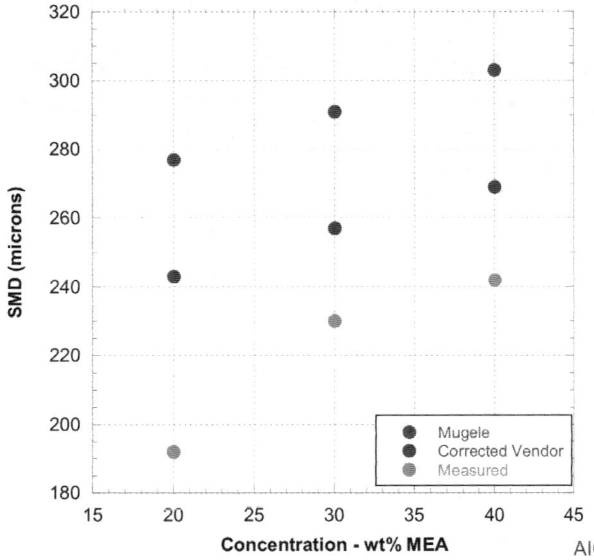

AIChE Meeting – Spring 2014

Radial Variation – T_S & T_V

20 wt% MEA

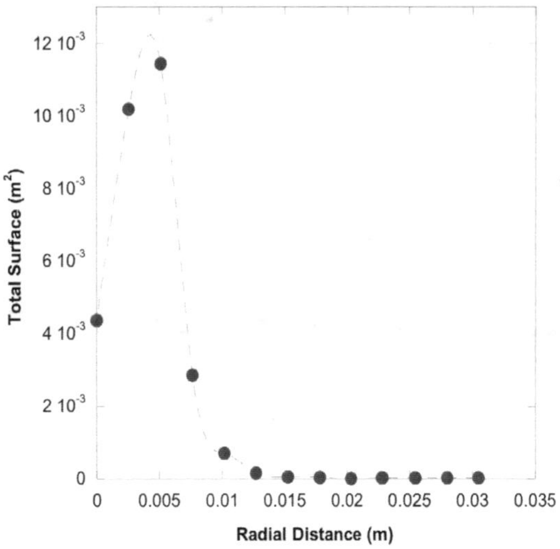

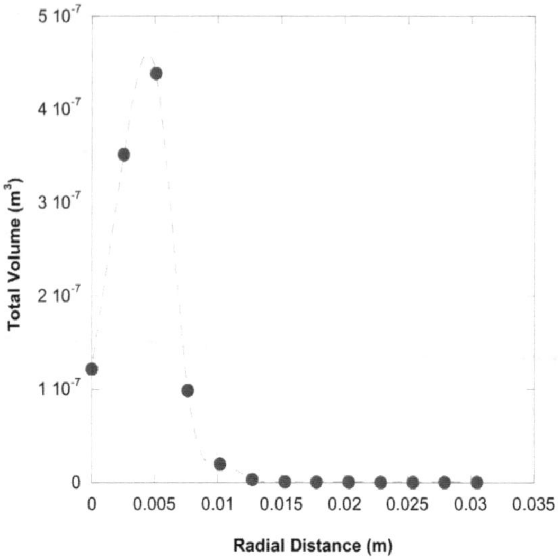

Surface Area Calculations

MEA (wt%)	Specific surface area, a_s (m^2/m^3) Method - I	Specific surface area, a_s (m^2/m^3) Method - II	Average drop size range (μm)	Specific surface area, a_{drop} (m^2/m^3)
20	1.46×10^4	4.97×10^6	54.9 – 124.3	$1.09 \times 10^5 - 4.83 \times 10^4$
30	3.02×10^3	3.64×10^4	114.2 – 153.3	$5.25 \times 10^4 - 3.91 \times 10^4$
40	5.03×10^3	2.59×10^4	130.9 – 158.9	$4.58 \times 10^4 - 3.78 \times 10^4$

Outline

- Research Motivation
- Set Up
- Methods & Matrix
- Results
- Conclusions & Future Work

AIChE Meeting - Spring 2014

Conclusions

- Increasing concentration of MEA produces larger drop size
  - Decreases specific surface area
- Measured SMD differs from published correlation and vendor data
- Limited value – singular measurements or estimates
- Two methods tested in quantifying surface area availability
  - Radial variation of $6/D_{32}$ (previously published method)
  - Radial variation of T_S and T_V (developed)
- Newly developed method appears to be more robust

AIChE Meeting – Spring 2014

Future Work

- Extend and test both surface area methods to other types of nozzles
  - Check the symmetry of nozzles
- Evaluate effect of axial location on surface area quantification
- Quantify surface area with counter current gas flow
- Extend procedure to other solvents
  - MDEA
  - NaOH

AIChE Meeting – Spring 2014

Acknowledgements

- Huntsman
  - Srini Godavarthy
  - Pat Holub
- Lake Innovation LLC
  - Chuck Lipp
- Artium Technologies
  - Chad Sipperly
  - Will Bachalo
- Swagelok, OK
  - David Fournoy

AIChE Meeting – Spring 2014

Questions?

AIChE Meeting – Spring 2014

Considerations in the Sizing of Power Generation Facilities for LNG Liquefaction Plants

Dragan Ristanovic, Neeraj Bhatia, Sunita Singh, Cyrus B. Meher-Homji
Bechtel Corporation, USA

ABSTRACT

High power demands of large scale LNG liquefaction plants have created the need for a careful and structured approach in selecting power generation facilities. Simultaneously, high thermal efficiency of LNG liquefaction plants has made it imperative to minimize feed usage and to reduce CO_2 emission. Further, in gas constrained situations, any savings in fuel auto consumption of the plant for liquefaction chilling and power generation can be converted into LNG production. Apart from optimal sizing of power generation, utilization of electrical infrastructure to support Spinning Reserve (SR) is one of the most important resources used by LNG plant operators to respond to unforeseen events such as generation outages and sudden load changes.

With focus on high overall plant efficiency, it is becoming increasingly popular to integrate process heat recovery and electrical power generation facility sizing. This paper delineates an integrated approach to process and power management, including optimal sizing of simple cycle gas turbine Generators (GTGs), configurations including cogeneration, power generation by expanders, essential service diesel generators and load management. Finally, an approach for evaluation of indices for GTG and project NPV calculations is presented.

NOMENCLATURE

AGRU	Acid gas removal unit
CC	Combined Cycle
CCR	Central Control Room
EPC	Engineering, Procurement, and Construction
ESDG	Essential Service Diesel Generator
FEED	Front End Engineering Design
GT	Gas Turbine
GTG	Gas Turbine Generator
HP	High Pressure
HRSG	Heat Recovery Steam Generator
HV	High Voltage
HVAC	Heat Ventilation Air Conditioning
LNG	Liquefied Natural Gas
MV	Medium Voltage
MW	Mega Watt
MTPA	Million Tons Per Annum
N_{Bla}	Number of Essential Service Diesel Generators required for Black start
N_{Exp}	Number of Expander generators
NPV	Net Present Value
SR	Spinning reserve
SC	Simple Cycle
SSTI	Sub Synchronous Torsional Interaction
ST	Steam Turbines
TIC	Total Installed Capacity
UPS	Uninterrupted Power Supply

1.0 INTRODUCTION

Natural gas currently supplies approximately 23% of the world energy production. According to the BP energy outlook [1], LNG represents a growing share of gas supply. Global LNG supply is projected to grow 4.5% per year to 2030, more than double the growth rate of total global gas production (2.1% per year). LNG contributes 25% of global supply growth 2010-30, compared to 19% for 1990-2010.

The size of LNG liquefaction facilities is gradually rising as shown in Figure 1 [2]. Currently most trains under development fall into the 3-6 MTPA range. A few mega-trains of 8 MTPA have also been built. As the size of the facility increases, and as multiple LNG trains are built within a single site, the selection of power generation capability becomes a significant challenge. The challenge includes managing performance at different ambient temperatures, step load changes, load drops, emissions, and other complexities such as the presence of sub-synchronous torsional interactions, when large VFD drives are present.

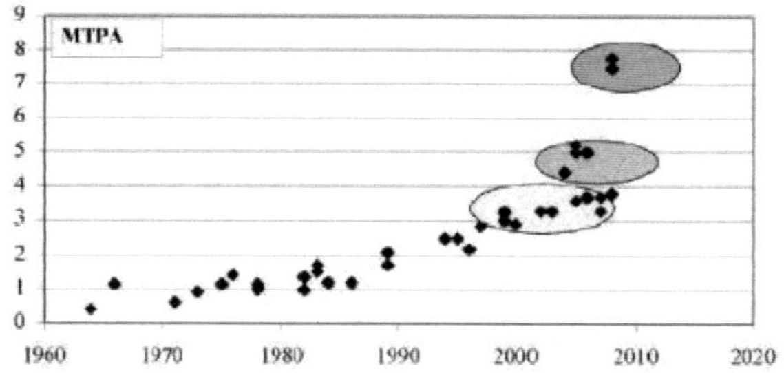

Figure 1 Growth in size of LNG Plants [2]

LNG train designs in the future will fall within four classes, having nominal capacities clustering around 1.5, 3.5, 5, and 8 million tons per annum (MTPA). These designs may coexist in the coming years depending on design selection by individual projects that match their gas supplies, sales, and other logistical and economic constraints.
The extent of electrical power needed in a LNG liquefaction process depends on multiple factors and is a strong function of the primary drive for the refrigeration compressors. The drives can be:

- **Single Shaft Heavy duty gas turbines.** If the drives are heavy duty single shaft gas turbines, then large VFD starter helpers are needed. Depending on the size of the single shaft gas turbine these can range from 6-40 MW per gas turbine. This results in relatively larger power generation requirements.
- **Aero-derivative Gas Turbine Drives**. In the case where two shaft (eg. Aero-derivative) engines are used, startup torque from the free power turbine is adequate, meaning that no starter motors are required. Power generation requirements for such facili-

ties can vary depending on the process, but are generally lower than if single shaft units are used [3].

- **All Electric LNG Facilities (eLNG).** If the facility is an all-electric facility (eLNG) then the power requirement is very large as all the main refrigerant trains will be driven by variable speed motors. In the case of an eLNG facility, the power requirements could be as large as 400-500MW for a two 5 MTPA facility.

Additional factors that will define the electrical load include:

- The liquefaction process- some processes require additional compression equipment, such as flash gas compressors.
- Site and project specific considerations that may include special domestic gas facilities, or additional power generation for local consumption.

LNG liquefaction plants are characterized by operational challenges including varying ambient conditions, heat integration and product quality constraints. Engineering, Procurement and Construction (EPC) phase of LNG plant projects have its own set of challenges. An important challenge in FEED stage is careful estimation of the plant electrical load. An additional challenge is the selection of the ratings and numbers of either the generators in an isolated system, or main incoming feeders from a power utility company. Occasionally the LNG plant requires a combination of generators and incoming feeders where generators operate either in isolation (island mode) or in synchronism with the utility company [4].

Future expansion requirements of the plant dictate its power block design. The electrical supply capacity normally is composed of two elements, one to match the known or initial consumption, and the second to account for keeping a spare generator or feeder ready for service in case of abnormal conditions.

Preliminary Screening of Power Generation Options

The earliest task in the plant development phase is the selection of power block configuration. Depending on the plant load requirements a wide range of options exist. These include:

- Simple Cycle Gas Turbine Generators- this is by far the most common approach taken for a majority of LNG Liquefaction plants.
- Cogeneration Cycles
- Steam Turbine Cycles

In very rare cases it may be possible to obtain power from a local utility. This is very rare as most LNG Facilities are in locations where minimal infrastructure exists. Even when cases exist of low cost hydro power being available or where large utilities are willing to sell across the fence power, experience has shown that plant often revert to self-generation.

The fundamental configuration for power generation is usually defined in the early phases of the project (during Pre FEED). The decision is based on CAPEX, user preference, operational complexity, and the strength for thermal efficiency, emission criteria and electrical requirements.

1. One of the set of configurations may be from a portfolio of standardized plant designs for each of the following Combined Cycle (CC) configurations:

   1 x 1 (one GTG, one HRSG, and one ST)
   2 x 1 (two GTGs, two HRSGs, and one ST)
   3 x 1 (three GTGs, three HRSGs, and one ST)

This approach allows the LNG plant's Engineering, Procurement and Construction (EPC) contractor to build a plant that best suits the owner's operating goals. A single ST for multiple GTs is less costly than a separate ST for each GT. However, multiple 1x1 configuration trains offer some significant advantages like (1) Phased construction flexibility - Owners can add units as dictated by market conditions, (2) Speed to market - As the construction duration of a 1x1 configuration is shorter than a 3x1 configuration, the first train can be brought on-line more quickly and start generating revenues, while the remaining units are still undergoing construction, (3) Closer matching with dispatch demand - A site with multiple 1x1 trains using supplemental duct firing can better and more efficiently match dispatch requirements, (4) Greater plant redundancy - Each unit is completely redundant, (5) Optimized spare parts inventory - Identical components are used for all trains.

2. A popular configuration discussed in this paper is a combination of generation and co-generation using GTGs, ESDGs Expander Generators and HRSG.

This paper discusses the second option in detail.

This paper mainly deals with the discussion of the integral components of the LNG plant's technology, process and electrical load considerations while emphasizing on the NPV calculation criteria for the main power block selection. The selection and sizing is based on an assumption of isolated plant operation with no connection to the power grid. The configuration discussed in the paper is with GTG as main source in the power block, expander generator/HRSG/ESDG as a co-generation source where ESDG is for black-start/emergency shut-down/essential load power supply.

Section 2 talks about the generation and co-generation used in LNG plants and presents an idea about how co-generation is sized for a given plant load. Reactive power requirements of the plant are also discussed in this section. Succeeding it is the Section 3 presenting the plant load list selection, plant load profile estimation, discusses typical LNG plant facts about contingency/growth allowance/load diversity, and concepts of normal operating load/black start load/essential load. It also briefly talks about the plant load shedding philosophy and power management system. Section 4 discusses generation configuration philosophy while Section 5 presents equipment selection criteria keeping thermal efficiency as a focus. Section 6 details main power source or generator sizing. Following all these conceptual sections is the Section 7 which utilizes information from all preceding sections and finally presents NPV analysis approach and GTG selection criteria for an LNG plant project.

2.0 POWER GENERATION

The number/size and manufacturer of the generators are preliminary considerations and are finalized during the FEED phase of the project. A variety of cogeneration configurations integrated with GTGs to improve overall plant efficiency can be utilized and in some cases the GT used for mechanical drive can be used for generation of steam that can be used for process and power generation purposes.

2.1 Gas Turbine Generator (GTG)

Gas Turbine Generators (GTGs) are typically the primary power generation system of the plant's real power and reactive power. The generation capacity of 'N' GTGs at site rated ambient condition should be equal to or larger than the final plant design operating load. In addition, during the largest motors starting at each voltage level with the other operating loads, the voltage drop criteria should be met.

Normally a $N_{GTG}+1$ criterion is applied. Most operators require the N+1units to be operating and sharing load.

2.2 Expander Generator

Within the Philips Optimized Cascade LNG Process, there are potential applications of flashing liquid expanders and also traditional gas expanders. The expanders drive electrical generators that augment the site power . An expander generator uses the energy (a pressure let down) from the process to generate real power. Depending on the size of the expander, the generator can be induction generators which consume reactive power from the electrical system or synchronous generators which generate reactive power [5, 6].

Each LNG train may have one expander generator. When the LNG train is in production, the expander generator will generate electric power and supply part of the load demand in that train area.

The sizing of the expander is a function of the enthalpy drop available within the process and the energy derived is a function of this enthalpy drop and the efficiency of the generator and gearbox (if one is needed). Typically the expanders are radial inflow turbine designs.

2.3 Combined / Cogeneration Cycles for LNG Liquefaction

There are several approaches wherein steam derived from the mechanical drive gas turbines may be used to provide both power and heat for the LNG facility. The full utilization of heat enables a significant improvement of efficiency when low efficiency industrial gas turbine drivers are utilized. Factors that have to be considered include [7]:

- Plot plan impact- the location of the HRSGs, steam piping and condensers if condensing steam turbines are used.
- Piping considerations- integrating the liquefaction section and the power generation section will call for piping routing to the power block.
- Design complexity- the design complexity has to be traded off with the increased thermal efficiency
- Transient operation- consideration must be made of transient scenarios such as a liquefaction unit trip, or half plant trips[1]
- Electrical transient considerations
- Startup considerations- very often additional boilers or stand by gas turbine generators may be needed.

2.3.1 Levels of Heat Integration

There are multiple levels that can be considered for heat integration:

- Use of HRSGs to provide process steam (no steam turbines)
- Use of HRSGs to provide process steam and also drive back pressure steam turbines. These turbines could be used for either power generation if aero-derivative engines are considered or as starter helpers if single shaft gas turbines are considered. If aero-derivative engines are used, then using the exhaust heat to produce power and process heat either from a back pressure or extraction condensing steam turbine, is a clean solution as the liquefaction block remains intact- driven totally by gas turbines.
- Use of HRSGs to provide steam and condensing extraction steam turbines which can be used for either power generation or as mechanical drivers. For example in the Optimized Cascade process, in a 2 Propane+ 2 Ethylene+ 2 Methane configuration, steam generated by the four Propane and Ethylene drivers, can be used in condensing steam turbines to drive the two methane trains.

2.3.2 Heat Requirements and Heat/ Power Ratio in LNG Liquefaction Facilities

The extent of process heat required is a strong function of the concentration of acid gas contaminants in the feed gas and the process heat required for the AGRU. The required process heat to power ratio for the facility is an important parameter and impacts the type of gas turbines selected and the heat recovery equipment and approaches used. Aero-derivative engines have higher thermal efficiencies around 40% and this means that the capability to raise heat in an unfired HRSG is lower than an industrial gas turbine commonly used in LNG mechanical drive service that has a thermal efficiency around 32-33%. Typical Heat/Power ratios for aero-derivatives are around 0.9- 1.0 while those for industrial engines are higher- around 1.4[2] . These values are without supplemental firing and can be significantly increased by the use of supplemental firing.

Large high-pressure, high-temperature reheat HRSGs burning natural gas and possibly some distillate are optimized for maximum electricity production and, in most cases, base-load service.

Compensating for the unexpected loss of an HRSG requires a quick re-distribution of the total electrical load onto the remaining generating sets without violating the voltage & frequency limits of the electrical system, or the speed limits of the turbines and compressors. With suitable steam reserves, and specific control actions in the power generation, and the compressor and drive systems, this situation can be coped with, generally without losing the refrigeration process [7].

A typical cogeneration configuration is indicated in Table 1and the thermal design of the cogeneration system is shown in Figure 2.

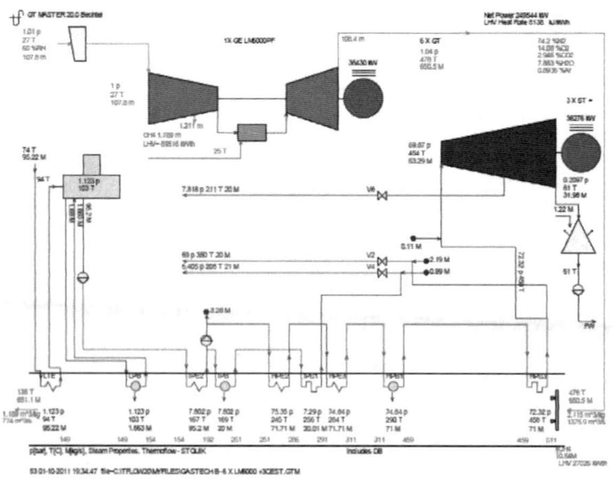

Figure 2 Cycle flow schematic for Cogen Type System [8]

[1] The Optimized Cascade Process can operate at part load with 50% of its refrigeration turbine drivers out of service due to the two trains in one configuration.

[2] The actual values are dependent on the number of pressure levels and process steam pressures.

Table 1 – Typical cogeneration configuration [8]

Configuration **6 LM6000PF engines in 2+2+2 Configuration** **Approximate MTPA = 5.x**	**6 x LM6000PF** **Propane – 2 x LM6000PF** **Ethylene – 2 x LM6000PF** **Methane – 2 x LM6000PF**	**All power delivered from Condensing Steam Turbines using heat recovered from Refrigeration Drivers.**	**Steam from HRSGs and extraction from steam turbines.**

2.4 Essential Service Diesel Generator (ESDG)

Essential Service Diesel Generators serve the basic purpose to black start the plant, emergency backup power and safely shut down of the plant. These generators are also used to provide essential power for selected loads that may be required in the event of an extended outage of the primary power generation system. The main steps in selection and sizing of ESDGs are documenting the essential/black-start load, location of the generator and controlling noise and vibration. 'N_{Bla}' ESDGs are selected to accommodate the Plant Design Black Start Load. Motor starting is also studied to make sure the voltage drop criterion is met during the largest essential motor starting on ESDG. $N_{Bla}+1$ criterion are also applied to maintain spare capacity. It is prudent to select 3 x 50% high voltage ESDGs for the process black start load. The 3 ESDGs also accommodate additional 'essential' loads which are not black start loads. If 3 x 50% ESDGs criteria can't be met due to essential load increase, a study should be performed.

For Operator Permanent Camp and Operation Center area, low voltage (400V) essential generator should also be sized to accommodate the essential load requirement. Centralized systems for controlling and managing, electrical power and process are installed at the Operation center. Operator Permanent Camp includes clinic, fire station, fire-water pump, canteen, kitchen, Refrigeration, etc.

2.5 Reactive Power Generation

Electrical load consumes both real power and reactive power. The lower the power factor between 0 and 1, the more reactive power is required. There are reactive power losses in power transformers and power cables. If the expander generators are induction generators, they also need reactive power supply as shown in Figure 3. GTGs are the main source of reactive power supply.

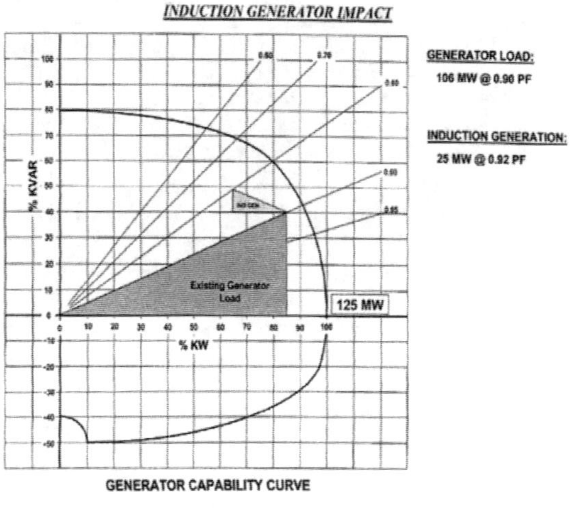

Figure 3 Generator capability curve

If the reactive power demand in electrical loads and reactive power losses in power transformers and cables are beyond the GTG reactive power supply capability, additional reactive power generation such as capacitors, synchronous condensers, etc., may be needed. This option has to be studied during the electrical system design [9].

3.0 DETERMINATION OF LOAD

3.1 Load List

An Electrical Load List is developed for the facility and maintained throughout the design which comprises of process and utility loads.

A load list should include the following as a minimum, so that Connected and Operating Loads can be derived:

- Voltage Rating, Amps, Motor or Utility System Load (kW), Power Factor, Efficiency, Service Duty, Demand Factor, Utilization Factor

Connected and Operating Loads should be derived keeping following factors in consideration.

a) Connected Load is the nameplate rating of a load.

b) Demand Factor is the calculated energy demanded by a load connected to the electrical system divided by its nameplate rating (connected load). For example, a connected load of 30 kW may actually require a 24 kW at design conditions, thus the demand factor in this case will be 24/30 or 0.80.

c) Utilization Factor is the percentage of a typical operating period that the load will be energized. Normally, a continuous load will be assigned a utilization factor of 100%, standby (or spare) loads will be assigned 0%, and an intermittent load will be assigned a utilization factor between 0 and 100%.

d) Efficiency represents the percentage of the electrical power delivered to the mechanical load.

The resulting load list should approximate the plant operating conditions at a given time.

3.2 Load Profile

A Facility Electrical Load Profile should be completed at the earliest possible time, to determine maximum facility loads and duration required for both normal and essential power supply. A load profile is a "time based" facility load requirement that can define minimum and maximum electric loads and duration for each based on available project data. Load profile showing generation capacity to facility load for the life of the facility is built.

In general, the maximum load should be used to determine size and quantity of Turbine/Generator system. Engineering judgment is necessary to determine whether:

a) It is economically feasible to add generation (size and/or quantity)

b) The maximum loads occur for a small duration throughout the facility life

c) Such an increase in size or quantity of generators is ultimately required.

3.3 Contingency/Growth Allowance/Load Diversity

Contingency and Growth Allowance applied to the power generation facilities should be determined to avoid shortages during operation, future modifications, and generator deterioration.

Contingency

In order to compensate for the uncertainty associated with the development and planning of a particular project, a contingency factor shall be applied. This factor should account for:

1. Typical loads associated with packaged equipment which has not been accounted for during the assessment period (lube pumps, cooling loads, etc.).
2. Facility non-process or utility loads that are unknown and only have past project experience as a guide (lighting, DC/UPS systems, space heater loads, etc.). These should not be confused with new loads or equipment as defined in the Growth Allowance.
3. Depending on the engineering knowledge of the uncertain loads, the contingency factor can be 0% to 10% and should be applied until the purchase of the generators is made.

Growth Allowance

a) Future growth of facilities or modifications to the processes should be limited by, but not limited to, the growth allowance of the power generation facilities. As facilities evolve over time, it is apparent that plants tend to adapt new processes to extend or improve the performance of facilities. To account for this unknown future development and for future changes in process conditions over the facility life cycle, a Growth Allowance should be applied. Historical data should be used to determine the level applied to the project in a way that is consistent with the current design practices and system developments.

b) The growth allowance should be 10%.

Load Diversity

Not all the electrical loads are coincident loads. For example, all the lighting loads are treated as continuous loads even though they are not operating all the same time. The maximum LNG train load and Domestic Gas (DomGas) load are used for load calculation. But in reality, with certain amount of inlet gas, LNG train load and DomGas load can't be the maximum at the same time. From LNG experience and historical data, load diversity of 0.95 is used for load calculation.

3.4 Normal Operating Load/Black Start Load/ Essential Load

The electrical load can be classified as normal operating load, black start load and essential load.

Normal Operating Load

For the normal plant operation, the electric load is the normal operating load. It can have minimum and maximum load conditions. For example, based on the process design, the minimum load can be one LNG train in production and there is no ship loading. The maximum load can be two LNG trains in production and with the ship loading. For the main power source generator sizing, the maximum load needs to be considered. Based on the maximal process design case and with ship loading, from the load list and the load profile, the maximum normal/plant design/final plant design operating load can be defined as:

Maximum Normal Operating Load = 100% x Continuous Load + 50% x Intermittent Load + 0% Spare Load.

Plant Design Operating Load = Maximum Normal Operating Load x (1 + Contingency + Growth Allowance) – Growth Allowance x Sum of Compressor Loads (1MW & above).

Final Plant Design Operating Load = Load Diversity x Plant Design Operating Load.

Black Start Load

Black start load is the electrical load required to start up the first main power source 'gas turbine generator (GTG)'. Based on the process input, the black start load can include the followings as minimum:

- Auxiliary load of one GTG, including the starter motor
- Electrical fuel gas heater, only the necessary partial load to start up one GTG
- Instrument air compressor, only the necessary partial load to start up one GTG
- Essential HVAC load for one GTG building
- Electrical driven fire water pump
- Jockey pump
- AC UPS
- DC System (Charger)
- Building essential lighting
- Essential load for the Central Control Room (CCR) or Operation Center etc.

Maximum Black Start Load = 100% x Continuous Black Start Load + 50% x Intermittent Black Start Load + 0% Spare Black Start Load.
Plant Design Black Start Load = Maximum Black Start Load x (1 + Contingency + Growth Allowance)

Essential Load

Essential load is the electrical load necessary to maintain the plant operation and for personnel safety. The essential load can include the following as a minimum:

1. Auxiliary load of all GTGs, including one GTG starter motor AC UPS and D.C. systems, 20% of the plant area lighting (exception street lighting and outdoor building areas)
2. Critical process load (such as fire water pump (if electrical driven), jockey pumps, electrical fuel gas heater, diesel system, etc.)
3. All lights in main control room (CCR)/Operation center and local control rooms at substations, 50% lights in electrical substations, 20% lights in administration building and other buildings, 50 % of local control rooms' and electrical substations' HVAC (exception of duct heaters), 100% of main control room (CCR) and administration building HVAC (exception of duct heaters), Laboratory HVAC (exception of duct heater)
4. Permanent Operator Camp Essential Load (such as clinic, fire station, fire-water pump, HVAC for canteen, kitchen essential lighting, Refrigeration, etc.)
5. Heat tracing load for LNG tank Instrument air compressor (only one), 100% Communications systems, 100% sea-navigation safety systems and air-warning light
6. Maximum Essential Load = 100% x Continuous Essential Load + 50% x Intermittent Essential Load + 0% Spare Essential Load.
7. Plant Design Essential Load = Maximum Essential Load x (1 + Contingency + Growth Allowance)

3.5 Plant load shedding philosophy/Power Management System

A power management system (PMS) is an integral part of LNG power plants these days. It maximizes system up-time and mitigates problems before experiencing an outage with proactive, high-speed load shedding, generation control, voltage control, and load management. It continuously assesses plant operating load, generation, and Spinning Reserve (SR) against the occurrence of unforeseen event such as generation outages and sudden load changes. For an automatic load shedding, PMS compares operating load and generation data to perform high speed load-shedding based on a pre-determined priority based load-shedding options matrix. PMS facilitates automatic load restoration once proper operating conditions are validated [9].

A load shedding options matrix must be developed in the EPC stage of the project during which PMS is configured. Non-critical loads are the first loads to be shed and typically include items such as permanent operator camp load, maintenance and warehouse building loads, incinerator and other miscellaneous loads. If these are not enough to maintain the generation/load balance, ship loading pump loads and /or of the Boil-Off-Gas compressor loads may be dropped. Alternatively in LNG plant production could be cut by going to half rates till recovery is attained.

Engineering studies, load shedding options, and economic comparisons must be performed including lessons learned and past experiences to get the optimal solution.

4.0 GENERATION CONFIGURATION PHILOSOPHY

Typical generation voltages for 50 Hz systems are 11 kV and 13.8 kV for 60 Hz systems. For example in a 50 Hz LNG facility, generation unit may have a step up transformer of 11/132kV, GTGs synchronize to the 132kV main distribution system through the 'synch bus' and supply the load demand of the LNG plant and Operator Permanent Camp though step down transformers.

Typical 11kV, 33kV, 69kV or 132kV 'synch bus' configurations can be adopted depending on the need of the generation & distribution system. 'N+1' GTGs are required if projects and clients choose to run all GTGs for some operational convenience. At all voltage levels, separate buses are for normal and essential load distribution. These buses are connected though two cascaded tie-breakers. At low voltage one of tie-breakers is closed at a time while MV and HV operate with both the tie-breakers closed.

The expander generators generate power, and supply the LNG trains load at the same distribution voltage (6.6kV) at the LNG train area. If the other voltage (i.e., 11kV) is required, a study needs to be performed.

The main 6.6kV ESDGs are on the 6.6kV essential service bus. To be able to black start the plant, especially to start the largest essential motor, the tie breaker of the 6.6kV essential service bus should be closed. After the largest essential motor is started, the tie breaker can be opened. The 400V essential generators in the Operator Permanent Camp and Operation Center (CCR) are connected to the 400V essential buses in those areas to supply the essential load.

5.0 EQUIPMENT SELECTION

Before selecting equipment from different suppliers, a thorough investigation is necessary to ensure that the owner's pro forma objectives are met for power output, heat rate, startup times, reliability and availability, etc.

The process includes an independent technology assessment of the equipment's operating history and quality control for the engineering and manufacturing processes. In addition, the performance offered by the original equipment manufacturers (OEMs) for a specific project is normalized and reconciled with past performance of the same equipment in a similar configuration, as achieved on other projects. It is a recommended practice to maintain a performance database of all past projects and routinely update with information from field tests.

With the increased concern related to global warming, and as fuel costs increase, the issue of overall thermal efficiency of the LNG facility has become more important. The thermal efficiency of an LNG facility depends on numerous factors such as gas composition, inlet pressure, etc. Gas turbine selection, the use of waste heat recovery, and the mode for power generation, all have a significant effect on the overall thermal efficiency of the LNG process. Process flexibility and operational stability are issues of paramount importance and must be incorporated into the considerations regarding thermal efficiency [10].

6.0 GENERATOR SIZING

'N' generators can meet the load requirement of final plant design operating load. The generation capability of N_{Exp}-1 expander generators is not discounted in the sizing of generation. Assuming that the total load requirement is of 100MW, the options can be of 10 x 10MW generators or 5 x 20MW generators.

In general, the bigger size of the generator, the smaller number of N. The total installed cost of the generators, associated step up transformers, breakers, power cables / bus bars, buildings and real estate will be less.

There are two disadvantages of the bigger size generator option. First one is that the generator load factor is low when N+ 1 generators are running. The generator efficiency must be checked to make sure it is not below its efficiency curve. Second one is that more loads need to be shed on the worst case scenario of only N generators running with one generator out of service suddenly. For the smaller size generator, less loads need to be shed for this worst case scenario.

Thus the proper value of 'N' and certain selection criteria must be defined. Engineering study, comfortable loads to be shed and economic comparison must be performed to get the optimal solution.

Transient stability study should also be performed to make sure that the electrical system can maintain the transient stability after the generator is out of service and enough loads are shed.

6.1 Generator Load Factor

Out of N+ 1 installed generators, output of N generators can meet the load requirement. The normal running philosophy is all N+1 generators are running. For the worst case scenario, one GTG is down for the maintenance and only N generators are running. When another generator is suddenly out of service, some load must be shed to maintain the balance between the generation and load. The following equations are used [11]:

P_i: Power Output required at each generator at site ambient condition
P_g: Rate Power Output of each generator at site ambient (driver known)
F_o: Overload factor in % when 1 generator is suddenly switched out of service
F_i: Initial Plant Generation Load factor in % before 1 generator is switched out of service
F_n: Normal Plant Generation Load factor in % when N+1 generators are running

Pre-disturbance Plant Generation Load factor is given by

$$F_i = \frac{100P_i}{P_g N}\% \qquad (1)$$

Post-disturbance Plant Generation Load factor is given by

$$F_0 = \frac{100P_i}{P_g(N-1)}\% \quad (2)$$

Normal Plant Generation Load factor is given by

$$F_n = \frac{100P_i}{P_g(N+1)}\% \quad (3)$$

Table 2 Number of installed generators Vs Plant Load Factor

No. of Installed Generators	Value of F to ensure no overloading
2	Not Practical
3	50.00
4	66.67
5	75.00
6	80.00
7	83.33
8	85.71

Table 3 Factors affecting plant Load factor

Power Consumption (MW) Pi	Initial Plant Load factor (Fi)%	No. of Installed Gen Sets	Generator rating at Site Conditions (MW) Pg (type)	Turbine ISO Rating (MW)	Final Plant Load factor (Fo)%
100	83.3	4	40(LM6000)	47.1	125
100	71.4	5	35(LM6000)	41.2	95.2
100	83.3	5	30.3 (FR6F)	35.3	111.1
100	66.7	6	30(LM2500P4)	35.3	83.3
100	72.7	6	27.5(LM2500)	32.4	91.9
100	80	6	25.0 (SGT600)	29.4	100
100	88.9	6	22.5 (FR 5)	26.5	111.1

In general, F_i between 85% and 90% and F_o between 105% and 110% are good factor numbers. Table 2 and Table 3 illustrate factors affecting plant load factor to ensure no overloading for given number of installed generators.

In practice if F_i is too high the operator of the plant will become nervous and will often switch into service the spare generator. If F_i is too low then there will be too many generators in service and it should be possible to withdraw one. Gas turbines have poor fuel economy when they are lightly loaded. High values of F_o should be avoided because of the risk of cascade tripping by the gas turbines. The margin of overload that a gas turbine can tolerate is relatively small and varies with the turbine design. The higher the normal combustion temperature within the turbine, the lower the tolerance is usually found to be available. A high overload will also be accompanied by a significant fall in electrical system frequency, caused by the slowing down of the power turbine and the relatively long time taken by the speed governing system to respond. Many power systems that use gas-turbine generators are provided with under-frequency and over-frequency protective relays, and these may be set to trip the generator when a high overload occurs. The initial rate of decline in frequency is determined by the moment of inertia of the power turbine, plus the generator rotor, and the magnitude of the power change seen at the terminals of the generator. If F_o is designed to be less than approximately 105% then the generators will be able to absorb the overload until some corrective action by an operator is taken e.g. puts the spare generator into service. However, it is also possible to introduce a high-speed load shedding scheme into the power system when F_o is found to be above 105%. Such a scheme will compute in an anticipatory manner how much consumption should be deleted in the event of a loss of one generator. The designer will be able to predetermine enough low priority consumers to achieve the necessary corrective action.

6.2 Rating of Generators in relation to their Prime Movers

Operation at low ambient temperature conditions

In some countries the ambient temperature can vary significantly over a 24-hour period, and its average daily value can also vary widely over a 12-month period. The power plant designer should therefore ascertain the minimum and maximum ambient temperatures that apply to the plant. The maximum value will be used frequently in the sizing and specification of equipment. The minimum value will seldom be used, but it is very important when the sizing of generators and their prime movers are being examined.

Prime movers will produce more output power at their shafts when the ambient temperature is low. The combustion air in the prime mover is taken in at the ambient temperature. Gas turbines are more sensitive to the ambient air temperature than the piston engines.

If the ambient temperature is low for long periods of time then the power plant can generate its highest output, which can be beneficial to the plant especially if a seasonal peak demand occurs during this period of low temperature. In some situations a generator may be able to be taken out of service, and hence save on wear and tear, and fuel. With this in mind the generator rating should exceed that of the prime mover when power is required at the low ambient temperature. A margin of between 5% and 10% should be added to the prime mover output to obtain a suitable rating for the generator. It should be noted that when the output of a prime mover is being considered, it should be the output from the main gearbox if one is used. Gearbox losses can amount to 1% to 2% of rated output power.

Upgrading of Prime Movers

Some prime movers, especially new designs, are conservatively rated by their manufacturer. As the years pass some designs are upgraded to produce more power. As much as 10% to 15% can be increased in this manner. If the power system designer is aware of this potential increase in rating then the generator rating should be chosen initially to allow for this benefit. At the same time the cables and switchgear should be rated accordingly.

Situations occur, especially with floating LNG, where no physical space is available to install an extra generator and its associated equipment. Sometimes the main switch-rooms cannot accept any more switchgear, not even one more generator circuit breaker. Therefore the potential for upgrading a prime mover without having to make major changes to the electrical system is an option that should be considered seriously at the beginning of a project.

7.0 NPV ANALYSIS AND GTG SELECTION

Net Present Value (NPV) plays an integral role in the selection of GTGs for LNG Liquefaction plants. Budgetary quotations for NPV analysis are obtained from a number of reputable GTG manufacturers for a variety of models available in the industry.

A budgetary quotation from various GTG manufacturers is based on the project specifications assuming no anomalies in the submission. The budgetary pricing is based on the required performance data and detailed scope of supply included in the Request For Quotation (RFQ).

The Bidders are evaluated based on a combination of lowest NPV Rank, lowest Total Installed Capacity (TIC) Rank and an acceptable GTG Load Factor Risk and Load Shedding. The results of the GTG Selection Study would culminate in a short-list of recommended suppliers and determine the suitability of a particular type of GTG. A final recommendation would be made after the selection of the gas turbine driver for the Liquefaction plant.

7.1 NPV and GTG Loading analysis

The NPV analysis is based on some notable evaluation criteria as listed below:

- Selection of a minimum number N of installed GTGs for reliable N+1 operation (N Running + 1 Installed Spare) as per the information in Section 6.0. This is normally a requirement based on the risk associated with high GTG Load Factors for an N-1 configuration.
- Apply escalation factors for fuel costs and O&M costs as per the project requirements. A de-rating factor of x% is applied for GT fouling and degradation.
- A load shedding options matrix for the LNG project is created based on the information in Section 3.0.
- An N-1 scenario is assumed where one (1) GTG is down for maintenance and a running GTG trips.
- Project load factor operational risk criteria are listed.

The GTG NPV Summary Tabulation is prepared based on Net Power @ x° C that may be de-rated from ISO conditions for site rated temperature, humidity, inlet/exhaust losses, fouling and degradation. Bidders' GTG Models that require more than N+1 installed GTGs based on Net Power are considered not acceptable.

7.2 GTG NPV Calculation

A model LNG project evaluation criterion is given in Table 4. The determining factors may differ for different LNG plants. The intent is to show the structure of such an analysis.

Table 4 Model Project Evaluation Criteria

Temperature	(°C)
Humidity	(%)
GTG Inlet Losses	(mm H2O)
GTG Exhaust Losses	(mm H2O)
GTG Fouling/Degradation	(%)
Emissions Requirement	DLE with NOx < 25 ppm
OPERATING PHILOSOPHY	N+1 with N GTGs Running with a minimum of N+1 installed
Design Running Load	(KW)
NPV Operating Life Cycle	(years)
NPV Discount Rate	(%)
Availability	(%)
EPC Schedule	(years)
EPC % Annual Expense Rate	(%)
Spare Parts Cost Factor	(%)
Installed Cost Factor for GTG Type 1	
Installed Cost Factor for GTG Type 2	
Installed Cost Factor for GTG Type 3	
Fuel Cost	(US$/MBTU)
Annual Escalation - Fuel	(%)
Annual Escalation - O&M	(%)

The above evaluation is as per model GTG NPV evaluation given in Table 5. The criteria are based on experience in the LNG plant projects for the past several years.

Table 5 GTG NPV Evaluation Criteria

GTG Manufacturer	
Model	
Type	
ISO Rated Power	(kW)
Power @ x°C Design Rated Case	(kW)
Efficiency @ x°C Design Rated Case	(%)
Number of Installed GTGs (N+1)	
Fired Hours per year	(Hours)
Equipment FOB Cost	(US$)
Annual Fuel Cost	(US$)
Annual O&M Cost	(US$)
Annual Operating Cash Flow	(US$)
Installed Cost Factor	(US$)
Total Installed Cost	(US$)
TIC Rank	
NPV	(US$)

Apart from the above factors, GTG performance is also evaluated based on ISO rated heat rate, ISO rated efficiency, site de-rated power loss, GTG F&D de-rate factors, net power at x °C, spinning reserve for N GTGs, and GTG load factor. Evaluations are done for minimum and maximum operating range of temperatures. GTG design running load for N calculated, N+1 calculated, N+1 installed and N GTGs running is an important criteria for evaluation.

The calculations in the Table 5 are detailed in Table 6 for each GTG for the complete life cycle of the plant.

Table 6 GTG NPV Calculations for the Design Life

Year	Ranging from first year till the last year of design life
Annual % Expense Rates	(%)
Equipment FOB Cost	(US$)
Installed Cost Factor	
EPC Cost	(US$)
Hours Per Year	(Hours)
Availability	(%)
Fired Hours per Year	(Hours)
O&M Rate per Fired Hour	(US$/Hour)
Heat Rate @ x °C Design Rated case	(BTU/kW-Hr)
Design Running Load	(kW)
Annual Fuel Cost	(US$)
Annual O&M Cost	(US$)
Annual Operating Cash Flow	(US$)
Total Installed Cost	(US$)
NPV	(US$)

In the Table 6, design life may vary from 25-30 years as per available data today. EPC cost is accounted for first few (2-5 years) only. It varies as per the rating of the LNG plant. Hours per year are 8,760. Availability is assumed to be 100%. Fired hours each year are same. All the indices in Table 6 are assumed not to change with years.

8.0 CONCLUSION

Power block selection and sizing criteria for an LNG Liquefaction facility should incorporate an integrated analysis for optimized performance, reliability and robustness. The process configuration, extent of process heat integration and power generation design philosophy has to be accounted for early in the design of the power generation system. Electrical load is the main driving factor for the sizing when simple cycle gas turbines are considered for generation sizing. If cogeneration type systems are considered, then process loads and their variability have to be taken into account. The selection and sizing of the power block should include adequate matching of the load profile while ensuring reliability, operability, efficiency and acceptable emissions. Transient behavior must also be analyzed to ensure electrical stability and recovery in the case of generator trips or sudden loss of electrical load. The design should take into account the contributions of expander generators if these exist in the plant. Essential service diesel generators sizing is dependent on the selection of plant's essential/black-start load. A life cycle NPV type calculation framework should be used to screen and select options. The final decision of configuration and equipment should be based on the NPV analysis coupled with practical factors such as operational experience.

9.0 REFERENCES

[1]. "BP Energy Outlook 2030", London, January 2012.

[2] Meher-Homji, C.B., Matthews, T., Pelagotti, A., Weyermann, H.P., 2011, "LNG Turbomachinery," Proceedings of the 1st Middle East Turbomachinery Symposium (METS-I), Feb 13-16, 2011, Doha, Qatar.

[3] Meher-Homji, C. B., Yates, D., Weyermann H. P., Masani, K., Ransbarger, W., and Gandhi, S., 2007, "Aero-derivative GasTurbine Drivers for the ConocoPhillips Optimized Cascade LNG Process—World's First Application and Future Potential," LNG15 Conference, Barcelona, Spain

[4] Liquefied Natural Gas: Enhanced solutions for LNG plants, GE Oil and Gas. Available online at:

http://www.ge-ner-gy.com/content/multimedia/_files/downloads/dataform_2047583989_2809802.pdf

[5] Vinod P. Patel, Hans E. Kimmel, "Fifteen Years of Field Experience in LNG Expander Technology," Proceedings of the First Middle East Turbo-machinery Symposium, February 13-16, 2011, Doha, Qatar. Available online at:

http://turbolab.tamu.edu/proc/pumpproc/P27/Fifteen%20Years%20of%20Field%20Experience%20in%20LNG%20Expander%20Technology.pdf

[6] Fritz Kleiner, and Steve Kauffman, , "All Electric Driven Refrigeration Compressors in LNG Plants Offer Advantages," GASTECH, 2005, 14-17 March, Spain. Available online at:

http://www.energy.siemens.com/us/pool/hq/energy-topics/pdfs/en/oil-gas/1_All_electric_driven_refrigeration.pdf

[7] Heat recovery steam generators: The essential solution to today's market conditions, ALSTOM. Available online at:

http://www.alstom.com/power/gas-power/heat-recovery-steamgenerators/

[8] Meher-Homji, C.B., Patel, H., Rajkumar, V., Messersmith, D., Rockwell, J., "Thermo-Economic Analysis of Combined Cycle Based Liquefaction Plants," Gastech 2011, Amsterdam Netherlands, 21-24 March 2011.

[9] Electrical power system studies for LNG plants, Mott MacDonalds. Available online at:

http://www.power.mottmac.com/projects2/transmissiondistributionprojects/powersystemstudieslng/

[10] Meher-Homji C.B., Messersmith D., Hattenbach, T., Rockwell, J., Weyermann. H.P.,, and Masani, K., "Aero-derivative Gas Turbines for LNG Liquefaction Plants-Part 1:The Importance of Thermal Efficiency, Part 2: World's First Application and Operating Experience," ASME paper Number GT2008-50839/50840, June 9-13, 2008, Berlin, Germany.

[11] Alan L. Sheldrake,"Handbook of Electrical Engineering for Practitioners in the Oil, Gas and Petrochemical Industry," John Wiley and Sons Ltd, 2003.

A HIGHLY ACCURATE DENSIMETER FOR CRYOGENIC LIQUID MIXTURES – FIRST RESULTS FOR LNG

Markus Richter, Reiner Kleinrahm, and Roland Span

Thermodynamics, Ruhr-Unversität Bochum, D-44780 Bochum, Germany

Natural gas is an important energy source today and for the future. According to the International Energy Agency the trade with Liquefied Natural Gas (LNG) will become more flexible and will increase significantly due to securing energy supply. In this context, the accurate determination of LNG densities is gaining importance for the custody transfer process when determining the transferred energy and also for process simulations, e.g., for modelling of economically and ecologically optimised liquefaction and evaporation processes. Today, the calculation models used for natural gas only offer uncertainties of 0.1 % to 0.5 % in the liquid phase. To reduce these relatively large uncertainties new accurate sets of experimental p-ρ-T-x-data for LNG are essential.

Within the scope of the presentation a new densimeter specifically designed for measurements of liquid mixtures at cryogenic temperatures will be described (also see *Richter, 2011* and *Richter et al., 2011*). A photo of this instrument in the laboratory at Ruhr-University Bochum is shown in Figure 1.

Figure 1 Photo of the densimeter for cryogenic liquid mixtures in the laboratory.

This apparatus allows the densities of LNG to be measured in the homogeneous liquid phase, along the saturated-liquid line and at supercritical states. Moreover, densities in the homogeneous gas phase can also be measured where required. Measurements can be carried out in a temperature range from 90 K to 290 K at pressures up to 12 MPa. The total relative measurement uncertainty in density is 0.02 % ($k = 2$), but this does not include the uncertainty associated with the gas analysis (which is required only once for the entire content of a sample gas cylinder). Taking the composition uncertainty into account the relative combined uncertainty is estimated to be smaller than 0.06 % ($k = 2$) in density. Specific problems involved in the measurement of LNG densities were effectively solved by using the state of the art single-sinker densmeter technique in connection with a magnetic suspension coupling and a new kind of VLE-cell. The single-sinker densimeter technique developed at Ruhr-University Bochum (by *Brachhäuser et al., 1991*) has proven successful for the accurate measurement of densities of liquids and gases over large temperature and pressure ranges. This type of instrument applies the Archimedes (buoyancy) principle to provide an absolute determination of density, i.e., a measurement that is independent of calibration fluids.

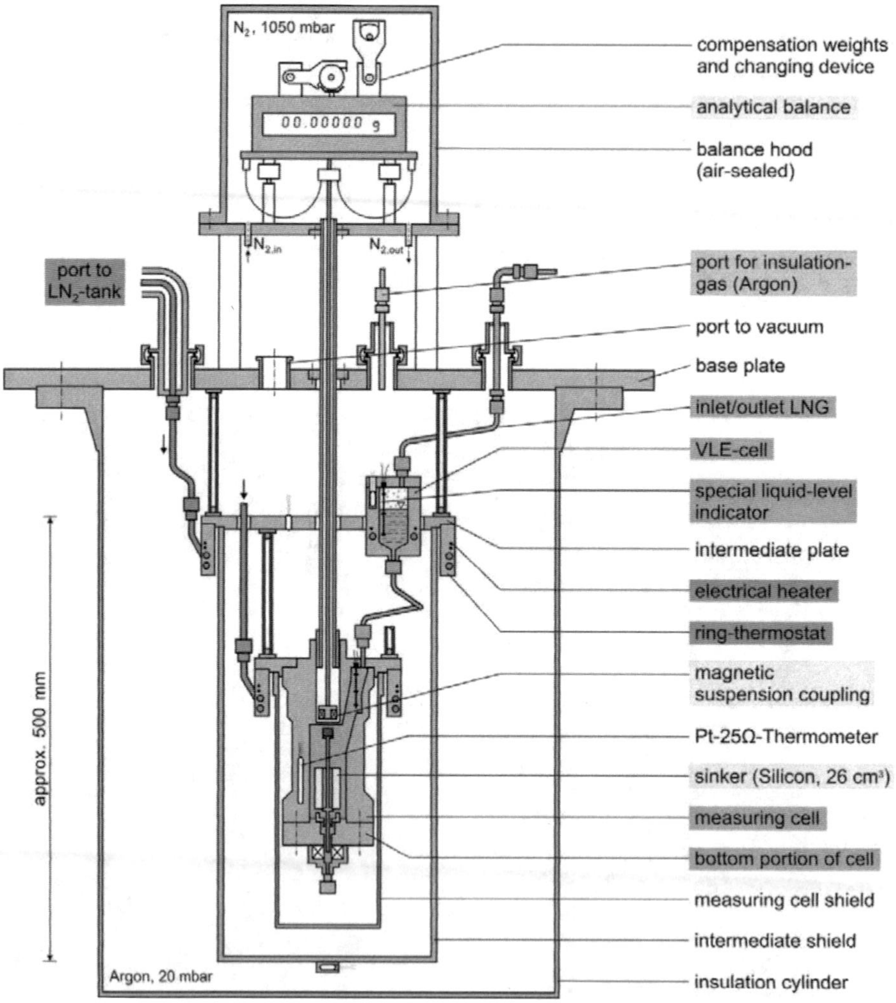

Figure 2 Schematic diagram of the densimeter for cryogenic liquid mixtures.

In addition to the technical description of the new densimeter, first results of LNG density measurements will be presented. The experimental data will be compared to the relevant thermophysical property models such as the GERG-2008 equation of state for natural gases of *Kunz and Wagner (2012)* and the Revised Klosek and McKinley Method of *McCarty (1980)* commonly used in LNG custody transfer (see LNG Custody Transfer Handbook transfer of *GIIGNL, 2011*).

The research leading to the presented results has received funding from the European Union on the basis of Decision No 912/2009/EC and the GERG (European Gas Research Group) partners E.ON New Build & Technology GmbH (Germany), Statoil ASA (Norway), Shell Global Solutions International B.V. (The Netherlands) and Enagas S.A. (Spain).

References

Brachthäuser, K.; Kleinrahm, R.; Lösch, H. W.; Wagner, W. (1993): Entwicklung eines neuen Dichtemeßverfahrens und Aufbau einer Hochtemperatur-Hochdruck-Dichtemeßanlage. Fortschr.-Ber. VDI, Reihe 8, Nr. 371, VDI-Verlag, Düsseldorf.

GIIGNL (2011): LNG Custody Transfer Handbook. Paris, Groupe International des Importateurs de Gaz Naturel Liquéfié, third edition, version 3.01.

McCarty, R. D. (1980): Four mathematical models for the prediction of LNG densities. Nat. Bur. Stand. Technical Note 1030.

Kunz, O.; Wagner, W. (2012): The GERG-2008 wide-range equation of state for natural gases and other mixtures: An expansion of GERG-2004, J. Chem. Eng. Data 57 (2012) 3032–3091.

Richter, M. (2011): Modifikation eines Normdichtemessgerätes für Erdgase und Entwicklung einer Präzisions-Dichtemessanlage für verflüssigte Erdgase (LNG). PhD thesis, Faculty of Mechanical Engineering, Ruhr-University Bochum.

Richter, M.; Kleinrahm, R.; Span, R.; Schley, P. (2011): A new apparatus for accurate measurements of the densities of liquefied natural gas LNG. Proceedings of the International Gas Union Research Conference, IGRC, Seoul, Republic of Korea.

353324 Flashing Two-Phase LNG Expanders

Abstract:

Conventional liquefaction processes for natural gas operate at high pressure through the condensation phase, after which the pressure of the liquefied natural gas, the LNG is reduced by expansion across a Joule-Thomson valve or a liquid expander before rundown to storage at near atmospheric pressure.

When the condensed fluid is flashed across a J-T valve with constant enthalpy an undesirable amount of LNG is vaporized. The aim of using an LNG expander rather than a J-T valve is to increase the amount of liquid and to decrease the amount of vapor at the expander outlet. By employing a flashing two-phase LNG expander an increased amount of liquid is produced in a near isentropic expansion process.

Flashing two-phase LNG expanders were developed and installed in 2001 and 2002. Since then they are successfully operating in the field for ten years without failures or interruptions except for routine maintenance shut downs. Flashing two-phase LNG expanders increase the LNG production output continuously by 3 - 5 %.

354015 The Past, Present and Future of LNG

Abstract:

In one decade, the gas prospects in North America have gone from dire to exceptional. In 2004, we were scrambling to build import capacity for gas as LNG. In 2014, we are scrambling to do the opposite - build export infrastructure for LNG. The shale gas revolution has led to a resurgence in petroleum engineering and gas utilization. The price of gas has dropped considerably, and has benefited those industries that traditionally depend on natural gas including power generation and chemicals manufacturing.

The low price and considerable environmental benefits over other fossil fuels have also made natural gas, in particular LNG, attractive to the transportation segment. New technologies are being developed for the use of LNG as a transportation fuel for trains and truck fleets. In addition the IMO (International Maritime Organization) environmental control area (ECA) regulations are scheduled to impose a maximum sulfur content of 0.1% on January 1, 2015 on bunker fuel for shipping in coastal North American waters, Hawaii, Puerto Rico and US Virgin Islands as well as Europe. This law has generated a significant order book for LNG fueled new build ships, and plans for the development of an LNG bunker fuel infrastructure.

These rapid developments do not come without challenges - both societal, regulatory and technological. Safety of LNG operations and the public perception are driving the development of innovative technical solutions. This presentation will review the current developments in the LNG value chain and discuss the safety and regulatory challenges associated with them.

Reliable Technology Selection Links Value Chains: LNG Transfer in FLNG

AIChE Spring Meeting

April, 2014

Joseph H. Cho, Ph.D., P.E.

Henry Ha, Team Leader

SK E&S

1

DISCLAIMER

SK

This presentation is not intended to provide the basis for any evaluation of SK E&S or its sister companies. Although SK E&S uses reasonable care to include in this presentation which it believes is up-to-date and accurate, SK E&S makes no representation or warranty as to the adequacy, accuracy, completeness or correctness of such information nor does not it warrant or represent that the presentation shall be complete in every aspect. SK E&S shall have no liability resulting from the use of the information provided in this presentation nor shall it have any liability for the absence of any specific information herein. The information may be changed by SK E&S at any time without prior notice.

2

AGENDA

1 Introduction

2 FLNG Economics and Status

3 FLNG Major Technical Risk & Mitigation

4 LNG Transfer System

5 Conclusions

3

1. Introduction

1. Introduction

SK

FLNG Concept: compacted and compressed an onshore LNG plant installed on the floating body

FLNG Lay-out

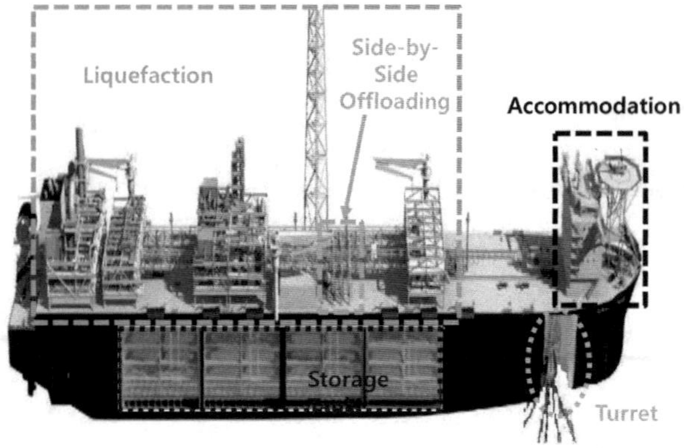

2 MTPA Basis:
LNG storage : 170,000 m3
Size : L 320m × Beam 55m
(Similar to VLCC 320,000 ton)

- **Topsides**
  Pretreatment, Liquefaction, Utilities, Accommodation Section
- **LNG containment system(Storage)**
  GTT Membrane: No. 96 and Mark III
  IHI SPB type
  Membrane & SPB
- **Mooring System**
  Turret mooring system:
  Internal turret or External turret
- **LNG Transfer System**
  Side by side loading system
  Tandem loading system
- **Accommodation**
  People on Board (POB): ~100 (Normal) to ~150 (Peak)

5

2. FLNG Economics & Status

SK

With mid-size stranded gas well, FLNG eliminates subsea pipeline, site preparation and marine facility

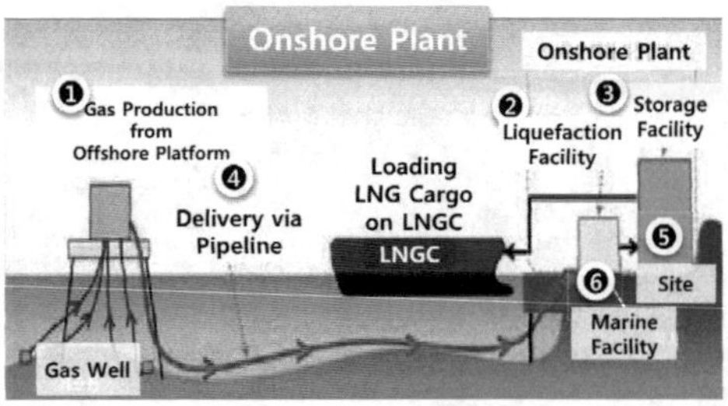

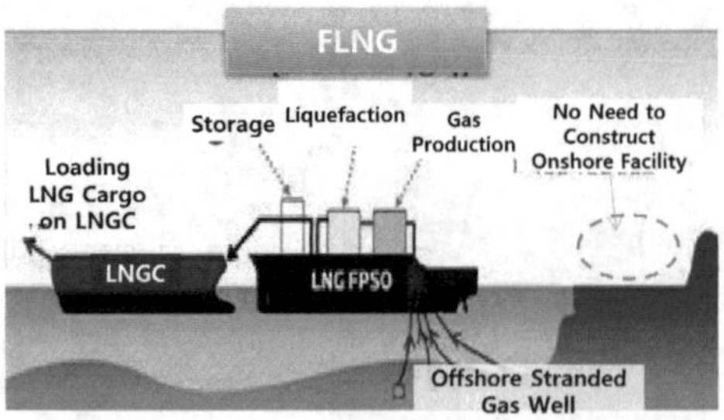

($/ton)

Main Item	Conventional*		FLNG
	NWS (AUS)	Pluto (AUS)	
① Gas Production Platform	849	373	1,290
② Liquefaction Facility	271	829	
③ Storage Facility	25	42	
④ Pipeline	58	259	-
⑤ Site Purchase/Preparation	45	46	-
⑥ Marine Facility	111	265	-
Total	1,359	1,813	1,290**

•Source : Wood Mackenzie, Pluto: '08 Data, NWS : '05 Data
Pluto : 4.3 MTPA, NWS : 4.4 MTPA (5TH Train)
** Pre-F/S Study with DNV and Prelude/Bonaparte/Petrobras Project Cost Analysis

6

FLNG Project Costs and Status

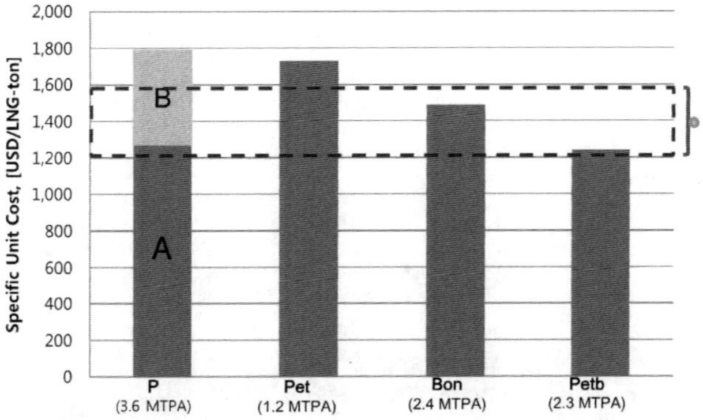

Source : Wood Mackenzie

※ Notes:
1. A specific cost is based on Liquid production (5.1 MTPA)
2. (A+B) specific cost is based on LNG production only
3. Specific cost is obtained from the TIC divided by overall liquid production capacity

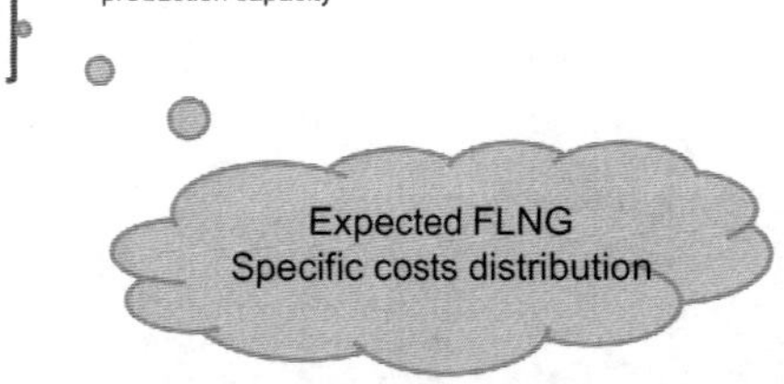

Subsea Pipeline Costs??

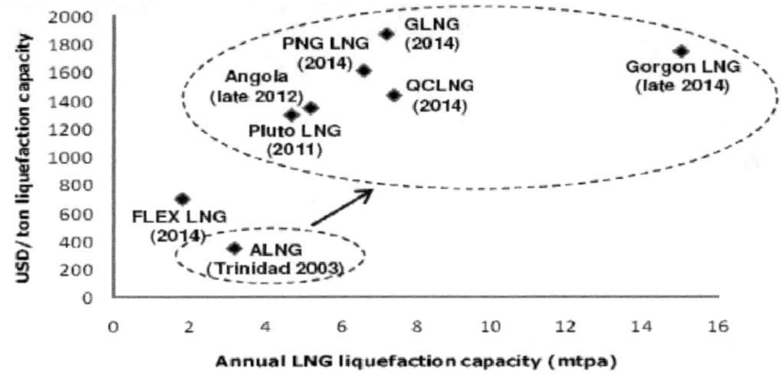

Source : FLEX LNG, Industry Reports

2. FLNG Economics & Status

#	FLNG	Location	MTPA	Sponsor	EPCI	Status	FID	COD	U$ /Ton	Remarks
1	Prelude LNG	NW. Australia	3.6	Shell	SHI + Technip	EPC	2011	2016	1,230	Hull, Topsides Manufacturing
2	PFLNG	Malaysia	1.0	Petronas	DSME+Technip	EPC	2012	2016		Hull Mfg
3	PFLNG2	Malaysia	1.5	Petronas	SHI+JGC	EPC	2013	2017		Hull Mfg
4	Petrobras LNG	Brazil	2.5	Petrobras	DSME+SBM/ Chiyoda SHI+Techinip/JGC HHI+Saipem	EPC contract Hold	Hold	2018	1,240	triple FEED
5	Bonaparte LNG	NW. Australia	2.0	GDF Suez	Doris	Pre-FEED	2015	2018	1,490	11.1 Pre FEED started
6	Lavaca Bay LNG	Gulf USA	4.0	Excelerate	SHI	FEED		2017		FLSO
7	Abadi LNG	Indonesia	2.5	Inpex	SHI+JGC HHI+Saipem STX+FW	FEED		2018		triple FEED
8	Cash Maple LNG	NW. Australia	2.0	PTTEP	Linde+SBM	Pre-FEED	-	2019		'Pre-FEED Completed('11.11)
9	Lavani	Tanzania		Statoil			-	-		Study hold until gas confirmed
10	Tamar FLNG	Israel	3.0	Höegh	KBR+DSME	FEED Started	-	-		Pre-FEED Completed
11	Scarborough FLNG	NW. Australia	6.2	Exxon Mobil		Feasibility				
12	Great Sunrise	NW. Australia	4.0	Woodside		Feasibility				
13	Browse	NW. Australia		Woodside		Feasibility				
14	Poseidon	NW. Australia	3.9	ConocoPhillips		Feasibility				

8

3. FLNG Major Technical Risk & Mitigation

SK

LNG transfer is one of the main concern of FLNG operation in case of severe weather. Failure in LNG transfer can cause both HSE and commercial matters.

	Risk	Mitigation	Assessment
Shutdown Caused by Severe Weather	• In case of Severe Weather, Disconnecting with gas well and taking shelter • It takes a week to resume the production after Disconnection	• Design and technology development to cope with severe weather occurring once in 10,000 years makes unnecessary disconnection and taking shelter - Design FLNG with basis of Australia (Wave Height 27m), Papua New Guinea (15m) - Development Turret technology makes unnecessary Disconnection ※ No Disconnection have happened to Oil FPSO in the North Sea for 13 years (Maximum Wave Height 15.5m)	• Even in case of Severe Weather like Cyclone, Possibility of shutdown is Low
Ship Rolling Lowering Productivity	• Ship Rolling caused by sea wave lower LNG Productivity of Topside	• Using existing Technology that makes it possible to maintain normal production up to 5 degrees of ship inclination caused by Rolling - CO_2 removal technology using solid absorbent - Minimizing the impact of sea wave using technology related to Turret	• Sea wave higher than 4m might diminish productivity (within 10 days per year)
Safety Issue caused by process layout in small space	• Safety Issue caused by facility installation in Small Space compared to Onshore Plant (50%~60%)	• Setting tighter standard of design and safety compared to onshore plant • Installing facilities Modularized on land	• Proven by Oil/LPG FPSO

9

3. FLNG Major Technical Risk & Mitigation

SK

	Risk	Mitigation	Assessment
Sloshing*	• Damage to Storage Tank Caused by Sloshing * Phenomenon of movement of LNG in tank damaging inside wall of tank	• Manufacturing Technology of Tank (Membrane, SPB, etc.) Reinforced weakness related to Sloshing is already developed	• Existing Technology is able to mitigate risk
Offloading Suspension caused by Severe Weather	• Offloading Suspension due to Severe Weather Condition	• Development of Flexible Hose improving safety of Side-by- Side method	• Side-by-Side method is applicable

10

4. LNG Transfer System

Offloading operations for FLNG could be carried out via the use of cryogenic LNG hoses or loading arms, and in either of the two offloading approaches:
1, Side by Side Offloading, 2. Tandem Offloading

Schematics	Description
Side by Side Offloading 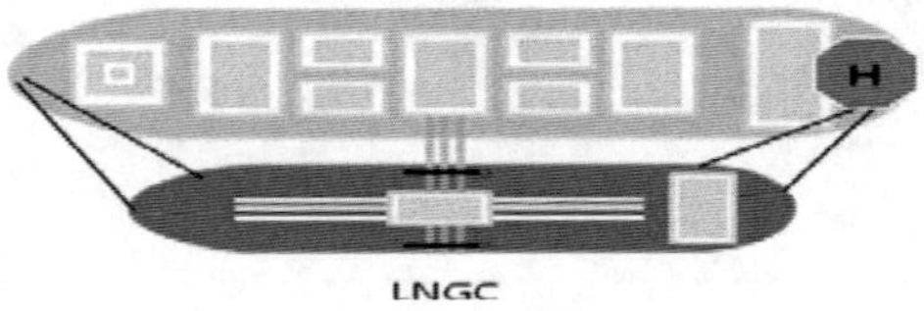	The LNGC is moored alongside the FLNG and offloading is carried out via loading arms or cryogenic hoses. **Movie**
Tandem Offloading (Via Floating Hose) 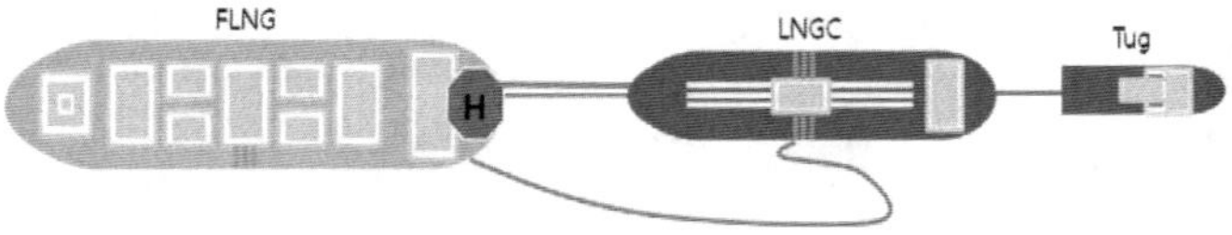	Tandem offloading is generally considered as a way to circumvent the limitation of side-by-side offloading by "separating" the LNGC and the FLNG **Movie**

11

Side by Side Offloading - Comparison

Type	*Side-by-side Offloading*			
	FMC Technologies	*SVT*	*Technip*	*Gutteling BV*
	Loading Arm	*Loading Arm*	*Hose*	*Hose*
Model	Chiksan DCMA-S LNG Marine Loading Arms	Atlantic Marine Arm	Amplitude LNG Loading System (ALLS)	Multi-LNG White Composite Hose

Maturity	• In use: yes	• In use: yes	• In use: no	• In use: yes
Classification Approval Status	• Couplers: DNV, RINA • Swivels, ERS: LR (in accordance with EN 1474, OCIMF)	• -	• -	• DNV, ABS, ClassNK, BV
Operating Life	• System: 25 years	• -	• -	• Hose: 5 years (weekly offloads)

12

Tandem Offloading - Comparison

Type	Tandem Offloading (LNGC to floating terminal)				
	FMC Technologies	SBM COOL	Trelleborg	Nexan	Bluewater
	ATOL	COOL Hose	Trelline Hose	CRYODYN Hose	Hose
Model	Articulated Tandem Offshore Loading System	Cryogenic Offshore Offloading Loading (COOL) system	Trelline LNG Transfer Hose	OCT Group System Cryodyn insulated corrugated metal pipe	Composite Cryogenic Hose
		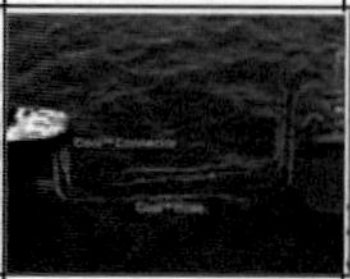			
Maturity	• In use (components): swivels, ERS, couplers • In use (technology): none	• In use: none • Gas test: none	• In use: none • Gas test: none	• In use: none • Gas test: yes • Cryoflex in use since 80s	• In use: none
Classification Approval Status	• Couplers: DNV, RINA • Swivels, ERS: LR (in accordance with EN 1474, OCIMF)	• Prototype qualification and fitness for service from ABS, DNV: per EN 1474 (hose and connector)	• None • Qualification by: June 2013	• DNV fitness for service tested and approved	• Lloyd's type approval for fabrication
Operating Life	• System: 25 years	• No details provided	• Hose: 5 years (≈ 500 offloads)	• System: 25 years • Flexible pipe: 5 years	• -

13-1

Max. System Operability

SK

Typical Maximum System Operating Sea State

Offloading Configuration	Shuttle tanker loading method	Shuttle tanker requirements	Approach / Mooring (Hs)	Offloading / Disconnecting (Hs)
Side-by-Side	Mid-ship manifold	Any tanker	1.5 – 2.0 m	2.0 – 2.5 m
Tandem	Bow	Non-DP Shuttle Tanker	3.0 – 4.0 m	4.0 – 4.5 m
	Bow	DP Shuttle Tanker	4.0 – 5.0 m	5.0 – 5.5 m

14

Conclusions

◆ The side-by-side loading arms offer the lowest risk. This is because in comparison to other technologies, it is the most proven technology, with the lowest schedule, CAPEX and OPEX risks

◆ For side-by-side and tandem offloading, the use of loading arms over hoses generally results in lower risk. This is largely due to better operability of the loading arms over hoses

◆ The smaller operating envelope associated with the side-by-side loading arm in comparison to the tandem offloading arm is indicated by the higher operability risk.

Membrane & SPB Tank

Item	Membrane Type (1 ROW)	Membrane Type (2 ROW)	IHI - SPB type (Self-supporting Prismatic shape IMO type-B)
General Shape	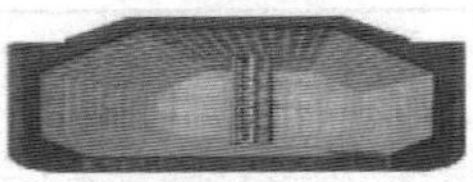	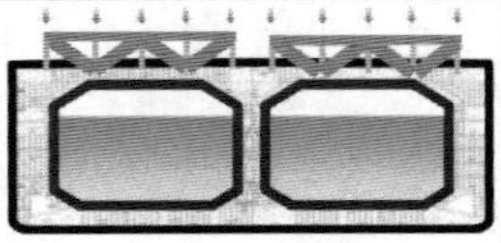	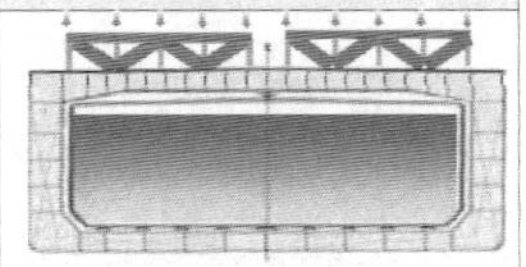
Sloshing	Limit	Half Limit	Free
Cost	Low	Medium	High
Current Application	All most all of the LNGC	Shell, Excelerate	Flex, SBM

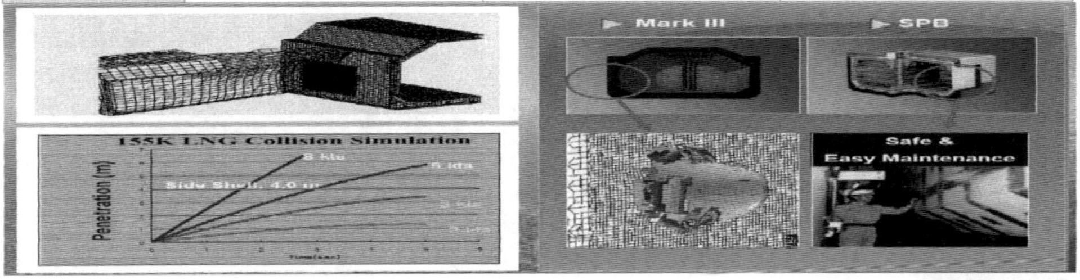

Return

THANK YOU

17

FLNG NEARSHORE COMMISSIONING METHODOLOGIES AND RECOMMENDATIONS

Herbert Moon, DSME, Korea
TH Kim, DSME, Korea
Paul Kim, DSME, Korea

Abstract

The normal Gas trial for LNG Carrier has been being carried out successfully for several decades and the purpose of this gas trial is to confirm integrity of the Cargo Containments and Cargo handling System in order to check the feasibility of LNG transfer in LNG terminal.

But, FLNG Gas trial is still necessary to be investigated more even through many studies have been done for long time. This is because there have never been any actual references up to now but it will not be long before to develop and establish the FLNG commissioning methodologies.

Offshore commissioning will be carried out for currently constructing two FLNG projects (Shell Prelude FLNG and Petronas Kanowit FLNG). But it is still considered as attractive method to perform gas trial and commissioning in nearshore because nearshore commissioning may be regarded as the solution which can reduce the overall project risks and prevent cost overrun.

It is acknowledged that nearshore commissioning may be the most effective and economic solution for FLNGs in Australia and Canada and its methodologies and recommendations for FLNG nearshore gas trial will be presented in this report.

Introduction

The purpose of normal LNG Carrier Gas trial is to evaluate the integrity of the CCS (Cargo Containment System) and CHS (Cargo Handling System) in the LNG Carrier. Gas trial is very important procedure and it shall be done before LNG Loading from LNG Terminal. In addition, Gas trial shall be considered in design and operation philosophy of LNG Carrier and LNG Terminal and it is carried out at open water because of safety. LNG Carrier shall be back to the shipyard and repaired if there are any defects or failures in LNG Carrier during gas trial.

Then, is it feasible to apply same procedures and methodologies of LNG carrier gas trial to the FLNG? In case of FLNG, she has special facilities, such as Gas processing system, Liquefaction Process, LNG offloading equipment, turret and etc., in addition to existing system of LNG Carrier and it makes FLNG bigger and heavier than LNG Carrier. Therefore, special attentions shall be paid to and additional matters shall be considered for the effective methodologies of FLNG gas trial. For application of gas trial to FLNG, possible location for mooring shall be implemented and risk assessment of cargo containment and mooring shall be prepared as well as studies of repair procedures and possibility of tests. Furthermore, permissions from authorities will be obtained for gas trial.

Comparison between nearshore commission and offshore gas trial regarding schedule, (dis)advantage and risk is shown in the report.

Proposed Commissioning Schemes for FLNG

Principles of FLNG Gas Trial

To establish the methodologies of FLNG Gas trial and nearshore commissioning, following matters shall be specially considered.

- ✓ Safety and Environmental Protection
- ✓ To comply with Relevant rules, code and standard
- ✓ To reduce the project overall risks and prevent cost overrun

Considering above matters, it is recognized that the gas trial and commissioning near shipyard where sufficient manpower and production & maintenance facilities are available is a reasonable solution for FLNG projects and performing methodologies are presented as follows;

A. Onshore Stage
- ✓ Utilities commissioned and ready for use and/or in operation.
- ✓ Fire water systems confirmed as fully commissioned and ready for use.
- ✓ Fire and gas detection systems confirmed as fully commissioned and ready for use.
- ✓ Desalination plant, potable water plant, service water plant and closed cooling water systems commissioned and ready for use.
- ✓ Sea water cooling system partially commissioned, to be fully tested upon arrival at location.
- ✓ Instrument air and nitrogen units commissioned and in use.
- ✓ Power generation systems are in operation as needed with the accommodation and central control room in full operation
- ✓ Emergency power generators, essential generator are in operation/standby depending on demand.
- ✓ Life boats and other LSA appliances in place and commissioned. This includes the serviceability of facility work boat and rescue boat.
- ✓ Accommodation, waste disposal treatment unit, drain and sewer systems commissioned and ready for use.
- ✓ Marine Growth Protection System (MGPS) commissioned, HVAC, domestic refrigeration and hotel services commissioned.
- ✓ Governmental permit, if necessary

B. Nearshore Commissioning Stage

When the activities with onshore commissioning are completed, the FLNG shall be towed to near shore a few miles away from shipyard by tug boats in order to implement the gas trial. Following are the commissioning activities nearshore.

- ✓ Nitrogen purging of lines for dry out purposes
- ✓ Import of LNG and/or LPG (if necessary) for tank cool down and start up
- ✓ Testing of the gas turbines with boil off gas
- ✓ Testing of boiler and WHRU operation on gas
- ✓ Agreement and permission when to leave Quayside
- ✓ Import LNG/LPG(if necessary) for tank test and boil-off gas for boilers steam system run
- ✓ Desalination unit full testing
- ✓ Tests of thrusters and ballast systems
- ✓ LNG loading / unloading using the loading arm or other loading systems
- ✓ Cargo handling for Hull
- ✓ Cryogenic test of liquefaction process
- ✓ Tests of BOG handling system
- ✓ Flare system
- ✓ Topside Process Modules
- ✓ Function testing of remained equipment

C. Offshore Stage

After arrival at the offshore site, many commissioning activities shall be done and main purpose of these activities is to check and proceed the all systems with feed gas from well and to confirm the performance of the FLNG.

This is the one of the most important stage in the whole procedures of FLNG EPCIC project and all the system and process shall be examined for LNG production. Followings are main activities in offshore stage.

- ✓ Installation and hook-up of the mooring lines, flowlines and risers
- ✓ Test and Commissioning of Liquefaction facilities with refrigerants and natural gas
- ✓ Test and commissioning of gas treatment facilities
- ✓ Test and commissioning of risers and flowlines

Solutions for Nearshore Commissioning

GTG (Gas Turbine Generator) to be fired with dual fuel

Main purpose of onshore commissioning is test and commissioning of GTG (Gas Turbine Generator) and PMS (Power Management System). Using gas for GTG commission is feasible onshore but the deployment of gas as fuel of GTG in the shipyard will impact seriously to the yard facilities and it is not recommended. Therefore, MGO (Marine Gas Oil) shall be considered as a fuel of GTG during onshore commissioning and fuel transfer system shall be applied at the GTG design. Dual fuel shall be applied in the specification of GTG.

If GTG is only available to gas firing, significant cost and schedule impact are anticipated because it is very difficult to deploy gas in the yard and additional facilities shall be implemented for GTG commissioning. Table1 is showing models for dual fuel operating GTG.

Table1. GTG Dual Fuel Models

GE Energy Aeroderivative	Siemens	Rolls Royce
LM2500	SGT-700	RB211
LM2500+	SGT-800	TRENT60
LM2500+G4		
LM6000PC		
LM6000PD		
LM6000PF		
LM6000PG		
LM6000PH		

SW (Sea Water)/CW (Cooling Water) System Design

A large amount of water is necessary as cooling medium for FLNG operation and relevant equipment and system are installed in FLNG hull and required water is supplied to the consumer in Hull or Topside. This cooling water is lifted from deep water through seawater lift riser as shown in Fig 1, normally 150meter depth of water, because the temperature and quality of water is very important for overall system in FLNG.

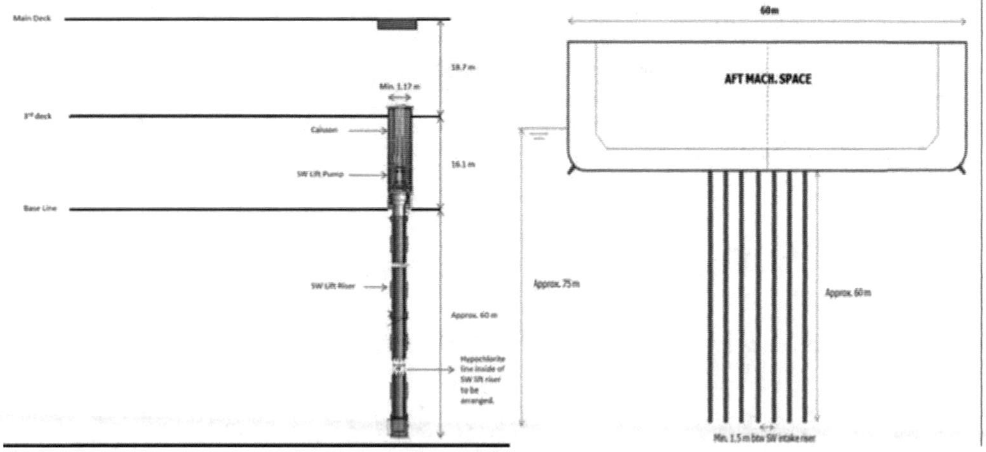

Fig. 1 SW Lift Riser (Section & Elev. View)

Seawater listing is generally applied to most FPSO and FLNG considering plant operation efficiency, maintenance and repair. This method is identically applied to essential cooling and fire water cooling system.

But it is not available to supply water using SW Lift Riser due to the shallow water depth and mud when commissioning is carried out onshore. In this case, Sea Chest will be applied for essential cooling system as shown in Fig.2 &3-1 and temporary spool pieces will be connected to topside cooling system for cooling water supply as shown in Fig.3-2.

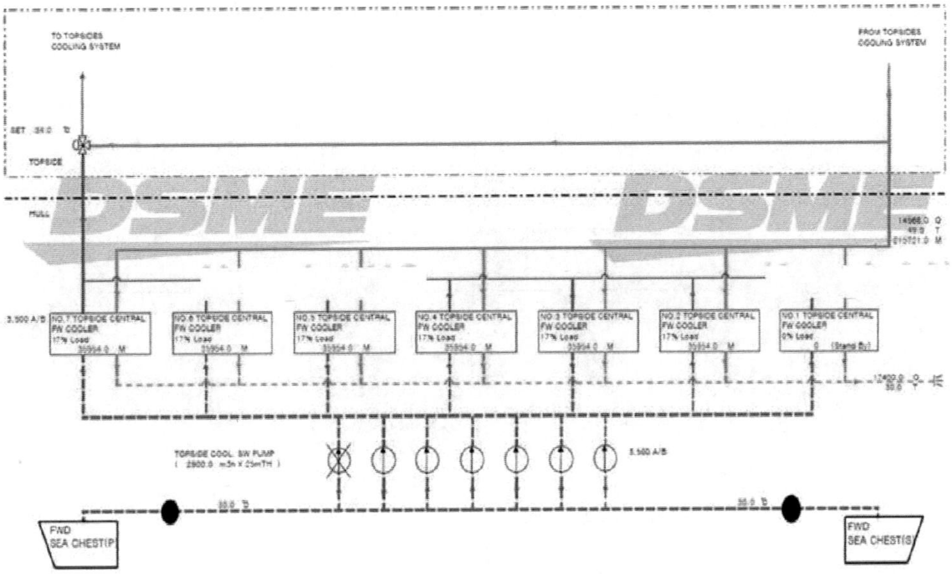

Fig 2. Sea Chest cooling system

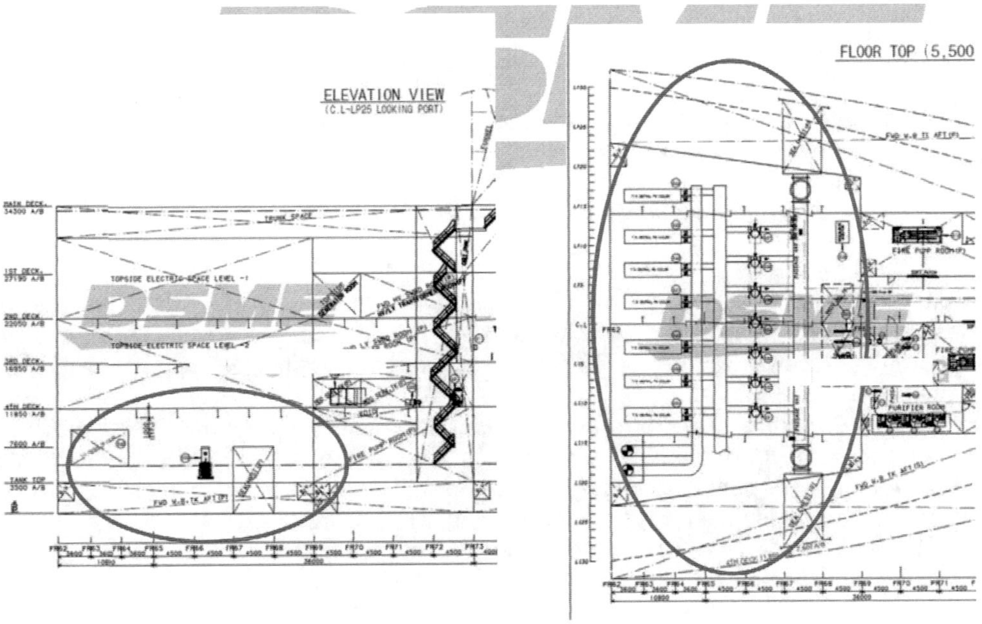

Fig 3-1. Sea Chest Arrangement (Elev. & Plan view)

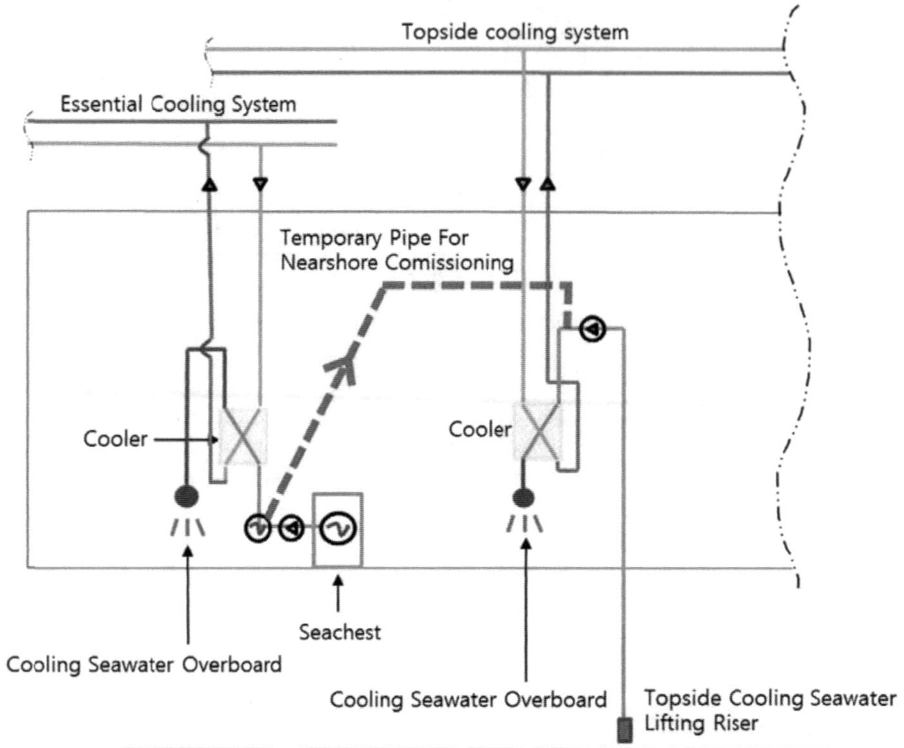

Fig 3-2. Sea Chest with temporary spool pieces for topside cooling

Temporary Mooring / Sea Keeping Facilities

It is not available for FLNG to receive LNG from terminal due to FLNG size and heavy displacement. Therefore, deployment of LNG carrier will be the most practicable alternative solution. In order to receive LNG from LNG Carrier, appropriate position keeping methodology should be provided.

It seems to take 2 months to perform above mentioned Nearshore commissioning whereas it takes 2 weeks to carry out ordinary LNG carrier gas trial. As the FLNG commissioning period is much longer than LNG carrier, countermeasures shall be established against the worst environment condition during nearshore commissioning.

Following Fig.4 is shown the track and frequency of typhoon in KOREA and Typhoon condition shall be considered on Temporary mooring system.

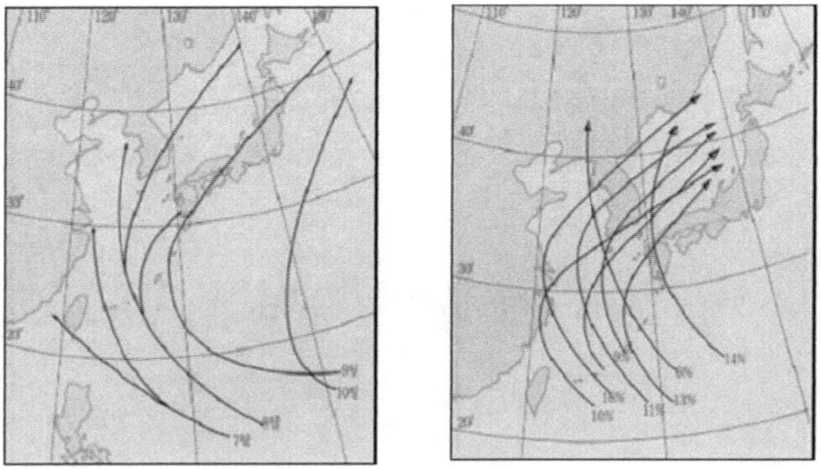

														1904~1995	
Month	1	2	3	4	5	6	7	8	9	10	11	12	Sum	Mean	%
Normal	-	-	-	-	1	14	68	92	68	7	-	-	250	2.7	88.7
Abnormal	-	-	-	-	-	2	13	15	2	-	-	-	32	0.4	11.3
Sum	-	-	-	-	1	16	81	107	70	7	-	-	282	3.1	100

Fig 4. Track and frequency of typhoon

It is difficult to determine the environmental condition because FLNG nearshore gas trial and commissioning has never been done before and it is necessary to determine environmental condition considering the period and time of gas trial and commissioning. In this report, 10 year returns Significant Wave Height (Hs) is applied but it shall be determined by agreement of COMPANY, Class and Contractor.

Several options for temporary moorings are studied and evaluated and it is found that most effective and safe mooring method is to utilize existing turret mooring system in FLNG. For the application of turret mooring system, mooring analysis shall be done and subsea facilities such as anchors and mooring lines are preinstalled at the location for temporary mooring. This may cause cost increase and design change in FLNG Project, so further investigation for temporary mooring system is necessary.

Followings are options for temporary mooring systems and detail design and procedure shall be further developed by 3rd party (Mooring Engineering Company)

Temporary Mooring using the FLNG Turret

Under the premise that the turret mooring system can be ready to be applied before gas trial as in Fig.5, turret mooring study shall be performed based on the information of specific FLNG configuration and environmental condition of the location for temporary mooring. For this solution, the configuration, number and speciation of mooring lines (diameter, length and grade) shall be investigated and provided. Also, the type and size of anchor shall be studied. However, existing windlass, fairlead and chain stopper in the turret will be utilized for this temporary turret mooring.

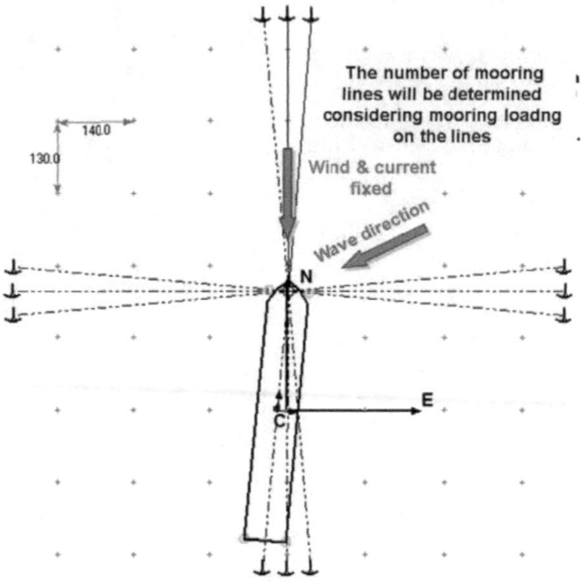

Fig. 5 Temporary Turret Mooring system

Single Point Mooring Tower

However, it cannot be assured that the FLNG Turret will match with the temporary mooring system and more over installation of mooring lines and anchors may be an expensive option. In this reason, as an alternative solution, temporary mooring tower has been investigated as in in Fig.6. For the application of this method, mooring tower shall be fabricated and installed on the location and the water depth of the location shall not be deep, otherwise the cost of construction and installation will be increased stiffly and some modification of FLNG for connection with mooring tower shall be necessary.

Fig. 6. Single Point Mooring Tower

Tug Boat Support

Tugboats with sufficient capacity will be deployed to keep positioning of FLNG during the gas trial as Fig.7. This method is simply considered if tug boats are available but this method seems to be very dangerous to personnel and asset. Therefore, this method need to be further investigated regarding risk and safety.

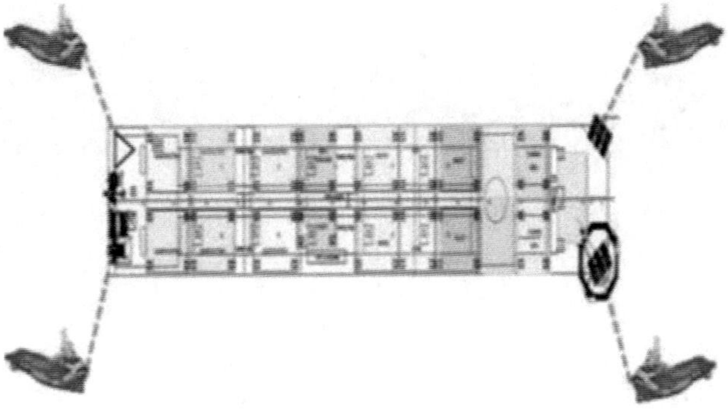

Fig. 7 Tug Boat Support

Provision of High Pressured Gas

The provision of high pressed natural gas is one of the most important prerequisites to nearshore commissioning. Ordinary method of fuel supply in FLNG is to deploy fuel supply compressor but it is not efficient because other system and equipment shall be operated in order to operate the fuel supply compressor. Therefore, it is necessary to develop alternative methods and there are two potential alternatives to supply high pressured natural gas. The first alternative is to deploy FSRU for high pressured gas supply and the other is to install temporary Regasification system on the FLNG and relevant basic engineering and schematic diagram have been already developed.

FSRU Deployment (First Alternative)

- ✓ Side by Side operation with by FSRU as below:

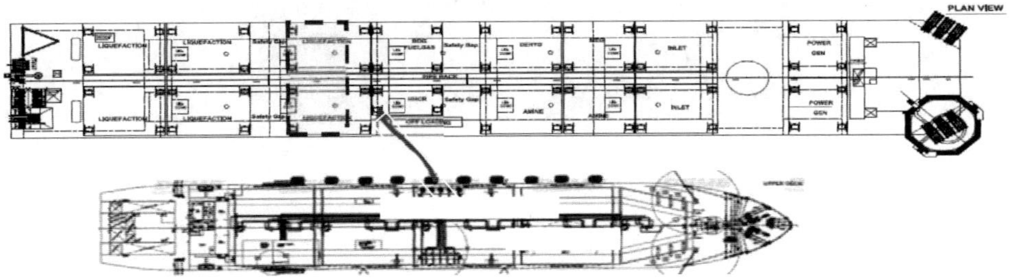

Fig.8 Pressured gas from FSRU

- ✓ Transfer of natural gas from FSRU skid outlet to compressor suction on Liquefaction process with qualified Jumper Hose
- ✓ Necessary pressure and flow rate out of FSRU skid is good match with feed gas rate of Liquefaction Process. (45-90 barg, 180 mmscfd @ 5 degC)

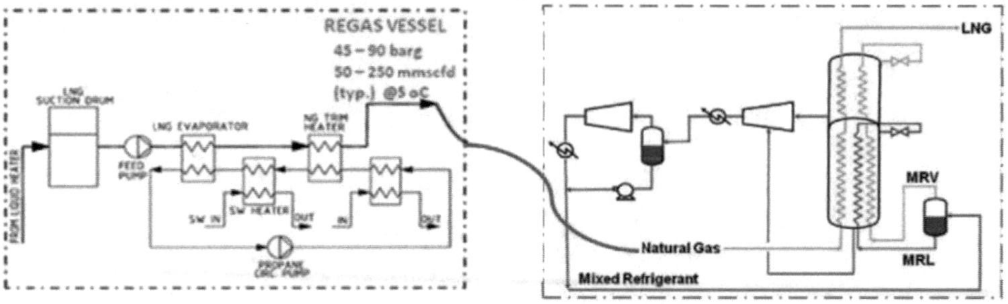

Fig.9. Schematic diagram

- ✓ 1-train tested with natural gas(close match with feed gas composition)
- ✓ Gas trial for FLNG finished, tanks and cargo systems to be passed acceptance criteria
- ✓ LNG in Regas vessel's tanks vaporized and subsequently liquefied into FLNG storage
- ✓ BOG, Flare, Utilities and PCS must be in operation

Temporary Regasification Skid (Second Alternative)

- ✓ Use of temporary regas. skid installed on FLNG
- ✓ Optimization of the test facilities of Liquefaction module on topsides
- ✓ Same as above base for schematic diagram and injection scheme

Fig.10 temporary Regasification

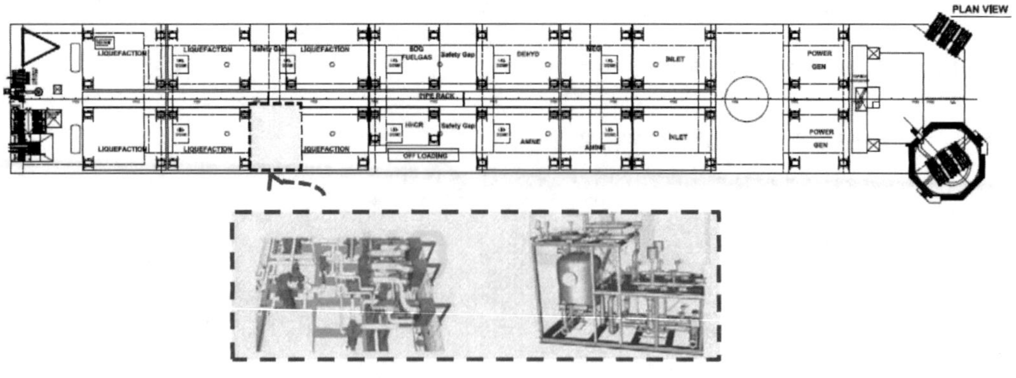

Table 2. Expected Risk and Cost Impact

	FSRU Deployment	Temporary Regasification Skid
Risk	✓Availability of regas vessel (FSRU) ✓Mooring method of FPSO (temporary, SPM or Ocean tug) ✓Location of gas trial ✓Governmental permit ✓Rule and Regulations in Korea	✓Purchase, design and construction of regas skid ✓Temporary installation of test facilities on FPSO (space available) ✓LNG feed stream to regas. Unit (main cargo line needs modification)
Cost Impact	✓Charter Regas vessel ✓Mooring cost ✓Jumper hose ✓Adaption to FPSO main gas stream piping ✓Additional fuel cost ✓Mobilization of start crew / test crew	✓Cost of regas skid ✓Cost of install. and adaption to main cargo line ✓Adaption to FPSO main gas stream piping ✓Additional fuel cost ✓Mobilization of start-up crew / test crew

LNG Loading/Unloading

Detail investigation shall be carried out in order to transfer LNG on nearshore. Table 2 is showing the information of LNG terminals in KOREA and it seems not to be feasible for FLNG to berth terminal due to the configurations of the terminals.

Table 3. Limitation of Korean LNG Terminal

Limitation Items		KOGAS						POSCO
		Incheon		Pyeongtaek		Tongyoung		Gwanyang
		No.1	No.2	No.1	No.2	No.1	No.2 (Future)	
Main Alongside		P & S	P & S	P & S	P & S	P	P	P
Water Depth (m) (1)		14.0	14.0	14.0	14.0	14.0	14.0	14.0
Tide Level	High (m)	9.84	9.84	9.93	9.31	2.21	2.21	4.46
	Low (m)	-1.01	-1.01	-1.04		0.00	0.00	-0.49
Acceptable Vessel at Terminal	LOA (m)	290.0	350.0	290.0	350.0	290.0	400.0	400.0
	Beam (m)	49.0	49.0	50.0	50.0	49.0	No Limit	No Limit
	Draft (m)	12.5	12.5	11.5	11.5	12.0	12.0	11.5
	Displ. (mt)	100,000	100,000	120,000	120,000	103,000	177,500	113,900
	Cargo (m3)		270,000		270,000	135,000	260,000	165,000
Turning Radius (m)		500.0	500.0	500.0	500.0	500.0	500.0	800.0
Allowable approaching Speed for berthing (m/s)		0.2	0.2	0.2	0.2	0.2	0.2	0.2

Alternative methods of LNG loading are to deploy LNG carriers. Depending on the LNG offloading system in FLNG, the type of LNG carrier will be determined. When the loading arm is applied in FLNG, ordinary LNG carrier will be deployed and moored with Side by Side mooring configuration. But, dedicated LNG shuttle shall be deployed with special LNG bow loading system and DP when LNG tandem offloading system is applied in FLNG. Fig.11 is showing ship to ship offloading using Side by Side mooring with ordinary LNG carrier and Fig.12 is presenting the tandem offloading with dedicated shuttle carrier which shall equip the requested bow loading system complying with FLNG tandem offloading system.

Fig.11. Side By Side LNG Offloading with Normal LNG carrier

Fig.12. Dedicated Shuttle LNG Carrier for tandem offloading in FLNG

Schedule Comparison between Nearshore and Offshore Gas Trial

Nearshore gas trial schedule is compared to carry out same test scope of work with the offshore gas trial. The periods of towing to field, Mooring, Hook-up and PAC test are same but other activities are different each other due to the test location.

Overall period of nearshore gas trial is longer than offshore gas trial by 2 or 3 months because additional activities such as temporary mooring and back-up offshore trial are necessary for nearshore trial. The period of "back-up offshore trial" is not determined yet and it can be determined considering remaining test items after nearshore gas trial.

Even overall schedule of near shore gas trial is longer than offshore gas trial, nearshore gas trial is still attractive solution because it can reduce overall project risks and prevent cost overrun.

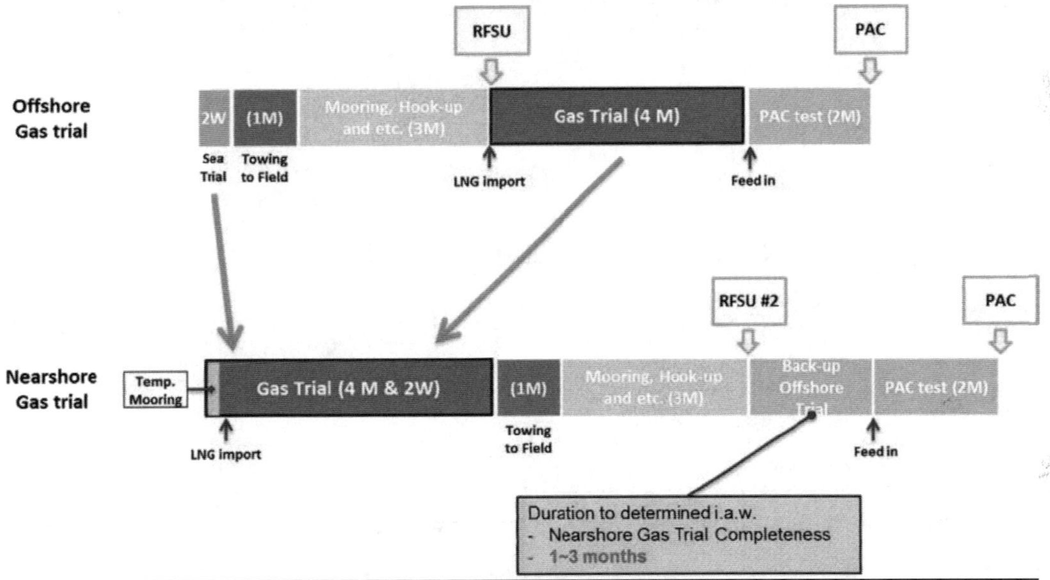

Fig.13. Schedule Comparison

Overall Comparison between Nearshore and Offshore Gas Trial

Table 4. Comparison Nearshore vs Offshore gas trial

	Nearshore Gas Trial	Offshore Gas Trial (No Nearshore gas trial)
Advantage	- Easy logistic with shipyard for maintenance work - Saving Offshore schedule and workload - LNG System and GTG tested before towing	- Short Schedule and low Cost
Disadvantage	- Additional cost for LNGC Deployment - Addition cost for Temporary Mooring system and installation - Overall schedule impact - the Safety Study for Nearshore gastrial	- LNG Tank, LNG Offloading and GTG are not fully tested before offshore
Schedule Impact	- 2 ~ 3 Month increase compared with Offshore gas trial	- N/A
Permission from Authority	- LNG Transfer - Temporary Mooring location - Flaring	- N/A
Risk	- Typhoon during gas trial - Permission for LNG Transfer and Temp. Moorng - Most System shall be RFSU at yard (tight schedule)	- Failure of LNG Tank in offshore - Failure of LNG Offloading system in offshore - Failure of GTG in offshore - Failure of Liquefaction system in offshore

Conclusion

Many studies and investigations have been being carried in order to perform nearshore commissioning properly and safely. Fortunately, technical solutions have been complete or developing but it is still necessary to investigate economic solutions and followings are necessary to be developed.

- ✓ Design of cost effective mooring design and installation
- ✓ Re-gasification facilities vs Deployment of FSRU
- ✓ Design philosophy of SW/CW system
- ✓ Manpower Mobilization
- ✓ Overall Project Risks and Cost Overrun

353142 Challenges and New Technologies for World Largest FLNG Facility

Abstract:

Growing rate of worldwide LNG demand is expected to be higher than that of the gas production, and such a higher demand has been driving forces for most of major oil companies to invest their resources in the offshore gas development. Especially FLNG concept will put vitality into approximately 2,400 marginal gas fields around the world, which were delayed because of excessive investment cost.

FLNG facility can provide attractive benefits by developing gas at the location of offshore gas field since it does not require conventional subsea pipeline to shore and additional upstream facilities. Based on the accumulated expertise in shipbuilding and offshore platform, it is possible to move LNG production facilities from onshore to offshore gas field.

As a world largest offshore LNG plant under construction, general overview of Prelude FLNG project will be presented. Although the state-of-the-art technologies applied for FLNG design have been obtained through past LNG-FSRU and oil FPSO projects, several technical challenges encountered during development and initial design stage of large scale FLNG concept will be introduced together with the implemented solutions.

Continuous efforts to improve the LNG related technologies by each party including oil & gas suppliers and shipbuilders will be the stepping stone to overcome any unrevealed challenge for the FLNG development. Some of latest design innovation devised for LNG containment system will be introduced.

Numerical Simulation of Rollover Phenomena for FLNG Application

Eun Sang Jung, Sangmin Park, Kiil Nam, Hyundai Heavy Industries, Ulsan, South Korea

Introduction

The demand of natural gas has been gradually increasingdue to the merit of clean fuel and cheap rice compared to oil. It is estimated that the global needs of natural gas will grow up to 345 million tons between 2010 and 2020, accordingly annual growth rate of 4.7% is expected. Because 20% of the world gas reserves are located at the offshore fields, LNG FPSO (Liquefied Natural Gas-Floating, Production, Storage and Offloading unit, called as FLNG) has become to be economical for exploring natural gas well in this area and has been recently spotlighted. [1] The FLNG is largely composed of two parts, i.e. hull and topside. While the FLNG is moored steadily above the well, the unprocessed natural gas is pumped from well to the FLNG's topside via riser. The reservoir gas is routed to several topside processings to obtain final products of LNG, LPG and condensates. Topside process includes pre-treatment and liquefaction. The products are sotrd at hull cargo tanks. FLNG is moored steadily, so the behavior of storage tank is similar with the on-shore tank.

The objective of this study is to inform FLNG storage tank about the issues surrounding rollover. This study contains the definition of LNG storage tank rollover and analysis of rollover phenomena.

- Definition of rollover
- Modeling and analysis of rollover phenomena
- Rollover phenomena scenario in FLNG
- Rollover mitigation method

Definition of Rollover

Rollover is one of hazards on LNG tank such as cryogenic hazard, pressure buildup, flammable vapor cloud/flash fire and rapid phase transition. [2] The meaning of rollover is the process of spontaneous mixing up of a similar or two different liquid cargos due to changes in the density of upper and lower layers level in the tank. This happens because of the boiling off of lighter fractions from the gaseous cargo, resulting in the liquid layer adjacent to the liquid surface to become denser than the layer beneath it. When this situation occurs, stratification develops and the unstable condition relieves itself with spontaneous mixing known as rollover.

The various reasons for rollover are:

- When there is density and temperature difference within the liquid
- Different initial grade of LNG (Chemical composition, Characteristics of LNG)
- Occurrence of stratification (N_2 content, Boil-off activity, Heat leak)
- If the cargo is stored for a long time
- Similar or Compatible cargo stored in one tank
- If LNG or condensate(NGL) of two or more tank is returned to one tank

Rollover is a phenomenon that can occur in systems containing stratified liquids. Rollover is a sudden mixing of existing layers of liquid. The stratified layers are characterized by different values of temperature and composition. Originally the layers were in mechanical equilibrium, but over time these densities can alter through changes in temperature and composition, driven by heat and mass exchanges between the layers. If the layer densities evolve in such a way that the density difference is zero but the temperature difference remains finite, the system can reach a point of hydrostatic instability, causing the liquid layers to mix rapidly. For LNG storage tank, rollover is most likely to occur following the loading of two or more LNGs from different sources into a single storage tank, each with a different temperature and composition. The potential for rollover occurs if the richer composition is on the bottom, and the lighter layer on top.

Modeling of Rollover

Rollover Assumption

The key quantities needed for the model and for solving rollover modeling differential equations are:

- Initial value for the temperatures, compositions, and total number of moles in each layer.
- Values for the physical properties of the LNG layers.
- External heat fluxes
- Heat transfer coefficient
- Mass transfer coefficient
- A criterion for the initiation of rollover phenomena

To validate the modeling of LNG rollover, the information of La Spezia case is used as reference data in this paper. [3],[4]

Governing Equation

Modeling of rollover can be applied to the tank, which is already filled and stratified. LNG of different compositions induce density difference and hence, it tends to be stratified in different layers. The LNG storage tank considered here is stratified into two layers: a lower layer and an upper layer, as shown in Figure 1. There also exists a thin film region at the top of upper layer, which is in thermodynamic equilibrium with evolving vapors, as shown in Figure 2. As LNG is stored below – 160 °C, there is a continuous heat leakage from the bottom and sidewalls of the tanks. The heat leakage from the bottom is represented as q_b, heat leakage from the top is represented as q_t, whereas that in the lower layer, upper layer and vapor space from the side walls are represented by q_L, respectively. The change in composition and temperature of LNG in each layer can be estimated by applying material and energy balance to the individual layers. The model gives flexibility of choosing any number of species up to a maximum of 10. It is assumed that there is no accumulation of mass in the film layer and LNG in the film region is in thermodynamic equilibrium with evolving vapors. Representative material and energy balance equations are discussed for lower layer and upper layer in this section.

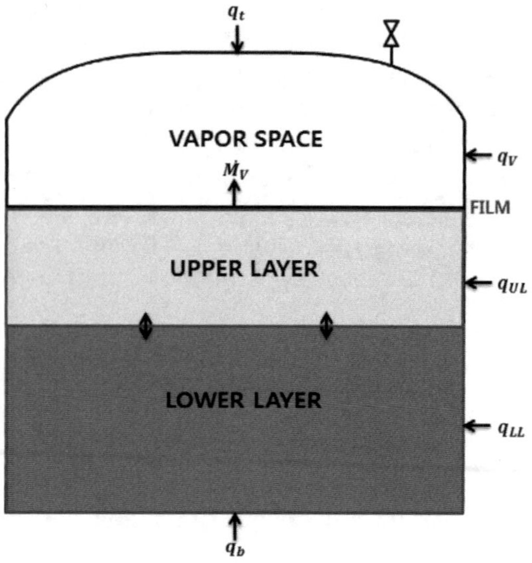

Figure 1 Schematics of LNG Tank

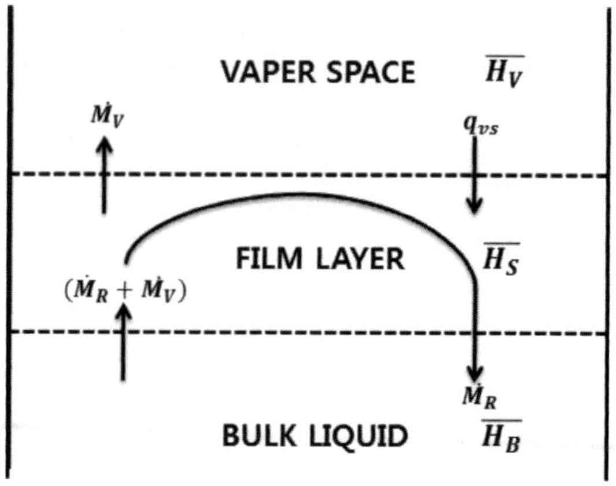

Figure 2 Schematics of Interfacial Film

- Material balance

$$C_l\delta_l \frac{dx_l(i)}{dt} = -\kappa \cdot (x_l(i) - x_u(i)) \quad \text{(Eq.1)}$$

$$C_u\delta_u \frac{dx_u(i)}{dt} = -\dot{M}_V y(i) - \kappa \cdot (x_u(i) - x_l(i)) \quad \text{(Eq.2)}$$

(C: Molar concentration, δ: Height, x: mole fraction, κ: mass transfer coefficient
$\dot{M}_V$: Molar vaporization rate, y: mole fraction of vapor phase)

In the above material balance equation for lower layer (Eq. 1), the rate of change in composition of species i is evaluated by mass transfer flux between lower and upper layers. Material balance for upper layer (Eq. 2) is written in a similar regime as for lower layer with additional term for molar flow rate of evaporation from upper layer to the vapor space, which is also called boil-off gas.

- Film layer

$$(\dot{M}_V + \dot{M}_R)x_B(i) - \dot{M}_R x_s(i) - \dot{M}_V y_s(i) = 0 \quad \text{(Eq.3)}$$

$$\dot{M}_R = 0.3276 \frac{\kappa}{C_{L,u}} \left(\frac{g\Delta\rho}{\nu\alpha\bar{\rho}}\right)^{1/3} \quad \text{(Eq.4)}$$

$$(\dot{M}_V + \dot{M}_R)\overline{H_B} - \dot{M}_R\overline{H_S} - \dot{M}_V\overline{H_V} + f_Q((q_t A + q_{UL}\pi D\delta_{vs})/A) = 0 \quad \text{(Eq.5)}$$

$$\dot{M}_V = \left[\frac{\dot{M}_R(\overline{H_B} - \overline{H_S}) + f_Q((q_t A + q_{UL}\pi D\delta_{vs})/A)}{(\overline{H_V} - \overline{H_B})}\right] \quad \text{(Eq.6)}$$

(M_R : Molar Recirculation Flow Rate ,f_Q : Fraction of Total Heat Transfer Rate)

The film layer is assumed that there is no accumulation of mass in the film layer and LNG in the film region is in thermodynamic equilibrium with evolving vapors. f_Q is the fraction of total heat transmitted to the vapor space, which is returned to liquid. Enthalpy of liquid and vapor phase is correlated in terms of temperature from which specific heat can be estimated. Correlations for enthalpy of liquid and vapor phase are obtained from NIST REFPROP 9.1. Rayleigh recirculation liquid flow rate between upper layer and the film is shown at Eq. 4. In addition to lower layer and upper layer, material balance (Eq. 6) is also applied to the film region, which is assumed to be in equilibrium with evolving vapors, to estimate the composition of LNG in the film.

- Energy balance

$$C_l \delta_l C_{L,l} \frac{dT_l}{dt} = q_b + (q_L \pi D \delta_l)/A - h \cdot (T_l - T_u) \quad \text{(Eq.7)}$$

$$C_u \delta_u C_{L,u} \frac{dT_u}{dt} = -\dot{M}_V \overline{H_V} + f_Q((q_t A + q_v \pi D \delta_v)/A) + (q_u \pi D \delta_u)/A - h \cdot (T_u - T_l) \quad \text{(Eq.8)}$$

($C_{L,l}$: Molar heat capacity, D: Diameter, q: heat flux, h: heat transfer coefficient)

The rate of change in heat content of the lower layer (Eq. 7) is calculated by considering the rate of heat coming in from the rate of heat transferred from the bottom of the tank, the side walls of the tank and the rate of heat transfer between lower and upper layers. Similarly, the rate of change in heat content of upper layer (Eq. 8) can be calculated by incorporating rate of heat transfer from fraction of total heat returned from vapor space to upper layer, rate of heat transfer from side walls and the rate of heat transfer between lower and upper layers. Specific heat of LNG in lower and upper layer is calculated from enthalpy correlations taken from REFPROP 9.1.

- Heat and mass transfer coefficient

Heat and mass transfer rates between the stratified layers are traditionally estimated from the empirical correlations. The empirical correlation of Globe and Dropkin (Eq. 9) is more appropriate to estimate heat transfer coefficients in this work, as it was proposed for heat transfer between the two horizontal plates heated from below and can be expressed as, [5]

$$h = 0.0597k(\frac{g\Delta\rho}{v^2\bar{\rho}})^{1/3} \quad \text{(Eq.9)}$$

The proportionality constant in the above correlation is quite significant, as Heestand et al. reported that the time to rollover is sensitive to this parameter. [4] Assuming turbulent conditions inside the tank, mass transfer coefficient can be obtained from Reynolds analogy (Eq. 10):

$$\frac{\kappa C_p}{h} = 1 \quad \text{(Eq.10)}$$

- Stratification forecast

Overall mass balance equations can be incorporated along with material balance and energy balance equations, in order to evaluate layer thickness of each layer. The evolution of an individual layer is strictly based on initial stratification and operating conditions.

$$\frac{d\delta_l}{dt} = \frac{[-\kappa \cdot \sum_{i=1}^{N}(x_l(i) - x_u(i))]MW_l}{\rho_l} \quad \text{(Eq.11)}$$

$$\frac{d\delta_u}{dt} = \frac{[\kappa \cdot \sum_{i=1}^{N}(x_l(i) - x_u(i)) - \dot{M}_V]MW_u}{\rho_u} \quad \text{(Eq.12)}$$

(ρ_l: Density, D: Diameter, q: heat flux, h: heat transfer coefficient)

The evolution of lower layer (Eq. 11) is estimated from molar flow rate from mass transfer rate between two layers, whereas for upper layer (Eq. 12) there is an additional term of mass lost due to boil–off. The thickness of vapor space is estimated from total height of the tank and lower and upper layer thickness.

Physical and thermophysical properties

For calculating the thermodynamic properties of the LNG an equation of state or equivalent thermodynamic model is needed. In this work the GERG-2008 equation of state was used to calculate all thermodynamic properties. This equation of state is an example of a new multi-parameter (~40) equation that has been regressed to all of the available high-quality thermodynamic data for natural gas and LNG and claims to represent that data within its experimental uncertainty. For a review of modern multi-parameter equation of state the reader is referred to Span et al. The GERG-2008 equation of state 1 is a formulation based on the multi-fluid approximation and provides a functional form for the reduced Helmholtz energy explicit in the fluid's density, temperature, and composition for a mixture. All thermodynamic properties can be calculated from appropriate derivatives of the Helmholtz energy function. The GERG-2008 equation of state has been adopted recently as the reference equation of state for natural gas and LNG by the Groupe Europeen de Recherche Gazieres (GERG) who commissioned its development. This equation of state has been implemented both as an add-in function for MATLAB and also in the software REFPROP 9.1.

The REFPROP 9.1 software allows the computation of transport properties for natural gas mixtures as well as pure components. For mixtures, model used by REFPROP 9.1 to calculate the thermal conductivity and viscosity is based upon the modified Ely-Hanley method, which is an extended corresponding states model.

Evaluation of Model

The model is applied to the case study namely La Spezia, Italy where rollover incidents occurred and which are well documented in the literature. The model predictions are subjected to various operating parameters and initial conditions for temperature, composition and level of stratified layers of the storage tank. Various heat leakage rates from bottom, top and sidewalls to lower and upper layer and physical properties of LNG such as thermal conductivity, thermal diffusivity and kinematic viscosity also contribute towards predicting time to rollover. All of the differential equations derived from governing equations are solved using MATLAB connected with REFPROP to calculate the characteristics of LNG.

Evolution of density

The density profiles of lower layer and upper layer, calculated using MATLAB model are shown in Figure 3. Density of the lower layer, as represented by dashed line, decreases with time, whereas that of upper layer, as represented by solid line, increases with time due to heat and mass transfer between the stratified layers and boil–off from the upper layer. Densities of the stratified layers eventually attain a uniform value. Density equalization is the criterion for prediction of rollover using governing equations. It can be seen that rollover occurs at about 32 hours, which is in a good agreement with the reported value of Sartsten and predictions by Heestand et al. The component of methane, ethane and nitrogen are volatile because of equilibrium with vapor space. There is mass transfer between both layers by mass transfer coefficient. The mass transfer coefficient is calculated by Reynolds analogy.

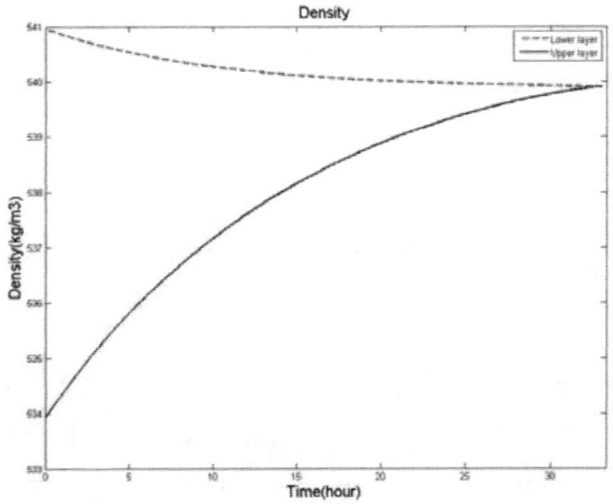

Figure 3 The density profile of lower and upper layer of LNG

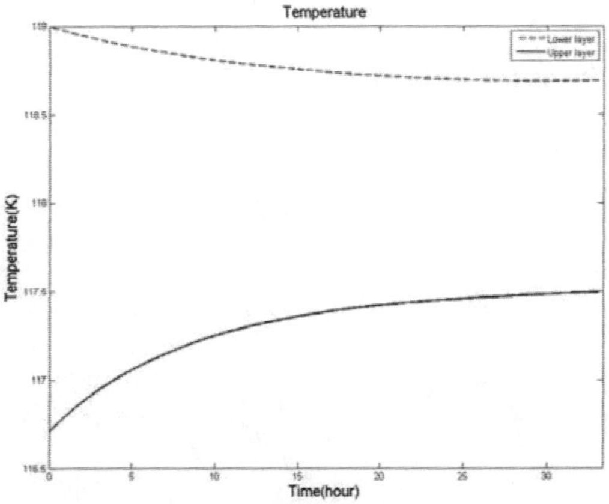

Figure 4 The temperature profile of lower and upper layer of LNG

Evolution of temperature

The calculated temperature profile of LNG in the lower and upper layers obtained is shown in Figure 4. It can be seen that temperature of lower layer decreases with time and that of upper layer increases with time by heat transfer. Although, the lower layer is getting significant energy through heat leakage from the bottom and side walls, there is considerable heat transfer between the stratified layers. It can also be seen that there exists a temperature gradient between the two layers, even just before rollover, which contributes to the higher boil–off rate at the time of rollover. Thus, the magnitude of severity of rollover due to higher boil–off rate is subjected to the temperature gradient, just before the rollover.

Evolution of Composition

The each composition profiles of LNG in the lower and upper layers obtained using the governing equations shown in Figure 5. Methane, ethane and nitrogen in upper layer are decreases with time. The high volatile compositions such as methane, ethane and nitrogen are evaporated to vapor space. The mass transfer between upper and lower layer is calculated by Reynolds analogy.

Rollover scenario case in FLNG

- Auto-stratification

In this scenario one of the tanks is full but is not used for an extended period of time (for instance during plant turnaround or because there is a problem with this particular tank). Stratification develops and the operator fails to take suitable action (ie. mixing of the tank content). Sea-state is calm and does not lead to significant mixing of the tank content. The dense bottom layer warms up and its density decreases. Eventually the density of the bottom layer and top layer gets very close to each other and rollover occurs.

- Off-spec LNG

In this scenario the tanks are partially filled with normal LNG. If a mechanical failure occurs which leads to the loss of NGL re-injection into the LNG, the LNG composition during LNG rundown

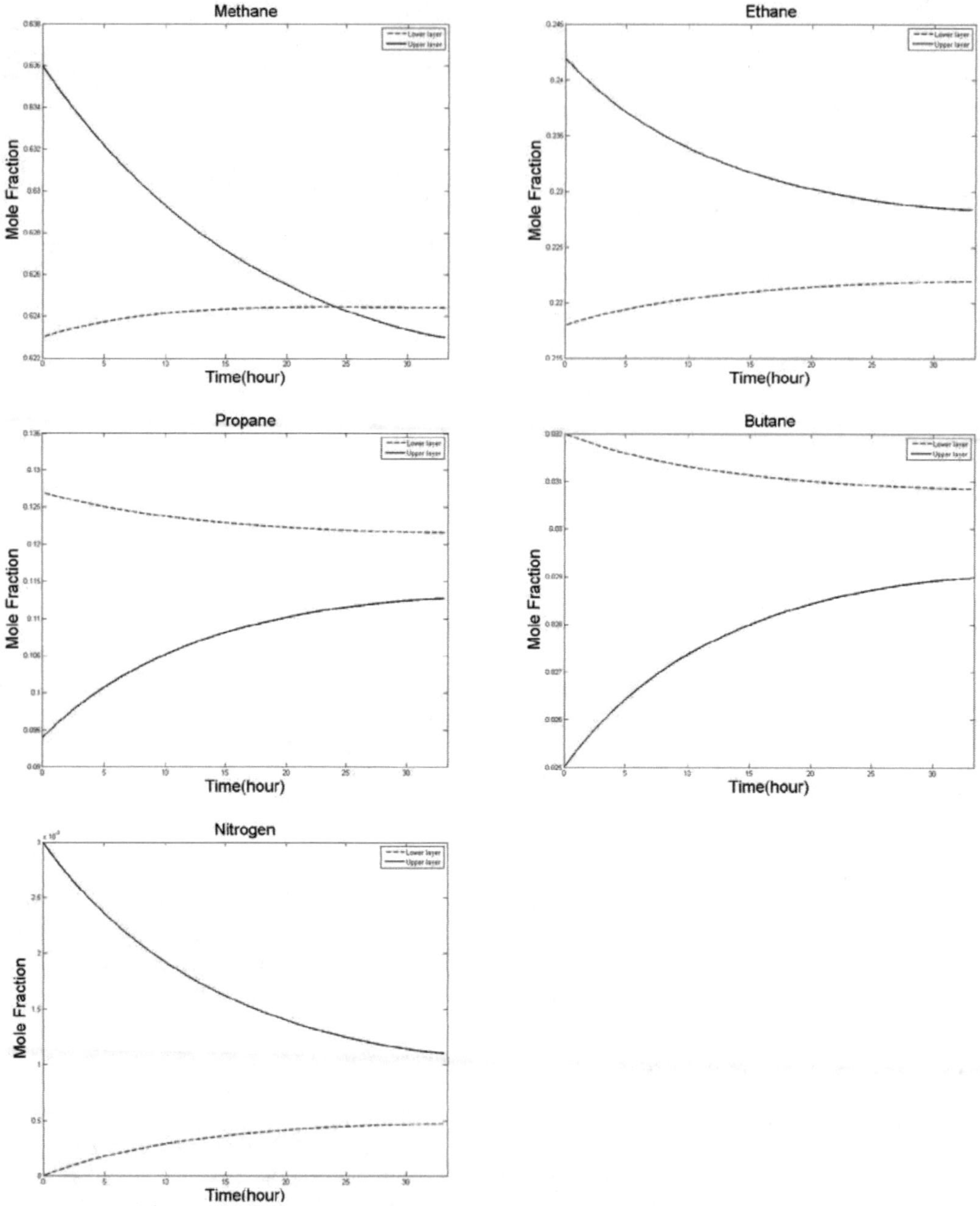

Figure 5 The composition profiles of lower and upper layer of LNG

becomes lighter and tends not to mix with the heavier "normal" LNG at the bottom of the tank. Stratification develops and the operator fails to take suitable action (ie. mixing of the tank content). Sea-state is calm and does not lead to significant mixing of the tank content. The dense bottom layer warms up and its density decreases. Eventually the density of the bottom layer and top layer gets very close to each other and rollover occurs. Worst case scenario would be if rollover occurs when the tank is full (probably with 50% light – 50% heavy LNG). Solving the worst case (50% lean LNG – 50% rich LNG), conditions are needed such as heat ingress, temperature of both layers, compositions of layers and layer height.

General method of mitigation of rollover

Detection of rollover

The first signal of occurance of stratification is increase of the temperature of the LNG in the bottom part of the tank, because the heat leaks in the bottom layer are not evacuated at the free surface by evaporation but contribute to that layer's temperature increase. The most obvious sign that conditions exist for a rollover event is the stratification of the stored LNG. This results in the development of two distinct layers of liquid densities within the product in the tank. Tank instrumentation can be included to detect these stratifications by measuring temperature and density at various levels. Software is also available to predict rollover incident, using storage tank data which maybe many days after a filling operation has been completed. LNG storage tanks should be equipped with a density measurement system to monitor the density of the LNG over the full liquid height.

There are main purposes of storage-tank instrumentation in connection with rollover

- Monitoring temperature and density gradient in the LNG tank and detection of stratification
- Obtaining data to investigate any incidents that may have occurred

A list of useful measurements for these purposes is as follow:

- Vertical temperature profile in LNG
- Vertical density profile in LNG
- Vapor withdrawal rate
- LNG level
- LNG filling and withdrawal rate
- LNG recirculation rate
- Composition
- Tank pressure

It is generally not necessary to make all these measurements: for example, stratification can be detected from the temperature profile, the density profile, analysis of composition and the vapor evolution. Also, not all parameters need to be monitored continuously. The choice of which measurements are made and when they are made ultimately depends on individual site conditions and requirements.

Mitigation of rollover

The present section discusses methods generally implemented in the industry to mitigate rollover phenomena. [6] Mitigation measures that are used are:

- Stratification inside storage tanks is avoided by top or bottom filling according to heel and fill LNG densities, bottom/top recirculation, mixing the liquids by filling using jet nozzles and distributed fill systems
- Different compositions of LNG are stored in separate tanks
- Specify LNG with nitrogen content less than 1%
- Monitoring of LNG density and temperature over height of tank
- Monitoring of total boil-off and heat balance to detect superheating
- Use of software based on LNG tank thermodynamic modeling to predict potential for roll-over
- Ensure LNG residence period in tank is not too long
- Process relief systems and safety valves are designed to handle roll-over effects

Conclusion

When rollover phenomena occurred, boil-off gas rate is much bigger than the normal boil-off gas rate. The possibility of a sudden release of large amounts of vapor and the potential over-pressurization of the tank leading to damage or failure is recognized by the major design codes. NFPA 59A - "Standard for the Production, Storage and Handling of LNG" requires this phenomenon to be taken into account at the phase of sizing relief devices. Emergency events induced from rollover can be risk to people, asset and environment have happened. Also the released LNG vapor is not only flammable, but a valuable commodity and a potent greenhouse gas and therefore venting should be avoided whenever possible.

Rollover phenomena occur at steady tank state to stratify the layers of LNG. For the FLNG, rollover is considered very unlikely due to the motions, low LNG composition fluctuations and continuously changing trim. If the feed gas has low nitrogen contents less than 1%, there is little possibility of rollover situation for FLNG.

References

1. Lee, S. (2012), "Technical Trend for LNG-FPSO," *The Korean Society of Marine Engineering*, Vol. 36, No. 1, pp. 23~39.
2. Society of International Gas Tanker and Terminal Operators (2012), "Guidance for the Prevention of Rollover in LNG Ships,"
3. Deshpande, K.B., Zimmerman, W.B., Tennant, M.T., Webster, M.B., Lukaszewski, M.W., (2011), "Optimization methods for the real-time inverse problem posed by modelling of liquefied natural gas storage," *Chemical Engineering Journal*, Vol.170, pp. 44~52.
4. Heestand, J., Meader, J. W., (1983), "Predictive Model for Rollover in Stratified LNG Tanks." *AIChE Journal*, Vol. 29, No. 2, pp.199~207.
5. Globe, S., Dropkin, D., (1959), "Natural Convection Heat Transfer in Liquids Confined by Two Horizontal Plates and Heated from Below," *Trans. ASME, J. Heat Trans.*, Vol.81, pp. 24~28
6. Zellouf, Y., Mariotti, J., (2010), "ON ASSESSING ROLLOVER UNDER OFFSHORE CONDITIONS," *The 16th International Conference & Exhibition on Liquefied Natural Gas*, Paper PS6-6.

353346 LNG Expanders for Small Scale LNG Liquefaction Plants

Abstract:

Small scale liquefaction plants for natural gas operate at low flow and at high pressure to produce LNG at competitive costs. The same type of turbomachinery that is used in regular liquefaction plants, like compressors, gas expanders, pumps and liquid expanders, is also installed in small scale plants. All turbomachinery with centrifugal design have a low efficiency if the pressure difference is high at low flow rate, increasing the costs for the liquefaction. To achieve competitive liquefaction costs, the design of the liquid and two-phase LNG expanders is of non-centrifugal design and is presented in this paper.

Increasing Reliability and Functionality of Molecular Sieve Dehydrators

A Case Study: Williams Ignacio 400MMSCFD Plant Durango, CO

March 2014

CONTACT INFORMATION

BiLFINGER WATER TECHNOLOGIES

Benjamin Schmitt, PE
Senior Applications Engineer
Phone +1 651 638 3274
Mobile +1 651 253 1258
Benjamin.schmitt@johnsonscreens.com

Johnson screens®

Hydrocarbon Processing
GLOBAL BUSINESS UNIT

PRESENTATION OVERVIEW

- Review Pre-Installation Operation
- Options Available to Address Known Problems
- Known Engineering Challenges
- Engineering Configuration Alternatives
- Installation Review
- Review Post-Installation Processing Operation

PRE-INSTALLATION: OVERVIEW DATA

Dryer Overview:

- 400MMscfd
- Fish Engineering (1956)
- (4) Adsorber Train
- 8' Diameter x 23' Tan-Tan
- Multitude of Conversions to Achieve Final Loading
- 34,200lb Mole Sieve / Vessel

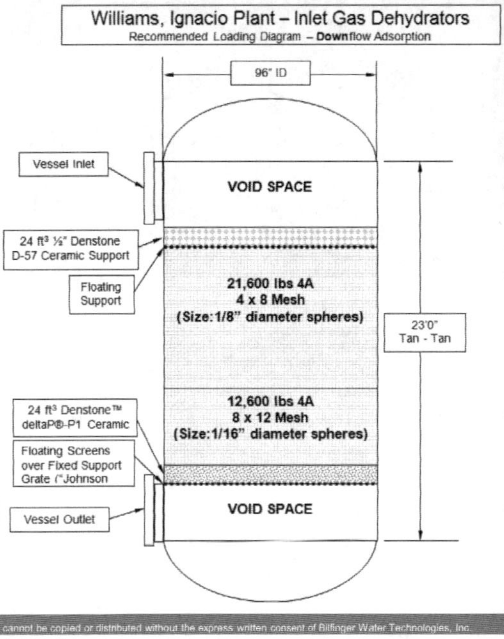

PRE-INSTALLATION: SUPPORT GRID RELIABILITY

BILFINGER WATER TECHNOLOGIES

- Multiple Unplanned Shut-downs Between 2010-2013
- Extensive Unplanned Down Time & Maintenance Costs
- Loss Of Production
- Loss Of Sieve Containment

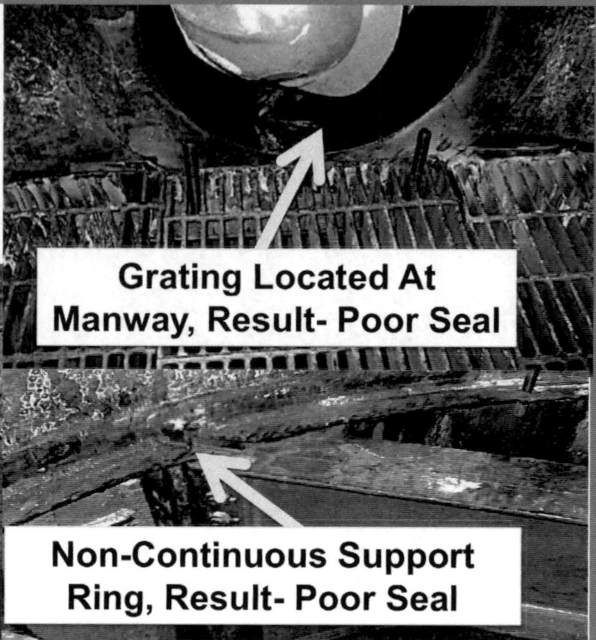

PRE-INSTALLATION: SUPPORT GRID CONFIGURATION

BILFINGER WATER TECHNOLOGIES

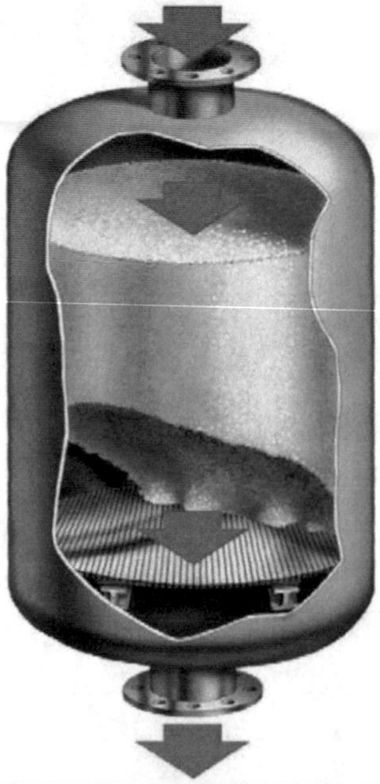

- Applicable in Mole Sieve Dryers
- Full Cross Sectional Catalyst Support
- Versatility of Design
- Ability To Support Catalyst In Direct Contact

JOHNSON SCREENS SSG™

BiLFINGER WATER TECHNOLOGIES

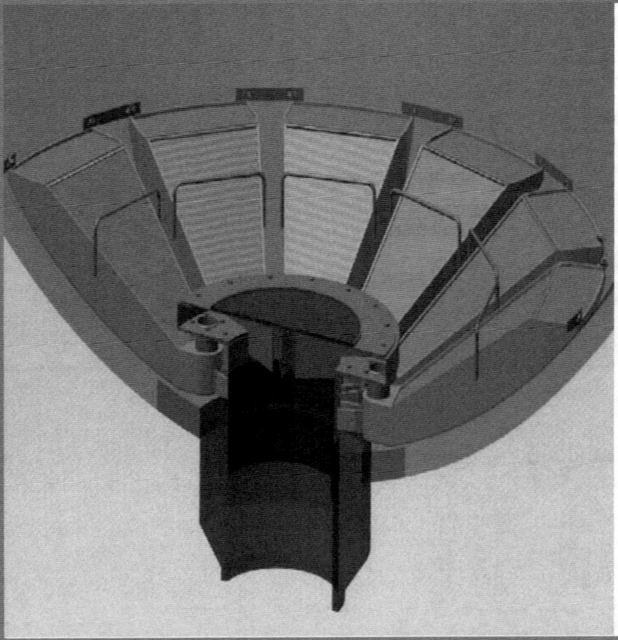

- Applicable in Mole Sieve Dryers
- Versatility of Design
- Ability To Support Mole Sieve In Direct Contact
- Increased Mole Sieve Weight

KNOWN ENGINEERING CHALLENGES

BiLFINGER WATER TECHNOLOGIES

- Mechanical Reliability
- Mole Sieve Containment
- Thermal Expansion
- 16 Inch Manway
- Space Limitations
- Late 1950's Vessels
- No Welding To Vessel

KNOWN ENGINEERING CHALLENGES

- Side Outlet / Inlet
- Unknown Head Condition
- Increased Bed Weight
- TA Schedule
- Internal Piping Connections

CONFIGURATION ANALYSIS:

	SUPPORT GRID	SSG™
Mole Sieve Weight		✓
Mole Sieve Usage	✓	✓
Reduced Regeneration Cycling		✓
Fluid Distribution/ Hydraulic Efficiency	✓	✓
Support Media Volume	✓	✓
Bed Containment Mechanical Reliability	✓	✓

©2014 All rights reserved. Bilfinger Water Technologies, Inc., New Brighton, MN USA Information contained in this presentation cannot be copied or distributed without the express written consent of Bilfinger Water Technologies, Inc.

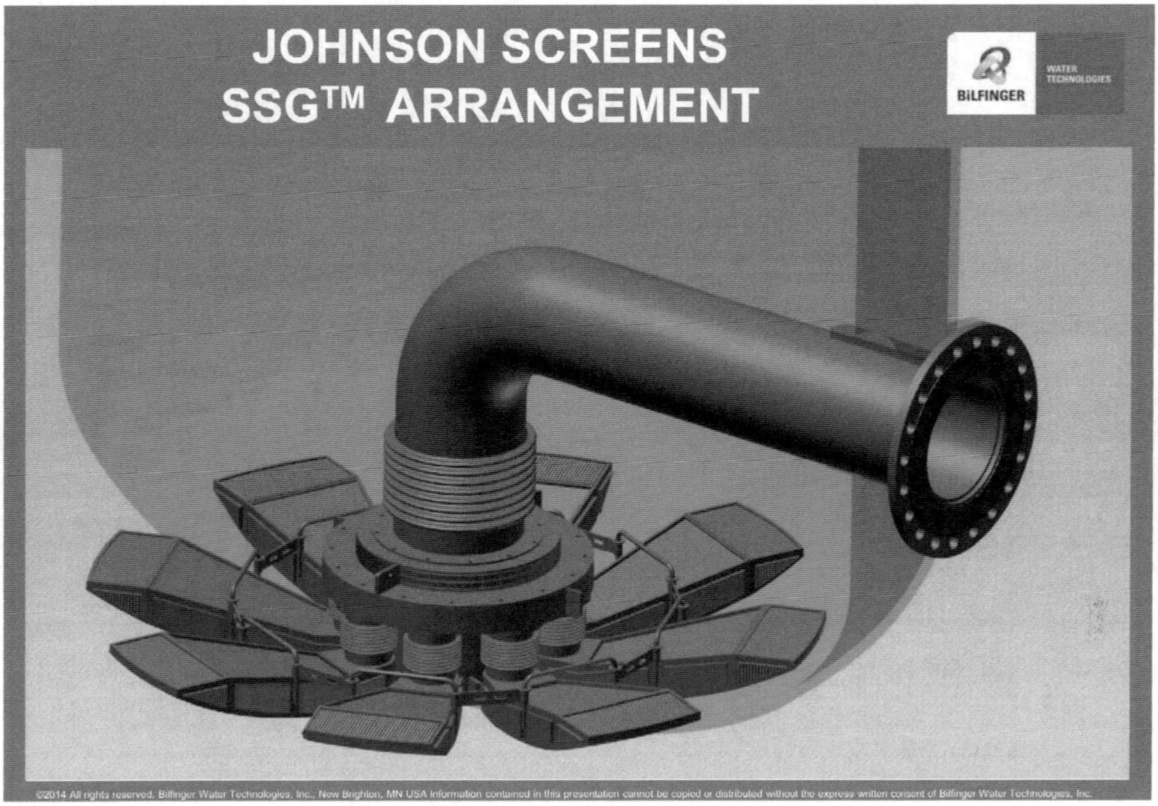
JOHNSON SCREENS
SSGTM ARRANGEMENT
BiLFINGER
WATER TECHNOLOGIES
©2014 All rights reserved. Bilfinger Water Technologies, Inc., New Brighton, MN USA Information contained in this presentation cannot be copied or distributed without the express written consent of Bilfinger Water Technologies, Inc.

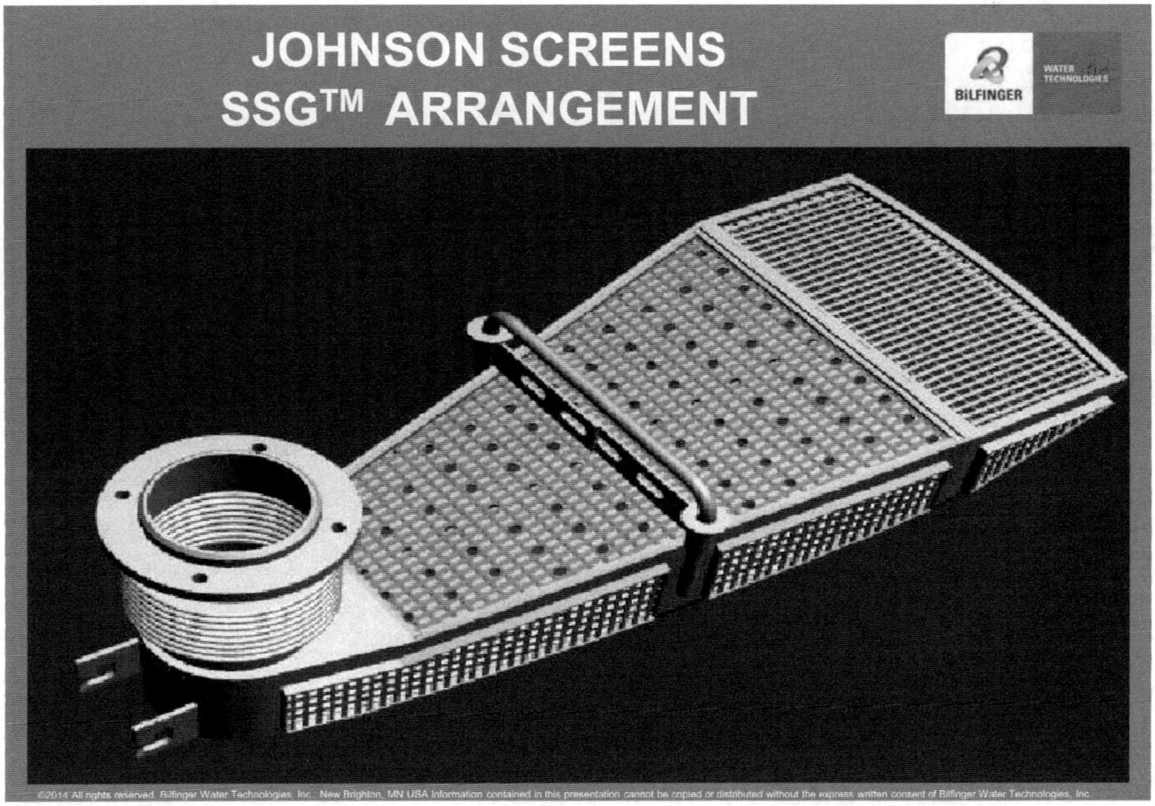
JOHNSON SCREENS
SSGTM ARRANGEMENT
BiLFINGER
WATER TECHNOLOGIES
©2014 All rights reserved. Bilfinger Water Technologies, Inc., New Brighton, MN USA Information contained in this presentation cannot be copied or distributed without the express written consent of Bilfinger Water Technologies, Inc.

PRE-INSTALLATION: OVERVIEW DATA

- **400MMSCFD (4 Vessels)**
- **8' Diameter X 23' Tan-Tan**
- **Installed Weight: (Each Vessel)**
  - **43,200lb Mole Sieve**
- **2' Unused Void Space In Head**
  - **4,500lb Additional Sieve Potential**
  - **42.3% Increase In Sieve Volume**

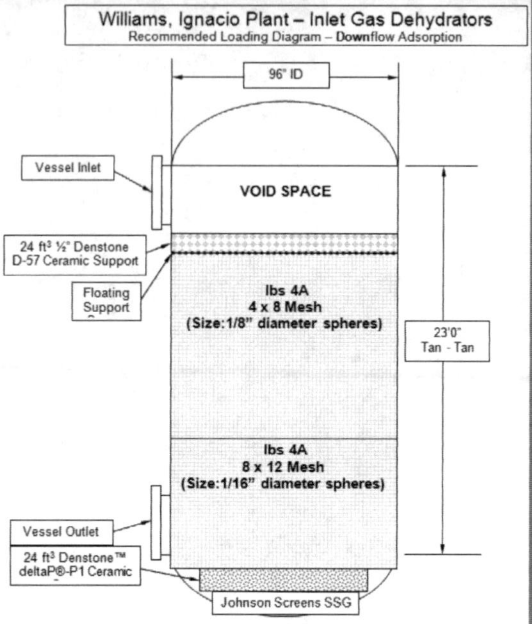

GENERAL CONFIGURATION LOADING DIAGRAM

Diameter: 8' Tan-Tan: 23'	Support Grid	SSG™ As Loaded	SSG™ Potential
Sieve Weight	34,200 lb	43,200 lb	48,700 lb
Inert Material Volume	48 ft³	48 ft³	48 ft³
Un-Used Volume	397 ft³	148 ft³	48 ft³
Sieve Weight Increase		26.3%	42.3%

CFD REVIEW: SUPPORT GRID

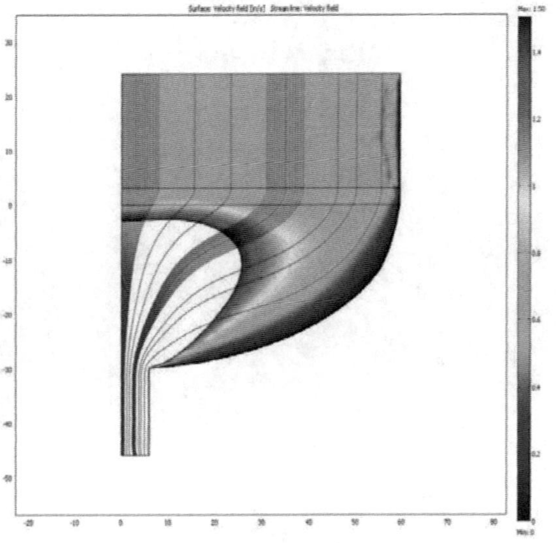

Modeling Assumptions:

- Down Flow
- Homogeneous Catalyst Bed
- Uniform Inlet Flow
- 2:1 Elliptical Head Configuration
- Support Grid Located At Tangent Line Of The Vessel
- Shadowing Effect Of Support Beams Not Included In Analysis

CFD REVIEW: SSG™

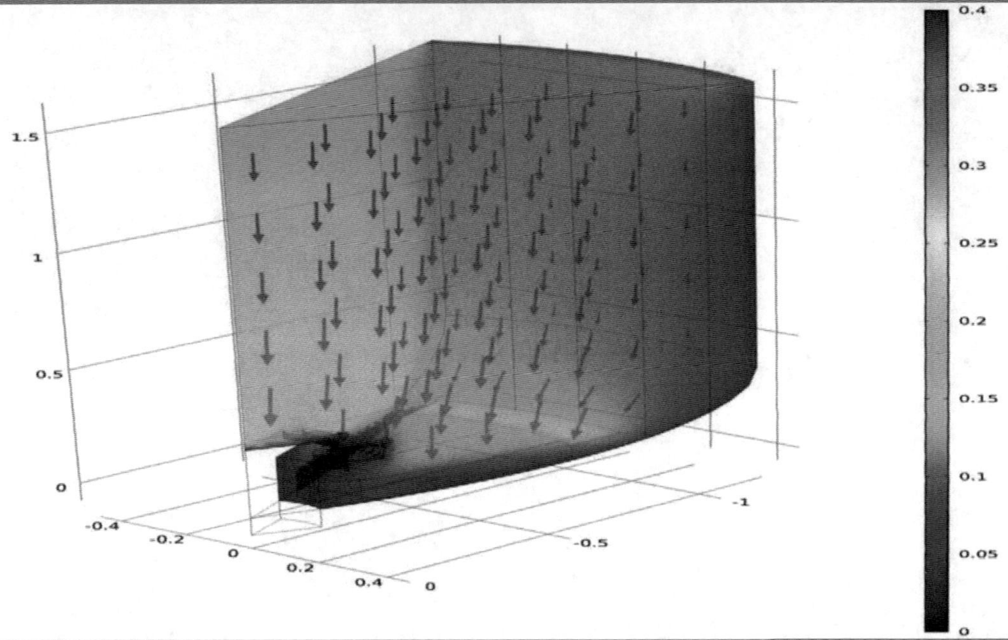

INSTALLATION REVIEW

BiLFINGER WATER TECHNOLOGIES

INSTALLATION REVIEW

BiLFINGER WATER TECHNOLOGIES

©2014 All rights reserved. Bilfinger Water Technologies, Inc., New Brighton, MN USA Information contained in this presentation cannot be copied or distributed without the express written consent of Bilfinger Water Technologies, Inc.

INSTALLATION REVIEW
BILFINGER
WATER TECHNOLOGIES
©2014 All rights reserved. Bilfinger Water Technologies, Inc., New Brighton, MN USA Information contained in this presentation cannot be copied or distributed without the express written consent of Bilfinger Water Technologies, Inc.

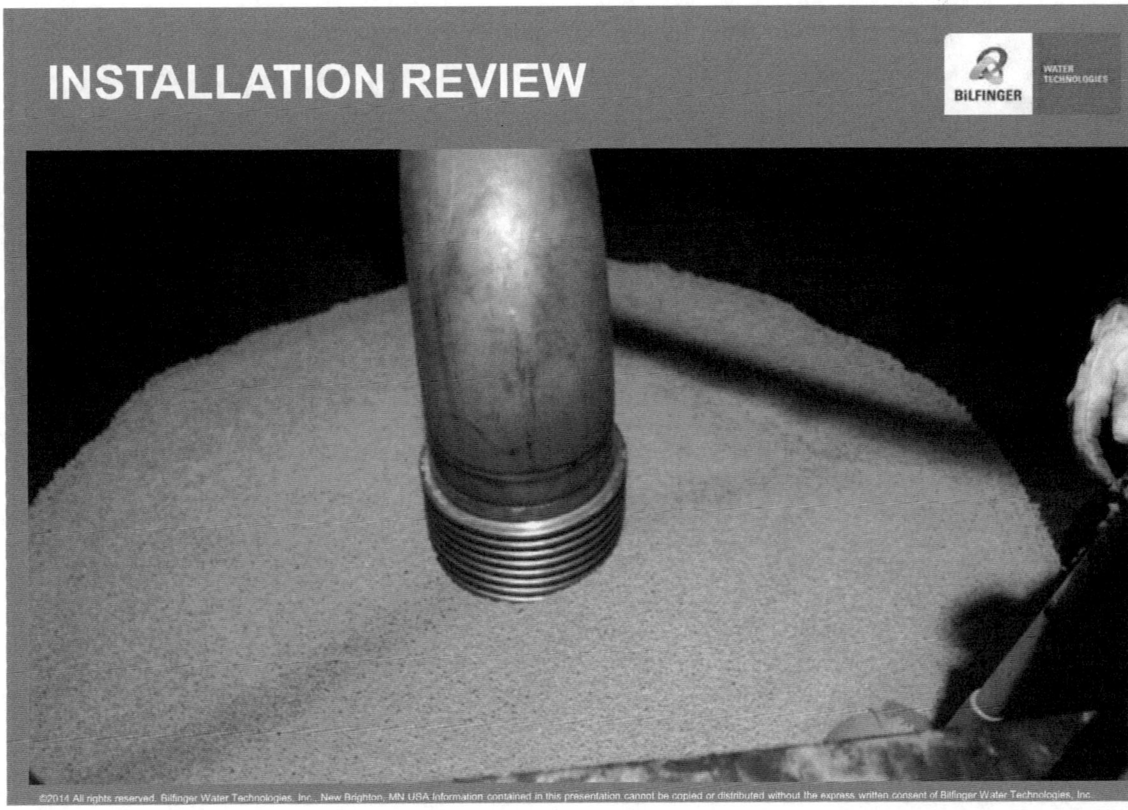
INSTALLATION REVIEW
BILFINGER
WATER TECHNOLOGIES
©2014 All rights reserved. Bilfinger Water Technologies, Inc., New Brighton, MN USA Information contained in this presentation cannot be copied or distributed without the express written consent of Bilfinger Water Technologies, Inc.

REVIEW POST PROCESSING DATA

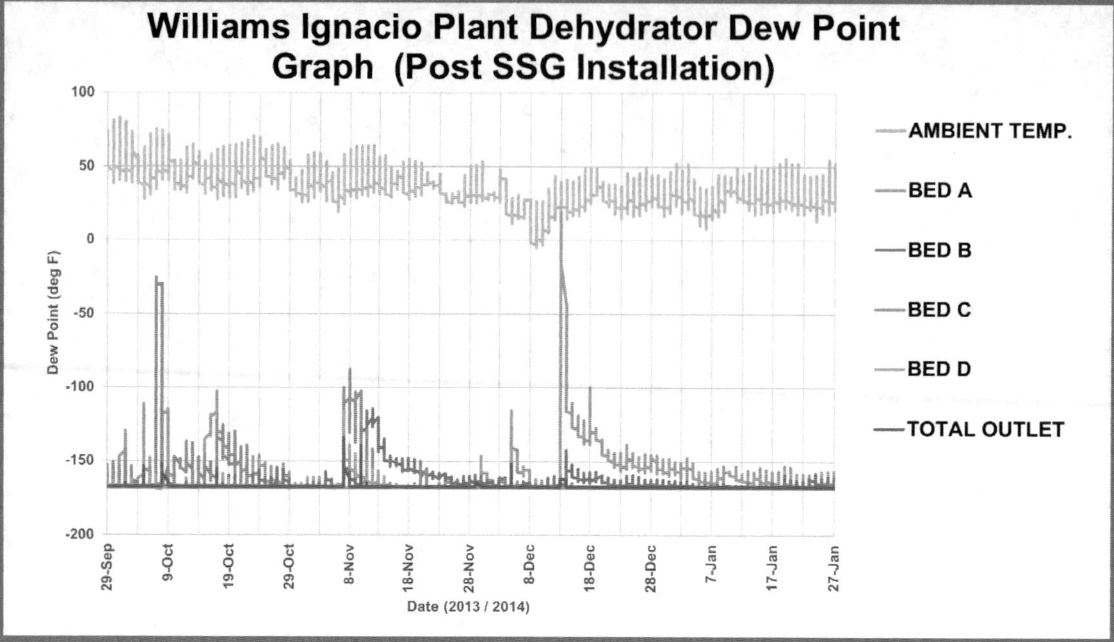

REVIEW POST PROCESSING DATA

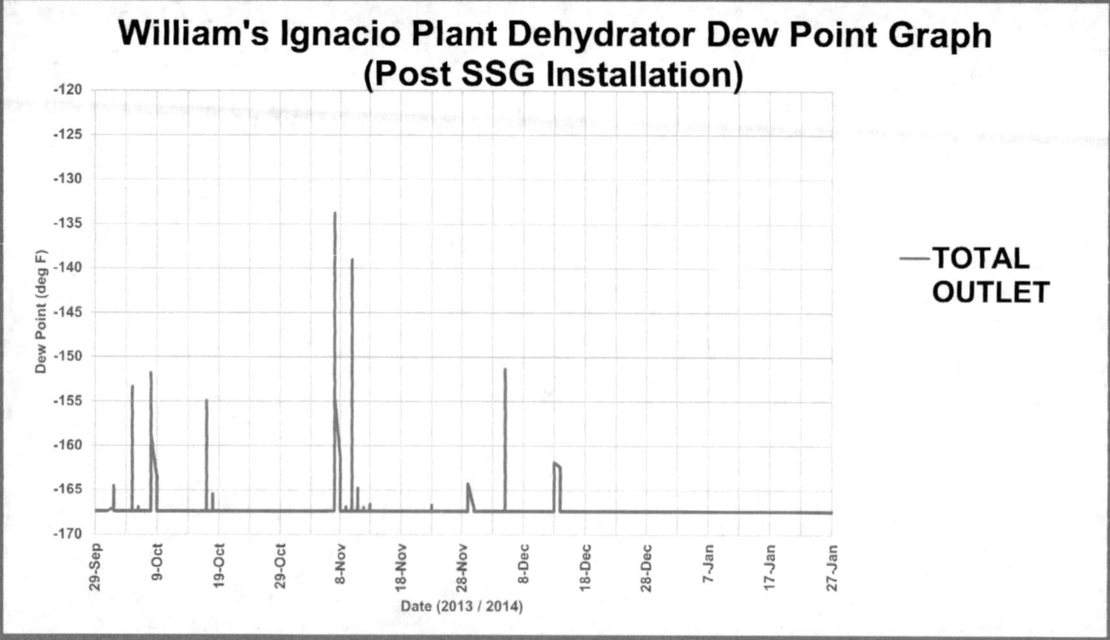

UNPLANNED ADDITIONAL BENEFITS

BiLFINGER WATER TECHNOLOGIES

INSIDE THE VESSEL ADVANTAGES

- **Bed Migration Eliminated - All Segments Are Sealed**
- **Increased Bed Capacity 26.3%, with 42.3% Possible**
- **Maximize Mole Sieve Utilization**
- **Fewer Regeneration Cycles Required**
- **Lower Fuel Consumption**

UNPLANNED ADDITIONAL BENEFITS

BiLFINGER WATER TECHNOLOGIES

OUTSIDE THE VESSEL ADVANTAGES

- **Bed Migration Eliminated - All Segments Are Sealed Removes The Need For Cleaning Down Stream Equipment**
  - **Lower Fuel Consumption**
    - **400,000 Cubic Feet Gas Used Per Bed For Regeneration**
    - **Fuel Cost $3.50 Per Million Cubic Foot**
    - **400,000 X $3.50 X 29 (Regen) X 4 (Beds) = $162,400 / 2 = $81,200 Savings Per Year**

©2014 All rights reserved. Bilfinger Water Technologies, Inc., New Brighton, MN USA Information contained in this presentation cannot be copied or distributed without the express written consent of Bilfinger Water Technologies, Inc.

UNPLANNED ADDITIONAL BENEFITS

OUTSIDE THE VESSEL ADVANTAGES

- Fewer Regeneration Cycles Required
  - Less Valve Actuations Required
    - 19 Valves Actuated Per Regeneration
    - 29 Fewer Regenerations
    - 19 X 29 = 551 Fewer Valve Cycles / Year (Potentially Doubles Valve Life)

JOHNSON SCREENS SSG™ Arrangement

CLOSING REMARKS

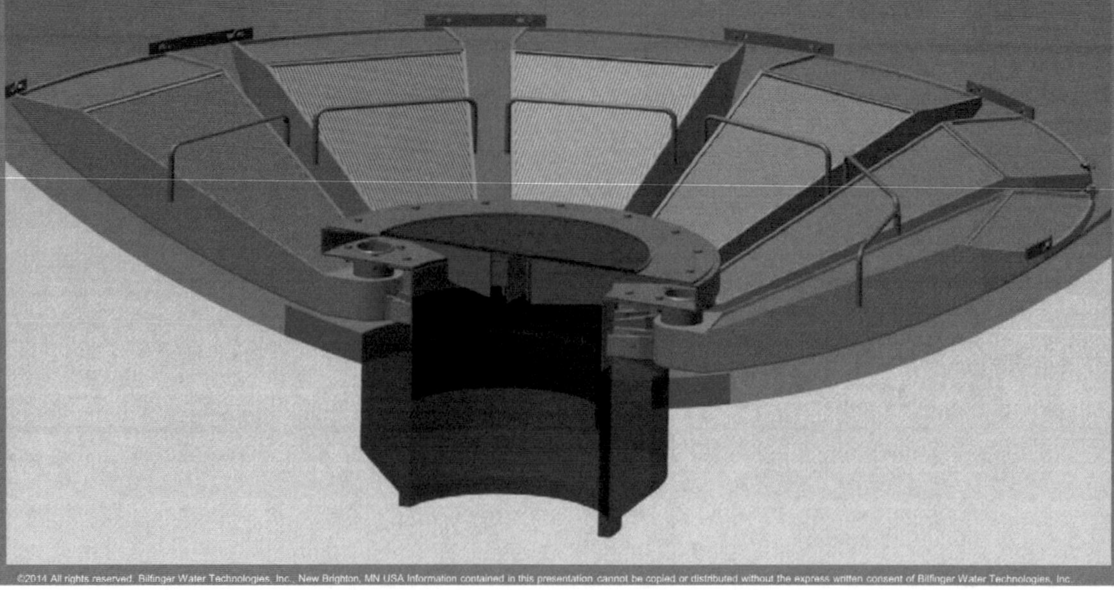

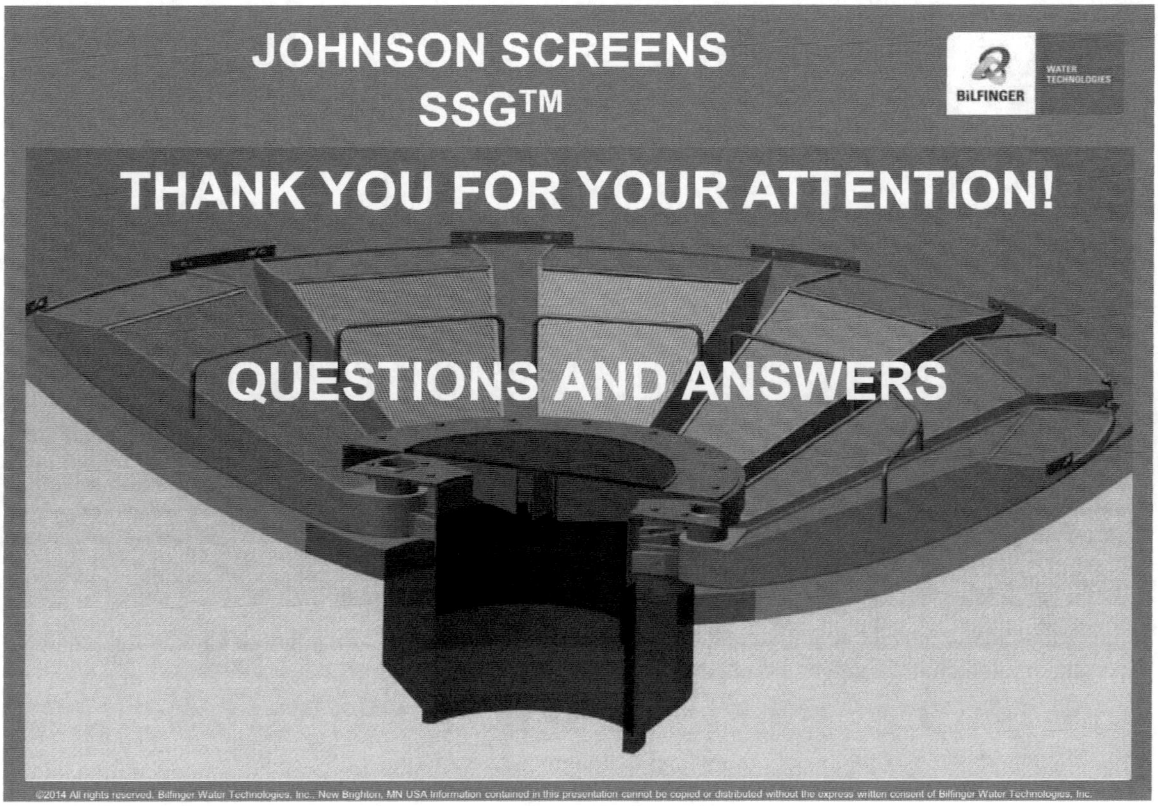
JOHNSON SCREENS
SSGTM
BILFINGER
WATER TECHNOLOGIES
THANK YOU FOR YOUR ATTENTION!
QUESTIONS AND ANSWERS
©2014 All rights reserved. Bilfinger Water Technologies, Inc., New Brighton, MN USA Information contained in this presentation cannot be copied or distributed without the express written consent of Bilfinger Water Technologies, Inc.

ZEOLITE ADSORPTION STUDIES FOR CONDITIONING OF HIGH-PRESSURE NATURAL GAS FLUIDS

Behnaz Hojjati
Robert A. Marriott
University of Calgary
Calgary, AB

AIChE Spring Meeting, New Orleans, LA, March 30- April 3, 2014

Dr. Robert Marriott's research group

Optimization of high-pressure conditioning requires advanced understanding of solid-gas phase behavior.

How much water can we tolerate and still avoid hydrate issues?

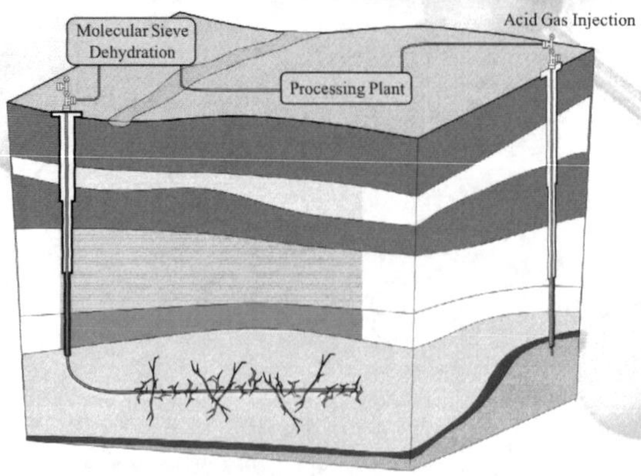

Can we inject sulfur or sulfur with viscosity modifying additives?

Why is there unexpected H_2S in some shale reservoir fluids?

2

NATURAL GAS

- In order to upgrade produced natural gas to sale gas, some unwanted components must be removed H_2O, H_2S, CO_2, and COS

- Dehydration of natural gas
  - *Absorption (glycol)*
  - *Adsorption (molecular sieves)*

- Selective high-pressure adsorption : *Molecular sieves*
  - can reduce water to very low dew-point concentration
  - small pressure drop
  - dehydration of natural gas at low cost (depending on location)
  - relatively free from problems of corrosion

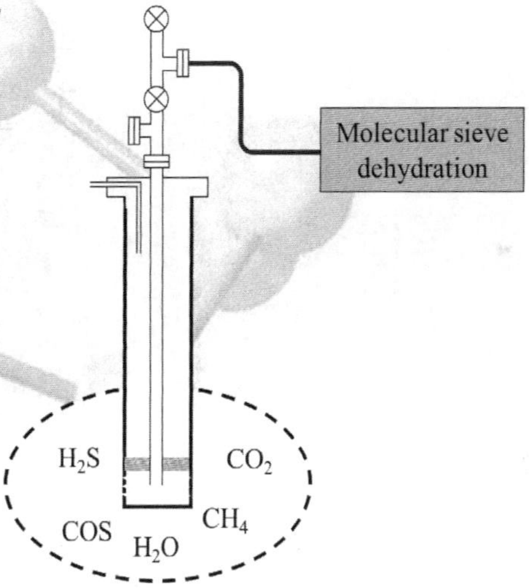

3

ADSORPTION OF H_2O AND CO_2 ON THREE MOLECULAR SIEVES

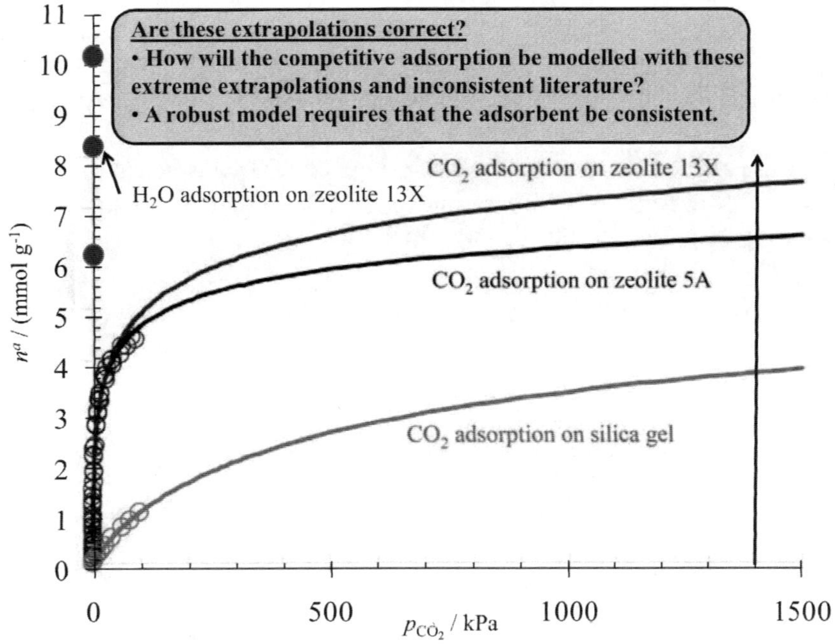

Data from Wang and LeVan *J. Chem. Eng. Data* **2009**, 54, 2839-2844.
Fugacities calculated using Span and Wagner, J. Phys. Chem. Ref. Data **1996**, 25, 1509-1596.

4

OPTIMIZATION HIGH-PRESSURE SOUR GAS ADSORPTION

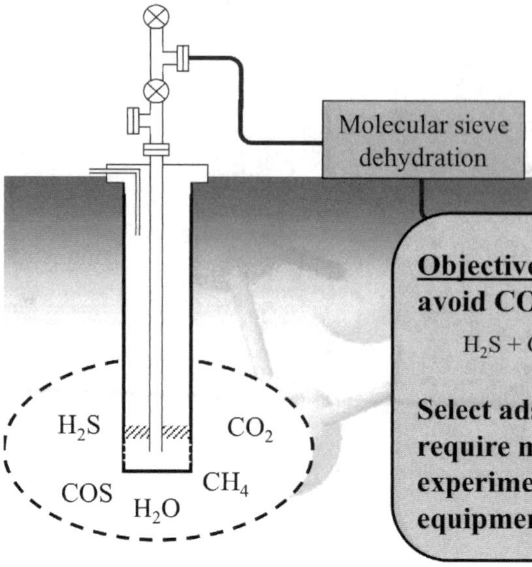

Objective: **optimize regeneration conditions to avoid COS production.**

$$H_2S + CO_2 \Longleftrightarrow COS + H_2O$$

Select adsorption relationships/equations require multi-component high-pressure experimental data from wet sour serviceable equipment (method development)

5

WHY BEGIN WITH ZEOLITES?

$M_{2/n}O{\bullet}Al_2O_3 \bullet ySiO_2 \bullet wH_2O$

- Zeolites are inorganic microporous materials having a **highly regular** pore structure.
- Aluminosilicate zeolites are easily synthesized under hydrothermal conditions from reactive silica-alumina gels in alkaline media at temperatures between $T = 80$ °C and 200 °C.

• Low silica zeolites (Si/Al:~1–1.5)	A, X
• Intermediate silica zeolites (Si/Al: ~2–5)	Y, L
• High silica zeolites (Si/Al: ~10–100)	ZSM-5, Silicate

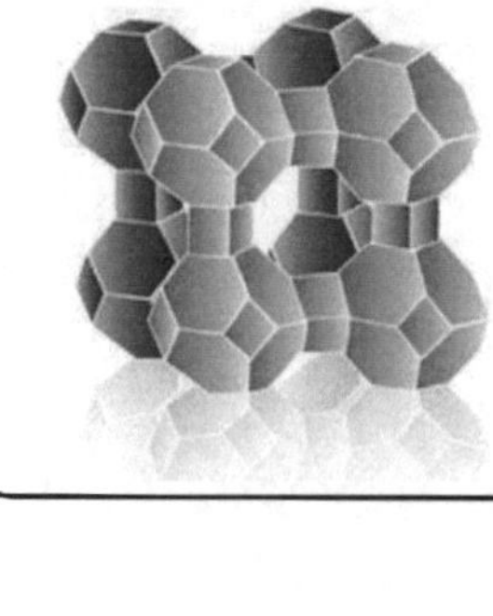

Zeolite A

$M_{2/n}O{\bullet}Al_2O_3{\bullet}2SiO_2$

Zeolite 4A: M= Na
Zeolite 3A: M= K
Zeolite 5A: M= Ca

6

www.zeoliteguide.com

SCANNING ELECTRON MICROSCOPY AND ENERGY DISPERSIVE X-RAY SPECTROSCOPY

Electron Dispersive X-Ray Spectroscopy:

O	50.79
Na	17.78
Al	15.48
Si	15.95

Si/Al : 1.14-1.15

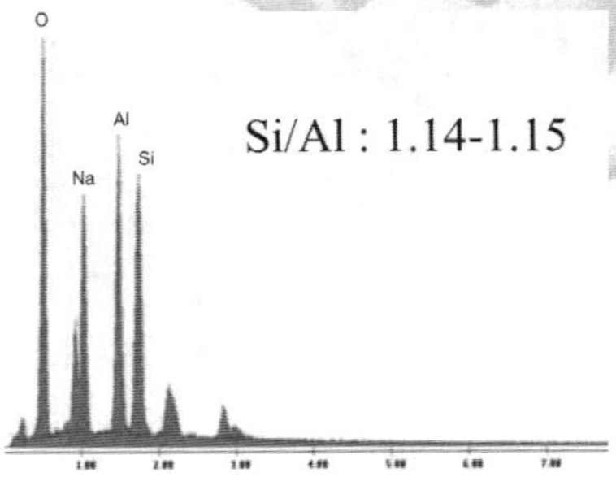

This Zeolite 4A — Commercial Zeolite 4A

Dynamic Light Scattering, $\bar{d} = 4.45$ μm

7

POWDER X-RAY DIFFRACTION

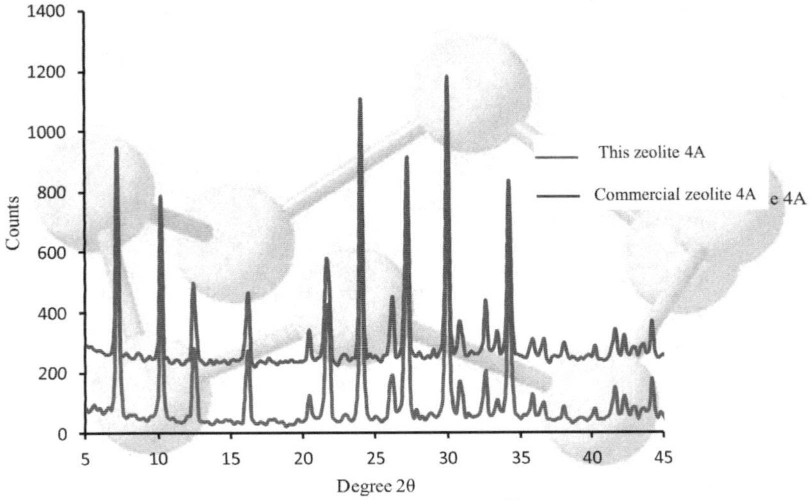

ρ **= 1.5174 g cm^{-3}**

IZA reports ρ = 1.5182 g cm^{-3} for fully dehydrated Zeolite 4A

8

In-situ ATR-FTIR &FT-RAMAN ADSORPTION OF CO_2ONTO ZEOLITE 4A AT T = 25°C

FT-RAMAN

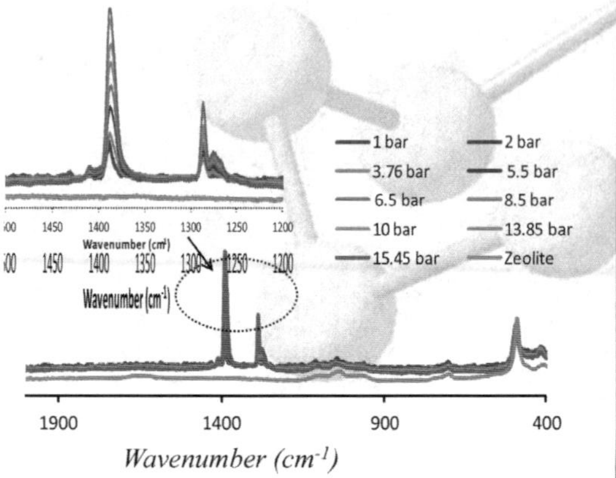

ATR-FTIR

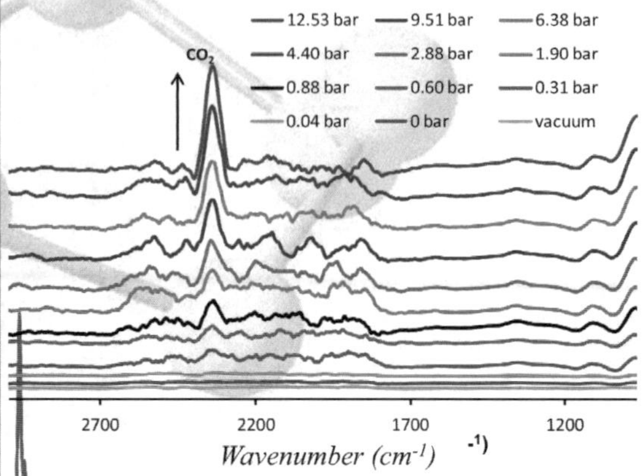

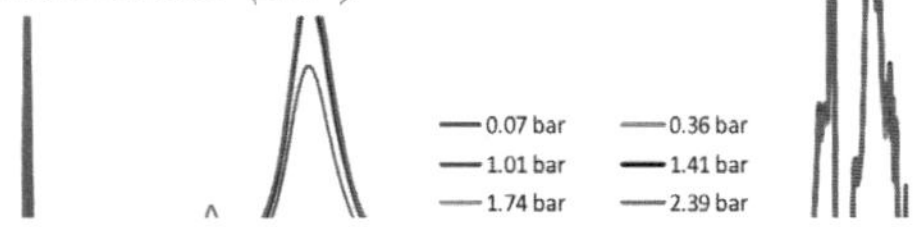

9

In-situ FT-RAMAN ADSORPTION OF H_2S AND CH_4 ONTO ZEOLITE 4A AT T = 25°C

CH_4

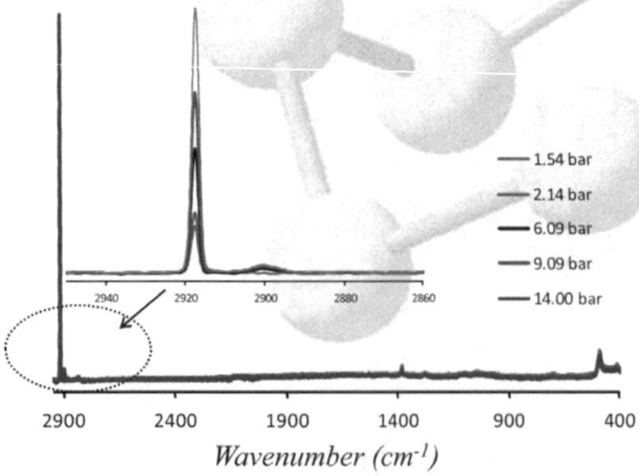

H_2S

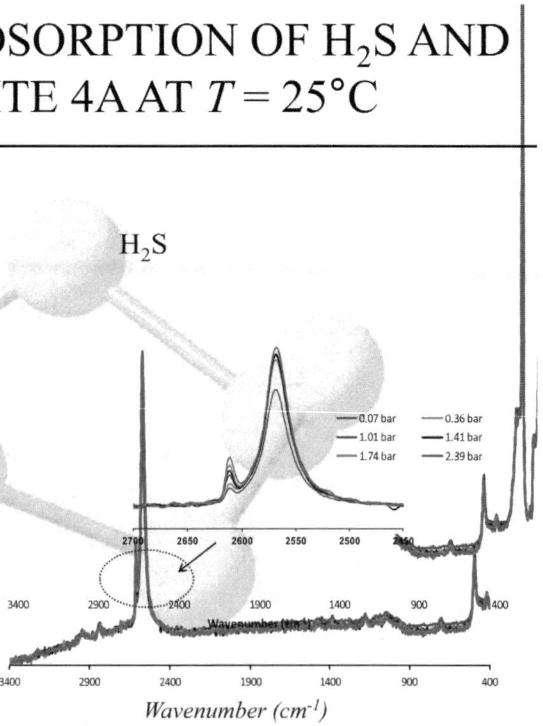

10

A SMALL CELL ADSORPTION APPARATUS FOR MULTICOMPONENT SOUR GAS ADSORPTOIN

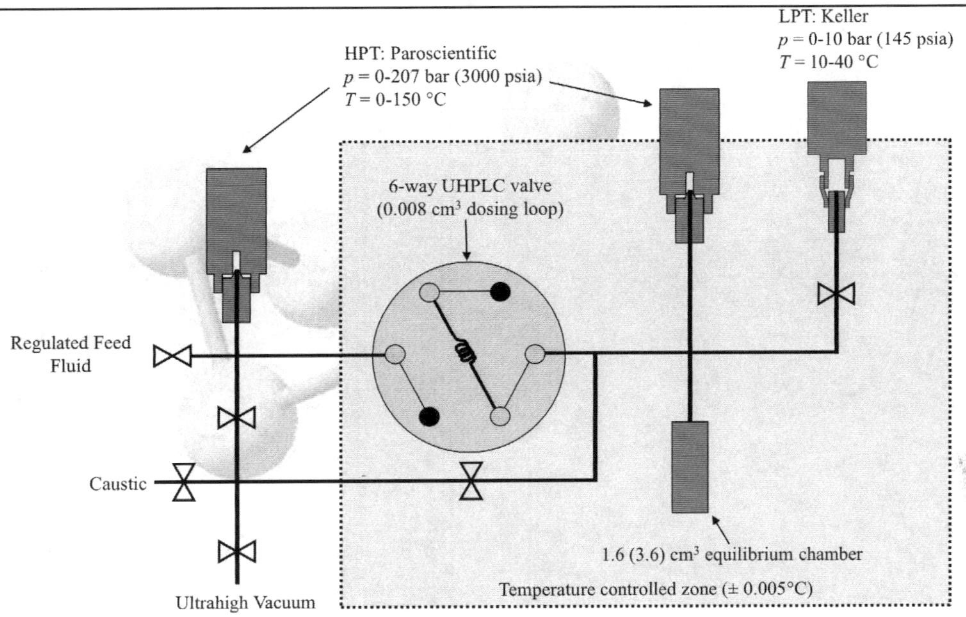

11

A NEW SMALL VOLUME ADSORPTION APPARATUS

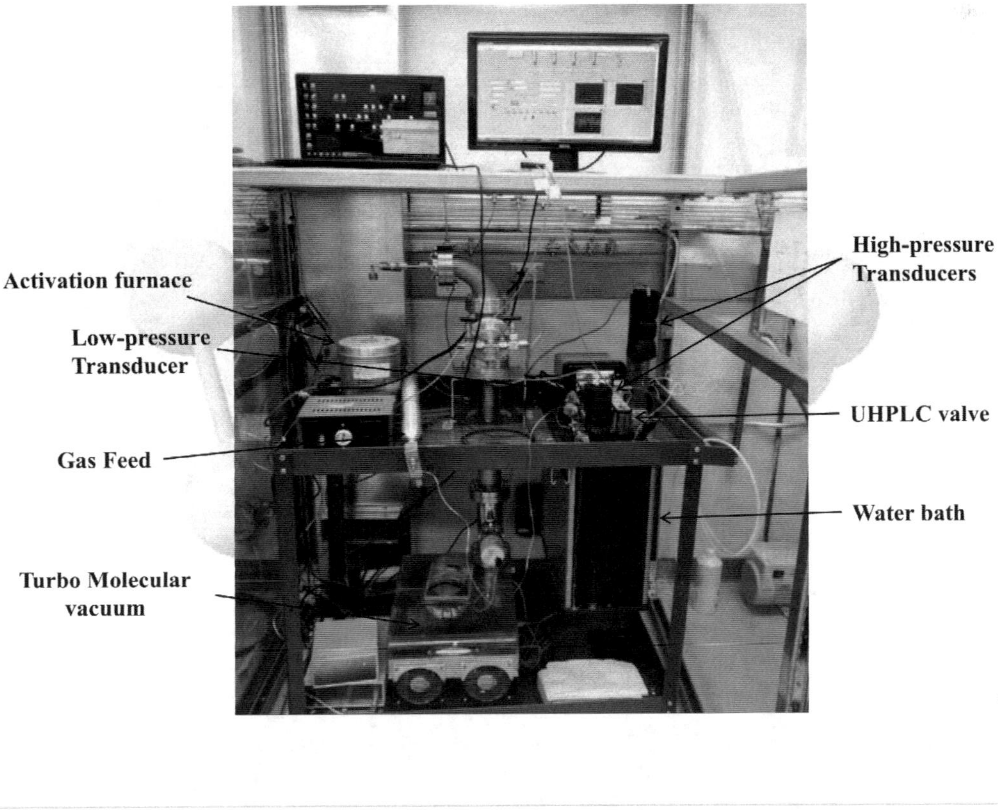

12

ADSORPTION CONDITIONS FOR PURE HETEROGENEOUS AND HOMOGENEOUS POROUS SOLIDS

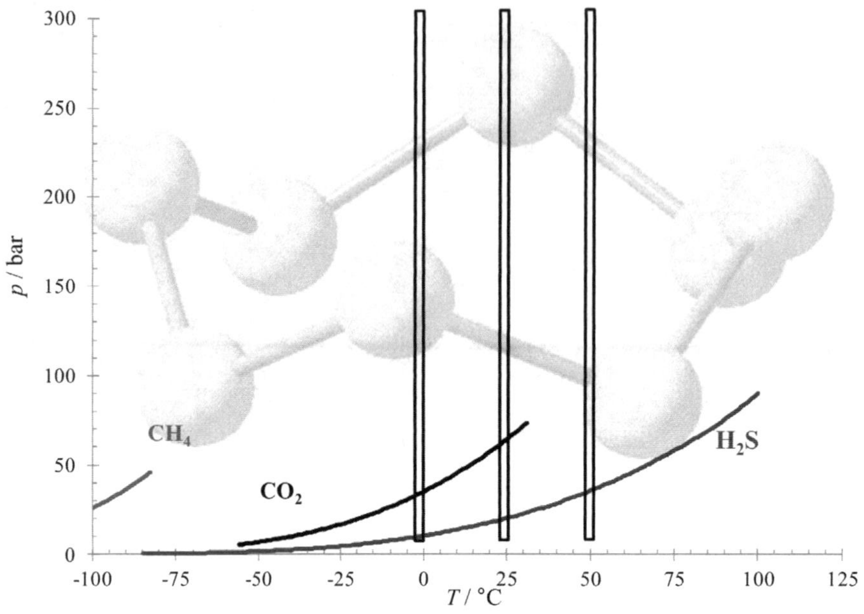

13

THE EXCESS ADSORPTION OF CO_2 ONTO ZEOLITE 4A

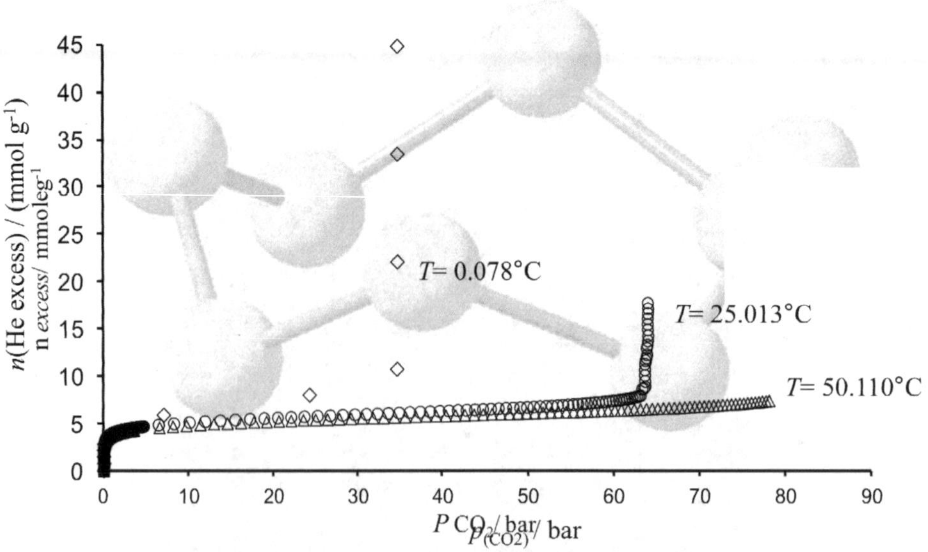

14

PRELIMINARY CO_2 ADSORPTION DATA

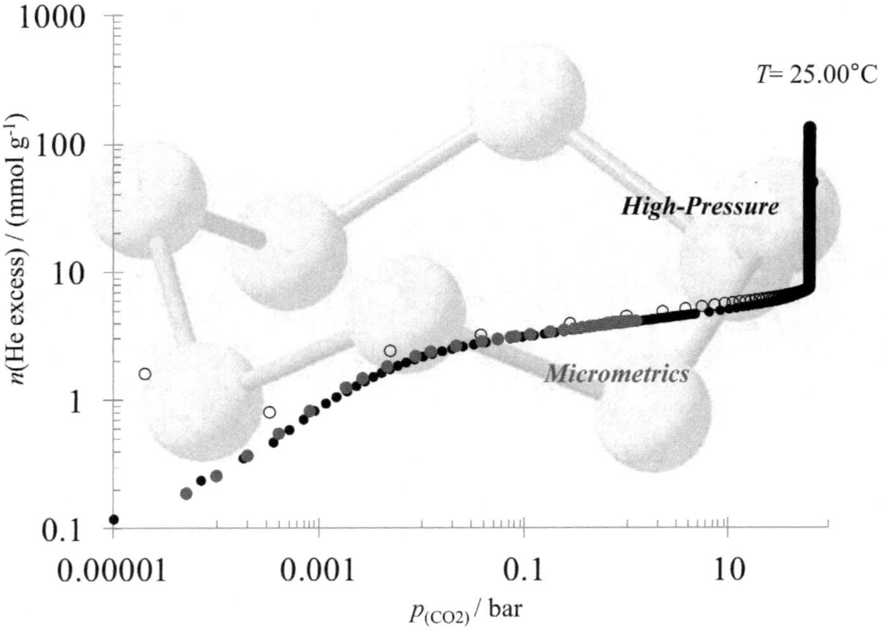

15

THE EXCESS ADSORPTION OF H_2S ONTO ZEOLITE 4A

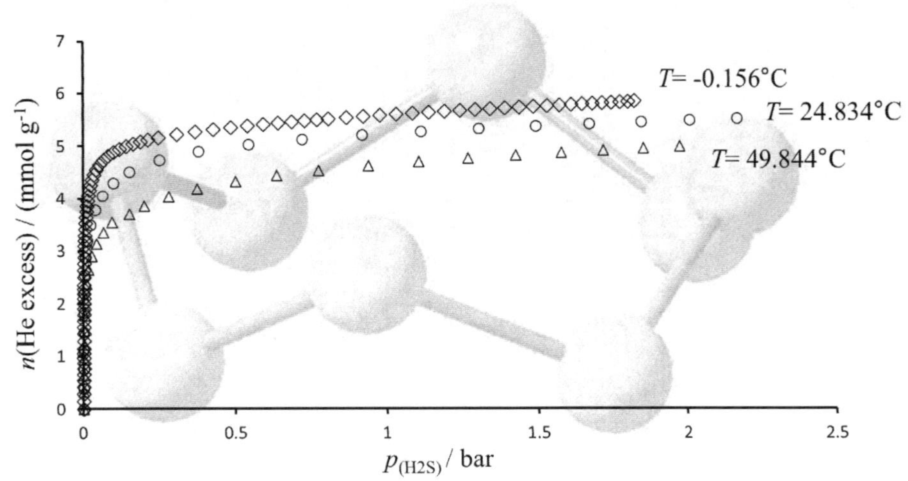

16

THE EXCESS ADSORPTIONS OF CH_4 ONTO ZEOLITE 4A

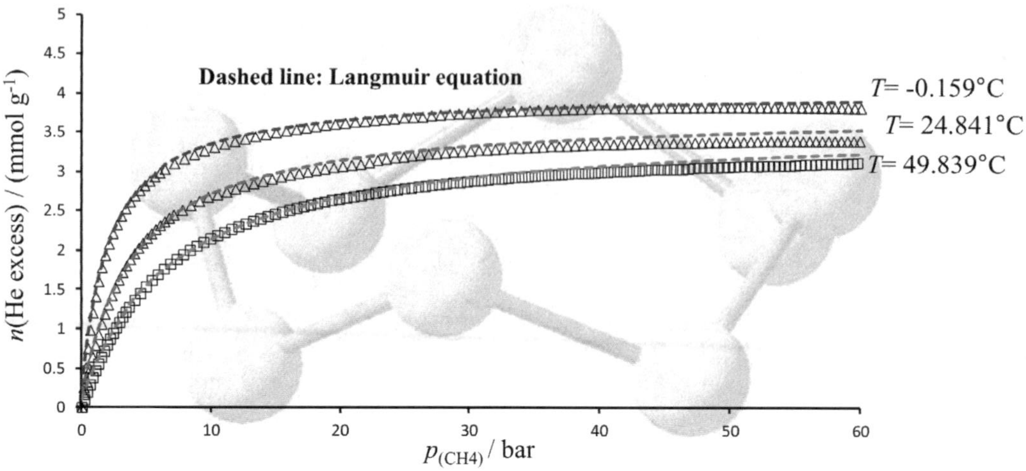

17

THE DIFFERENCE BETWEEN ABSOLUTE AND EXCESS ADSORPTION

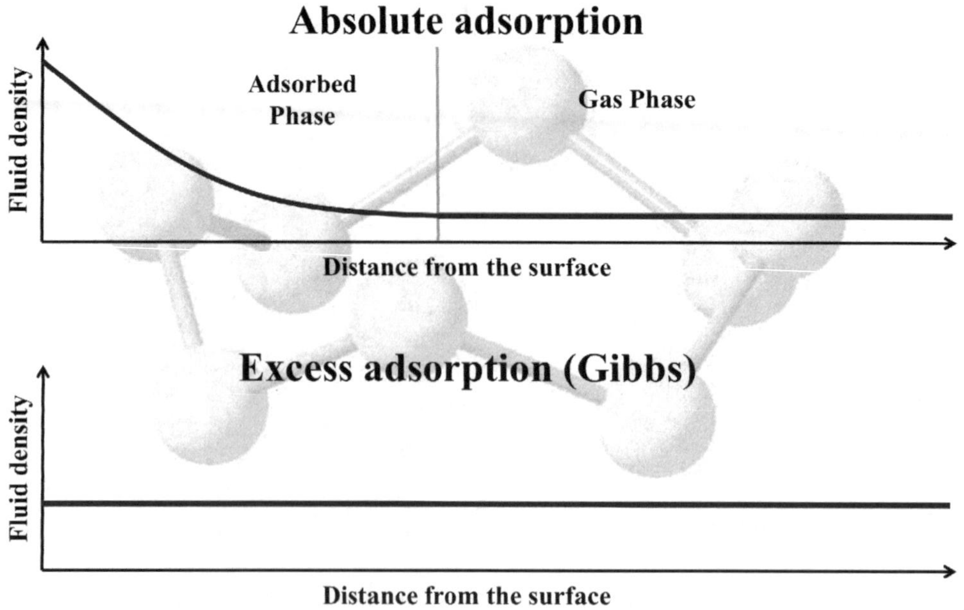

18

THE EXCESS & ABSOLUTE ADSORPTIONS OF CH_4 ONTO ZEOLITE 4A AT T = -0.159°C

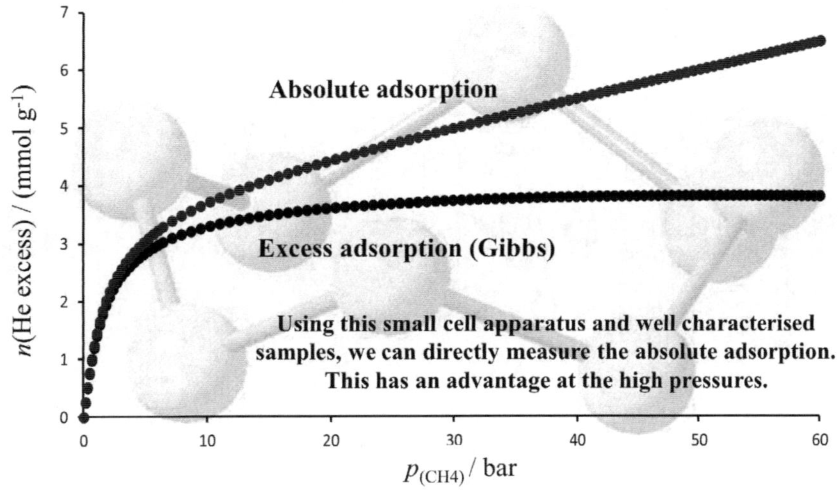

19

THE HEAT OF ADSORPTION OF METHANE ONTO ZEOLITE 4A AT T = 25°C

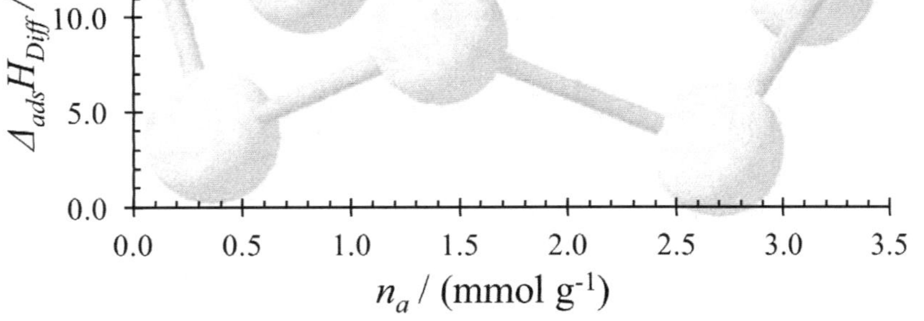

$\Delta_{ads}H_{diff} = 21 KJmol^{-1}$ "Adsorption of Methane on Titanosilicate ETS-10 Zeolite", *J. Chem. Eng. Data* 2005, *50*, 843-848

$\Delta_{ads}H_{diff} = 20.9 KJmol^{-1}$ "Adsorption of Methane on MFI", *Order Porous Solids*, 2011, Science

20

SUMMARY AND FUTURE WORK

- High-quality Zeolites 4Awas synthesized *via* hydrothermal techniques and characterized by XRD, SEM/EDX, and DLS.
- High-pressure ATR-FTIR and FT-Raman were used to check the purity of the zeolite 4A by the adsorption of CO_2, CH_4 and H_2S at different pressure (room temperature).
- High-pressure adsorption isotherms of CO_2, CH_4 and H_2S on the zeolite 4A at 0, 25, and 50 °C using a home-built adsorption apparatus were studied.
- The results can be used to benchmark and compare materials and instruments used for measuring multicomponent sour gas adsorption.
- Future adsorptions models to be parameterized with internally consistent data.
- Future work will include high pressure adsorption isotherms of CO_2, CH_4, H_2S, COS, SO_2 and elemental sulfur on the synthesized zeolite 5A, 3A and 13X at different temperatures.

21

Thank you

Natural Sciences and Engineering Research Council of Canada
(NSERC)
Alberta Sulphur Research Ltd. (ASRL)
The University of Calgary

352233 Design of Units for Mercury Removal from Hydrocarbon Gas Streams

Abstract:

Design of Units for Mercury Removal from Hydrocarbon Gas Streams

Robert W. Soffel, Selective Adsorption Associates, Inc.

John Markovs, Adsorption Solutions, LLC

Abstract:

The hydrocarbon industry became aware of the detrimental effects of mercury some 40 years ago when the aluminum heat exchangers were damaged by mercury at the Skikda LNG plant in Algeria. Since mercury is widely present in natural gas, subsequent plants have chosen to install Mercury Removal Units (MRU's) to protect heat exchangers and other plant operations. The predominant means for removing the mercury has been the use of fixed bed adsorption.

Review is made of the process design criteria for achieving dependable long-term operation of mercury removal units. Essentially total mercury removal is required, since even part per trillion concentrations of mercury in the gas can damage aluminum heat exchangers. The effects of gas residence time in the MRU as it affects the attainable purity are examined, as well as the effects of the adsorbent particle size. Examples of MRU designs are presented for a variety of feed gas conditions, including wet and dry feed gases. The physical and chemical properties of the four types of commercially-available mercury removal adsorbents are compared. The design of the MRU vessel and its relationship to the process gas pressure drop is examined. The importance of designing single-phase flow protection into the mercury removal process to attaining the desired design performance is discussed. Experience-based recommendations for achieving reliable mercury removal performance and predictable mercury adsorbent life are described.

Estimation of Calorific Value and Grindability of Colombian Caribbean Coals by Multiple Regression and Artificial Neural Networks.

Yamid Alí Gómez Rueda, Junior Researcher, Universidad del Atlántico, Barranquilla, Colombia.

M.Sc. Ricardo Angulo Mercado, Associate Professor, Universidad del Atlántico, Barranquilla, Colombia.

Abstract.

Coal is one of the most important raw materials in the world; it is used in diverse applications such as water treatment, steel production and molecular sieves; however coal is mainly used as fuel in power generation plants. In power industry the coal must have specific properties to make an affordable, efficient and environmentally responsible generation process, therefore a lot of instrumental analysis are required to determine the value of those properties, spending time, equipment and qualified personal. Between many properties measured, there are two that are fundamental for combustion and grinding processes: Gross calorific value and Hardgrove Grindability index.

Mathematical models of coal properties – including Gross calorific value and Hardgrove grindability index– have been developed, mainly, to reduce the number of instrumental analysis, determining the mathematical relations of certain properties with parameters that are taken from more simple analysis. Although mathematical models generalizing performance of properties are the goal, in empirical models of Coals this is very difficult, due to coal heterogeneity; hence empirical models are often developed for coals of specific locations, where the estimation of properties becomes the main goal.

Due to the above, added to the lack of studies of Caribbean Colombian Coals, in this study we aim to find through multiple regression and artificial neural networks an empirical mathematical model to determine Gross calorific value and Hardgrove grindability index for Colombian Caribbean Coals using the data from proximate and ultimate analysis as regressors.

In the development of this research, we collected data from three different coal mines, we applied parallelly the regression method and the artificial neural network and finally we compared the two mathematical models to determine which of two methods showed a better performance. After that we made a comparison of our final model with another mathematical models published in the international literature applied to studied coals. Finally we established some criteria to determine which procedure – of the two utilized in this study - use when somebody wants to predict certain property.

Introduction

The behavior of coals are often dependent on the abundance of C,H,N,O,S, also fixed carbon, volatile matter and, in low range coals, humidity and ash content. It is possible to predict a large number of coal parameters using these components. Even if there is no a causation link, useful correlations are allowed by the coalification processes, thus many studies have been published in the last 50 years. Many of them are plots which coal scientists still use, like Seyler's plot. In recent years, the trends are accompanied by equations and artificial neural network models that made significant predictions. In this paper we aim to use traditional multiple regression and Artificial Neural Network (ANN) regression to predict Gross Calorifc Value (GCV) and Hardgrove Grindability Index (HGI) of three Caribbean Colombian Coals for GCV and two for HGI, and the models obtained are compared with others of the scientific literature.

Methodology

Step 1. Database building. Proximate and ultimate analysis data was collected from a variety of coals, covering a diverse range of volatile matter, mineral matter, and humidity. The data was divided in two groups randomly: the first one, containing 80% of the data, was used to build the model; the second one, containing 20% of the data, was used to validate the model.

Step 2. Variable Selection. Here we aim to minimize the number of variables of the model, using the principle of Occam's razor. We use here correlation analysis and some theory of causation to select the variables.

Step 3. Selection of the equation's relationship and ANN structure. Here we select the type of equation (linear, quadratic, logarithmic) and the ANN architecture and learning method.

Step 4. Choose of the best model. To choose the best model we used tow criteria: Root Mean Squared Error (RMSE) and mean absolute error (MAE)

Step 5. Comparisons with scientific literature models. Here we aim to see the differences between predicting power of published models and their ones.

Results.

Gross Calorific Value.

A filtering was made and only 9000 BTU/lb and higher Calorific Power Coals were chosen; 773 samples were used to perform the analysis. For GCV we used the data provided by proximate analysis (Fixed Carbon, FC; Volatile Matter, VM; Ash, A) and their combinations as product. The next graph shows the correlations between proximate analysis variables and GCV.

Graphic 1. Correlation plots between proximate analysis and GCV

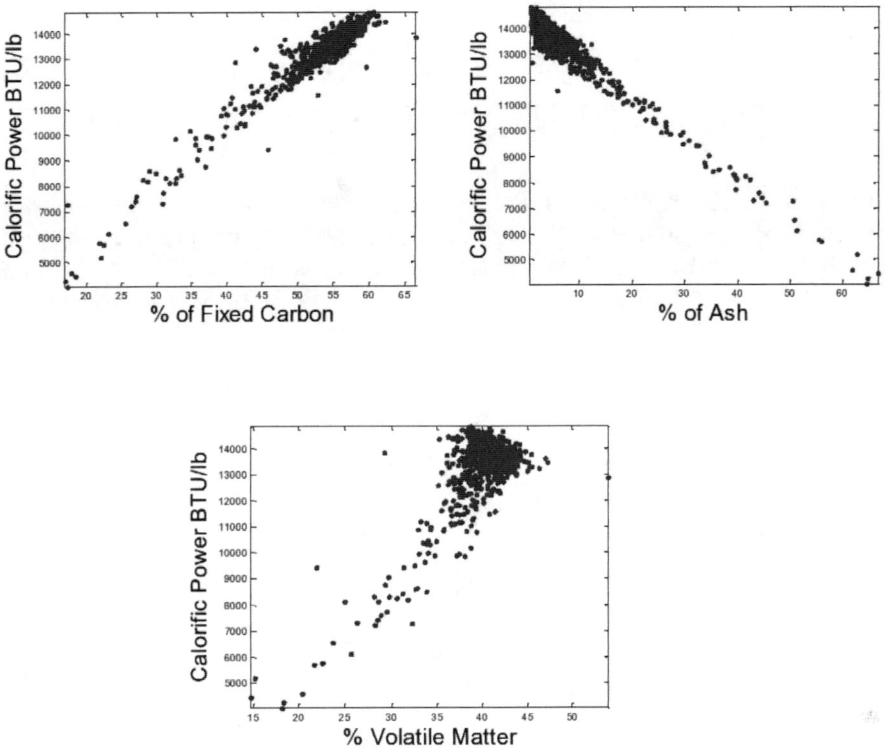

The regression equation that showed the best performance for global data is stated below with r^2=0.96

$$GCV(BTU/lb)=-22212{,}9-250{,}13*FC+56{,}908*VM-4{,}641*FC*VM-1{,}229*A*VM$$

On the other hand, two ANN architectures were used. The first ANN is a Multilayer Perceptron with 20 neurons in the hidden layer, using the Levenberg-Marquardt learning algorithm. The second one was a GRNN with a smoothing parameter optimized with a value of 2.

The performance of the regression and ANN models are presented in the Table 1.

Table 1. Performance of regression and ANN models

	r2			RMSE		
Data group	**Regression**	**ANN 1**	**ANN 2**	**Regression**	**ANN 1**	**ANN 2**
Guajira 1	0.938	0.954	0,965	271,98	4,50	205,59
Guajira 2	0.984	0.988	0,985	270,24	16,96	150,98
Cesar	0.988	0.991	0,990	122,69	3,82	106,23
Global	0.959	0.971	0,965	294,53	14,74	168,76

According to the table, there is no significant difference between the performance of ANN models and regression models. The best models were obtained on individual mines instead of the Global data of the three mines. In all cases, the best model was ANN 1 which corresponds to a model of Multilinear Perceptron with Levenberg-Marquardt learning algorithm.

Hardgrove Grindability Index.

In HGI the regression models shows an almost null relation between HGI and proximate and ultimate analysis data ($r^2 = 0{,}164$). Then only ANN models were used. We use 4 ANN models which are described in the table 2.

Table 2. ANN models for HGI

	Neurons in hidden layers		
MODELS	**I**	**j**	**k**
Model 1	20	0	0
Model 2	32	0	0
Model 3	10	10	0
Model 4	10	10	12

The performance of ANN models are presented in the next table.

Table 3. Performance of the ANN models for HGI

	Global	Guajira_1	Guajira_2
ANN Model 1	0,61663	0,69	0,92893
ANN Model 2	0,64086	0,63	0,97
ANN Model 3	0,78322	0,81635	0,95
ANN Model 4	0,70392	0,79216	0,94541

The performance is similar to the GCV, showing a better performance for individual mines than Global data. However in the three dataset the best model was the model 3, which corresponds to a model with two hidden layers

Comparison with other models.

Comparing the results of our study and other published empirical models we can conclude that the models have the same capacity of prediction, even when they were calculated to coals from different sources.

Table 4. Comparison of GCV models

BTU/lb	Banerje et al (2008)	Chelgani (2009)	Patel et al(2006)	This paper
r^2	0,9783	0,97	0,996	0,971

For the HGI we chose the Maximum Absolute Error (MAE) as comparison criterion. Due to our prediction parameters it can be said that the maximum error of 10 is admissible, although other empirical models showed maximum errors less than 10, they used prediction parameters more elaborated than proximate and ultimate analysis.

Table 5. Comparison of HGI models

HGI	This paper	Chelgani(2007)	Li et al(2005)	Ozbayoglu et al(2006)
MAE	10	4.5	5	17,46

Conclusions.

A lot of predictive models that relate proximate and ultimate analysis with properties of coal have been developed in the last 70 years. These models have been developed for coals of specific locations and in this study we create two models using multiple regression and artificial neural networks to obtain them. In this study, the ANN models show a better performance than regression models. In the case of GCV, the Multilayer perceptron with 20 neurons, one hidden layer and Levenberg-Marquardt learning algorithm, although the regression models showed a good performance, if is not required much precision, the regression models can be used; otherwise, ANN model can give the best performance.

The HGI prediction was harder, and no regression model was obtained, due to the show practically any correlation between proximate and ultimate analysis and the HGI. The ANN models worked better, and although they did not show an excellent performance, it is enough to make calculations in some processes that required the data quickly. The ANN model that performs the best was a Multilayer Perceptron with two hidden layers and ten neurons in each layer.

References.

1. AKYILDIZ, Mustafa; URAL; SUPHI. Studies of the relationship between mineral matter and grinding properties for low-rank coals. 2004.
2. ASTM Standards of Coal and Coke. ASTM D121, ASTM D167, ASTM D388 ASTMD409 ASTM D2013, ASTM D2234, ASTMD3172, ASTM D3174, ASTM3176, ASTMD3180, ASTM D3302, ASTM D4749, ASTM D5016, ASTM D5865 ASTM D6518.
3. BANERJEE et al. Development of a new proximate analysis based correlation to predict calorific value of coal. Elsevier. 2008
4. BAGHERIEH et al. Prediction of coal grindability based on petrography, proximate and ultimate analysis using multiple regression and artificial neural networks.
5. CANNAVOS G. Probabilidad y Estadística Aplicación y métodos. Ed. en español Mc Graw-Hill. México.1995.
6. CHANNIWALA; GHOSAL; PARIKH, Jigisha. A correlation for calculating HHV from proximate analysis of solid fuels.Fuel. 2004
7. CHELGANI, Chehreh; HOWER, James; JORJANI; MESROGHLI,Sh. Simultaneous prediction of coal rank parameters based on ultimate analysis using regression and artificial neural network. International Journal of Coal Geology. 2010.
8. CHELGANI, Chereh; JORJANI; MESROHI . Estimation of the gross calorific value based on coal analysis using regression and artificial neural networks.Elsevier. International Journal of Coal Geology. 2009
9. DUNXI Yu et al. Prediction of grindability with multivariable regression and neural network in Chinese coal.
10. KUCUKBAYRAK, Sadriye. Estimation of calorific values of Turkish lignites. Fuel. 1991.
11. HAXHI, Hajri. MARTIN, Duncan; LOLJA, Saimir. Correlations in the properties of Albanian coals. Correlations in the properties of Albanian coals. Fuel. 2002.
12. LI, Peisheng et al. Prediction of grindability with multivariable regression and neural network in Chinese Coal. Elsevier. Fuel Journal. 2005.
13. MATHEWS, Jonathan et al. A review of coal properties with elemental composition.Fuel process technology. 2014, pp. 104-133
14. MONTAÑO MORENO. Redes Neuronales Artificiales aplicadas al análisis de datos. Universidad de les Illes Balears: Tesis Doctoral. 2002.
15. OZBAYOGLU, Gülham. Estimation of Hardgrove Grindability index of Turkish coals by neural networks. International Journal of mineral processing. 2008
16. PATEL, Shagufta. Estimation of gross calorific value of coals using artificial neural networks. Elsevier. 2006.
17. SPECHT, Donald. A General Regression Neural Network. IEEE, transactions in neural networks. 1991.
18. SPEIGHT, James. Handbook of Coal Analysis. Wiley &Sons. 2005
19. The Mathworks Inc. Matlab Help, Neural Network Toolbox, Version 7.9.0.
20. WALPOLE; MYERS. Probabilidad y Estadística para ingeniería y ciencias. Prentice Hall 2000.

Safety Study for Optimum Safe Design of FLNG

Sung-eun Kim, Kiil Nam, Jinsang Park, Tae Jin Kim and Yoon Choon Kim,
Hyundai Heavy Industries, Ulsan, Korea

Introduction

As increasing the demand of energy all over the world, the consumption of natural gas is also increasing. The offshore gas reserves are estimated to have significant portion of natural gas. In this sense, the FLNG is being introduced to develop the natural gas in offshore. The FLNG generally combines a turret, topside gas processing facilities, hull storages of LNG, LPG and condensate, a flare, an offloading system, living quarter and all the supporting facilities.

The offshore facilities have considered that the safety-criticality is the most important feature. Safety-criticality stems from the hazards due to treating the flammable hydrocarbon materials in the limited and congested space which has the potential for large gas clouds and long flame distance. This will tend to increase the potential overpressures and accidental loads. From a safety standpoint, the FLNG should be normally open between topside modules with good venting of potential flammable gases and ample explosion relief areas to limit the amount of flammable gas from releases and reduce the consequences of any explosions.

The objective of this study is to quantify DAL (design accidental loads), to identify potential requirements for additional risk reduction measures and to provide a summary of risks to personnel working on the FLNG. Safety engineering activities have been focused during FLNG FEED stage on the subjects of the topside facility layout, since the safety consideration governs the topside process module layout to minimize hazards. This is to help the development of guidance for engineers of the FLNG in order to produce inherently safe design.

Background of FLNG Development

HHI (Hyundai Heavy Industries) developed its own model, Hyundai FLNG. The production capacity is 2.5 MTPA (Million Ton Per Annum) for generic design approach [1]. The development has been progressed by two design phase as pre-FEED and FEED. In the pre-FEED stage, the topside design of PFD level with essential control and preliminary 3D module design have completed in association with Linde, AG. Simultaneously HHI has conducted the interface and hull engineering as well as model test at Marine (Netherland). Based on the pre-FEED, FEED was conducted and P&IDs has been issued except liquefaction process which requires licensor's guarantee. The FEED development is initiated by the necessity to enhance the capability on new technology of FLNG as EPC contractor and give a fast feedback on the critical issue to execute the project. Topside module design was updated by lifting capacity of 10,000 ton and integrity with hull. Also, safety studies have been done to verify the FEED deliverables.

Main Features of Hyundai FLNG

Based on the survey of market needs, tentative target area of Hyundai FLNG is northwest sea of Australia. With regard to the design basis and technical review on gas processing and liquefaction, commercial class of 2.5 MTPA FLNG model is developed as below.

- LNG production capacity : 2.5 MTPA
- Liquefaction license : LIMUM (Linde AG)
- LNG tank type : Membrane Mark III (GTT)
- LNG tank capacity : 193,800 m3 (32,300 m3 x 6ea, 2row)
- Hull Dimension (L/B/D) : 355/ 70/ 35 m
- Mooring : Internal Permanent Turret
- Offloading type : LNG and LPG – Side by side, Condensate - Tandem

General Arrangement and Structural Design

Topside modules are arranged with regard to process flow and safety aspect. The final module design is based on the 10,000 ton lifting capacity. Structural calculation and reinforcement on module support are carried out. Safety gab of more than 15 m and fire wall for living quarter are designed to prevent accidental escalation. Hull is designed with the input of environmental loads, tank capacity plan, topside and weight distribution as well as a berthing vessel loads. LNG storage tanks are arranged with 2 rows to minimize the sloshing effect on the membrane tank.

Safety Study

It is the challenge to cope with the risks arising from the FLNG as a novel facility. So it is necessary to validate the process, estimate the risk level and obtain the solution for the risk reduction at early design phase. With assistance of well-authorized class, DNV, safety studies were conducted as the follows.

- HAZID workshop at the end of the Pre-FEED design
- HAZOP workshop at the end of the 1st P&IDs issue
- Risk and availability analysis at the final P&IDs issue
  - RAM (Reliability, Availability, Maintainability) Study
  - FRA (Fire Risk Analysis)
  - ERA (Explosion Risk Analysis)
  - QRA (Quantitative Risk Analysis)

Methodology

The design approach for safety engineering can be either prescriptive approach or performance based approach. The performance based approach is further divided into deterministic and risk based approach [2]. Considering the novel nature of the FLNG, the safety studies have been carried out in accordance with the risk based approach.

NORSOK Z-103 introduces the general procedure of the risk based analysis as shown in Figure 1. According to risk based approach, the risk of the facilities is estimated by the combination of the historical failure data and the corresponding accidental load for the event tree scenarios. In other words, the frequency analysis and consequence analysis are carried out to draw a risk picture such as exceedance curves. Finally, the risk picture is compared to the safety acceptance criteria. The comparison results can be used as the DAL specifications or the grounds for the enhancement of the risk reduction measures. Ultimately, it is important to demonstrate that the solutions of risk management should be within the ALARP (As Low As Reasonably Practicable) principle.

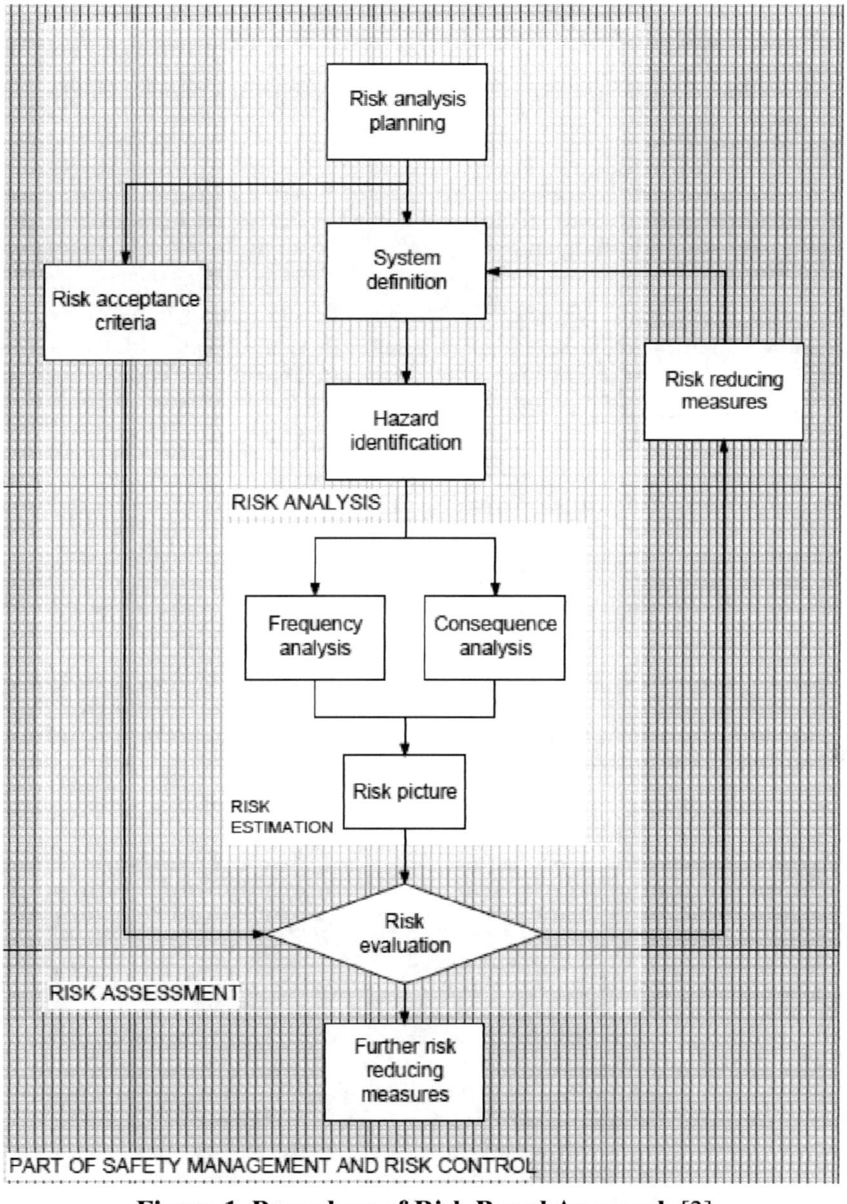

Figure 1. Procedure of Risk Based Approach [2]

Safety Acceptance Criteria

The risk analyses of FRA, ERA and QRA have been conducted on FLNG FEED. FRA and ERA present DALs for the safety critical elements by the potential accident induced from hydrocarbon leaks. QRA integrates the results from FRA and ERA and also accounts for the risk contribution from the hazards other than hydrocarbon release. The safety acceptance criteria are applied to the analysis as described at Table 1.

Consequence modeling of FRA is based on the flame length for the jet fire and the flame diameter for the pool fire considering time varying discharge rate. In ERA, local and global overpressures by explosion are estimated. Through the FRA and ERA, the exceedance curves for fire and explosion are obtained. The cumulative frequency of 10^{-4}/yr is applied to specify DAL for each topside module and structure. QRA can be conducted with the safety objective of personnel, environment and asset. In this study, the fatality of personnel working at the FLNG is a basis of QRA. As for individual risk criteria, the value of 10^{-6} < IRPA (Individual Risk Per Annum) < 10^{-3} is applied, which is generally accepted for oil and gas industry.

Table 1. Safety Acceptance Criteria

Study	Target	Value
FRA	DAL against fire	Exceedance frequency of 10^{-4}/yr
ERA	DAL against explosion	Exceedance frequency of 10^{-4}/yr
QRA	IRPA	10^{-6}/yr < IRPA < 10^{-3}/yr

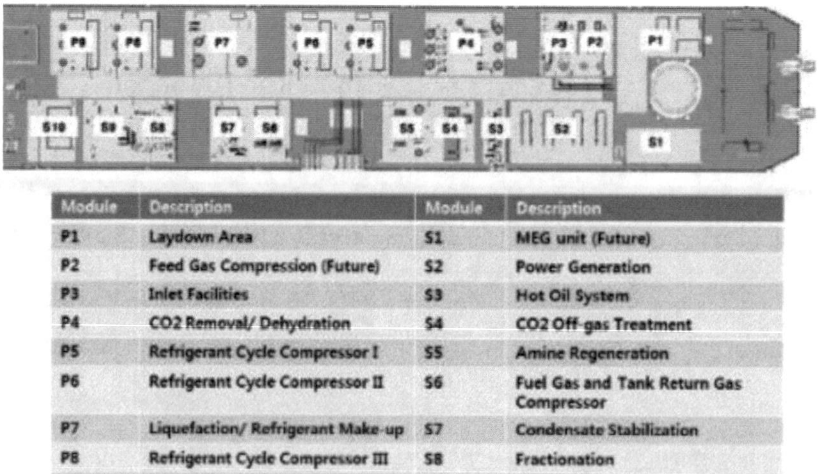

Module	Description	Module	Description
P1	Laydown Area	S1	MEG unit (Future)
P2	Feed Gas Compression (Future)	S2	Power Generation
P3	Inlet Facilities	S3	Hot Oil System
P4	CO2 Removal/ Dehydration	S4	CO2 Off-gas Treatment
P5	Refrigerant Cycle Compressor I	S5	Amine Regeneration
P6	Refrigerant Cycle Compressor II	S6	Fuel Gas and Tank Return Gas Compressor
P7	Liquefaction/ Refrigerant Make-up	S7	Condensate Stabilization
P8	Refrigerant Cycle Compressor III	S8	Fractionation
P9	Refrigerant Cycle Compressor IV	S9	Flare and Blowdown System
		S10	Laydown Area

Figure 2. Topside and Module Arrangement

Analysis Results

Risk based analysis is applied to estimate the risk of all hazard types which FLNG bears on duty. FRA and ERA give DALs specification (Exceedance frequency of 10^{-4}/yr) for the fire and explosion based on the exceedance curves, respectively. QRA includes the fatality of non-process accident types as well as that related to process leaks. Main results are described for each analysis result.

Fire Risk Analysis

Consequence of fire hazards is expressed as jet fire length or pool fire diameter depending on the released phase. The leak locations are assumed at the center of the same module. The one of the fire exceedance curves, for P7, the liquefaction module is shown at Figure 3. The DAL specifications are shown at Figure 4 for the jet and pool fire, respectively. According to results, DALs against the pool fire are not escalated to neighboring module. This is mainly due to the coaming around the modules. On the other hands, the DALs of the jet fire are expected to reach the other modules depending on the modules. This includes P7, S7 and S8. In case of module 7, the induced jet fire length is more than 40 m long after 30 minutes from the initial leak. This is mainly due to the large inventory and additional ESDVs are required.

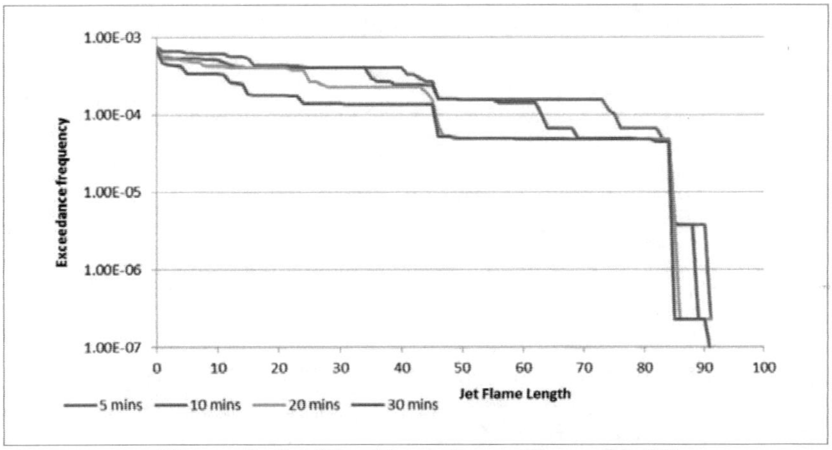

Figure 3. Jet Fire Exceedance Curve for P7

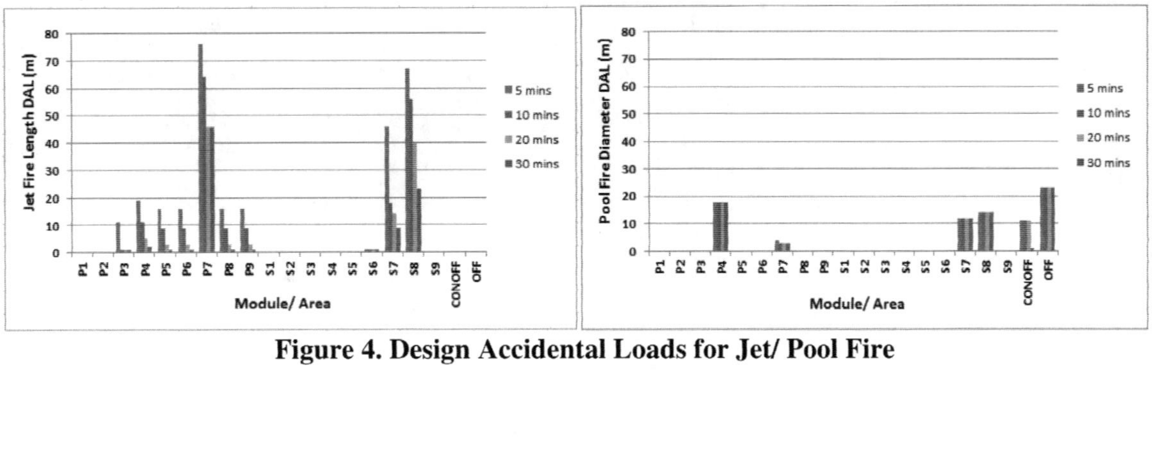

Figure 4. Design Accidental Loads for Jet/ Pool Fire

Explosion Risk Analysis

Explosion is the outcome of the delayed ignition after the flammable atmosphere is formed by the hydrocarbon gas leak. The overpressure can lead to structural damage as well as the fatality. The combined explosion exceedance curve for P7 is shown at Figure 5. Combined exceedance curve sums up the contribution of the overpressure from the neighboring modules as well as the target module itself. Explosion DALs are specified by the area averaged overpressure per 4x4 m^2 and 10x10 m^2 panels. The local DAL (4x4 m^2) can be applied to the large equipment and primary structures. The results of local and global DALs for each module are shown at Figure 6. In terms of local DAL, the highest DAL is generated at module P7. Also, significant DALs occur at P4, P5, P6, P8, P9, S6, S7, S8 and S9, and the modules are related to the fractionation and refrigerant compression except P4. From the results, it is found that the staggered module arrangement and safety gabs ensure good ventilation and also safety gab of 15 m has effectiveness to reduce overpressure less than 10%. The high inventory still requires additional ESDVs to reduce total flammable gas released.

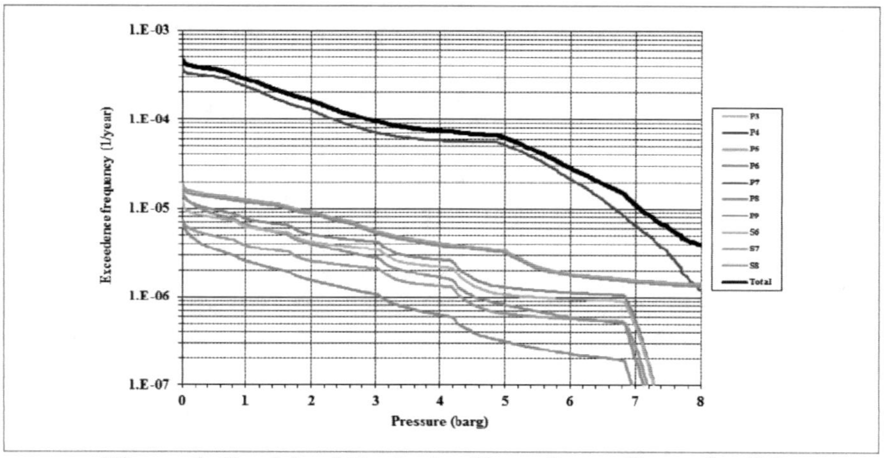

Figure 5. Combined Explosion Exceedance Curve for P7

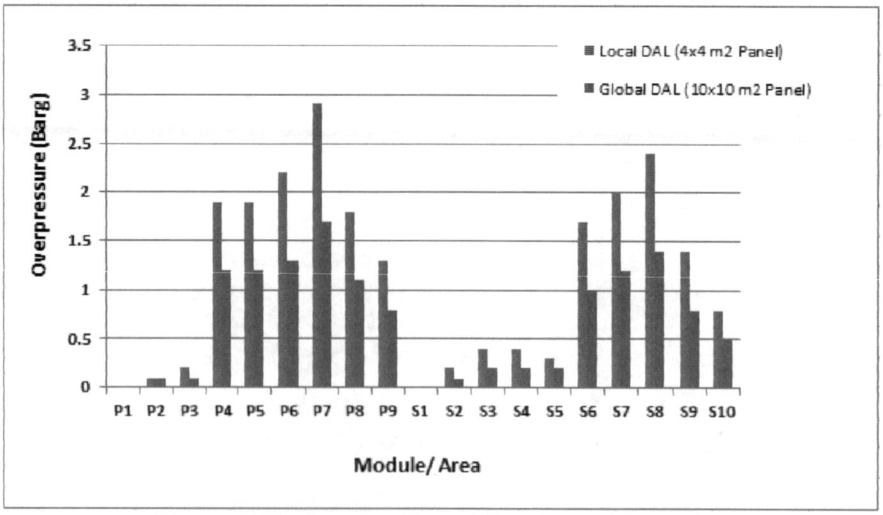

Figure 6. Design Accidental Loads for Explosion

Quantitative Risk Analysis

The safety objectives are personnel, environment and assets. QRA is carried out with the fatality assuming POB (personnel on board) is 125. The IPRA for the FLNG installation as per the categorized working group and the contribution of each hazard type is are provided as shown in Figure 7. The production working group has the highest IPRA of 7.0 x 10^{-3}/yr. However, the all the personnel on board are within tolerable region comparing to safety acceptance criteria. PLL (Potential Loss of Life) is the sum of all individual risk across all personnel at a facility. Overall PLL from the calculation is 1.5 x 10^{-1}/yr for FLNG. This corresponds to one fatality per 6.5 years and is high compared to offshore platform. The reasons for this high value are complex, large processes and piping routings and higher POB for FLNG. Figure 8 shows the relative contribution of each hazard type to overall PLL and hydrocarbon leaks take account of 51.4%. Helicopter transport accidents and ship collision are ranked after the hydrocarbon leaks.

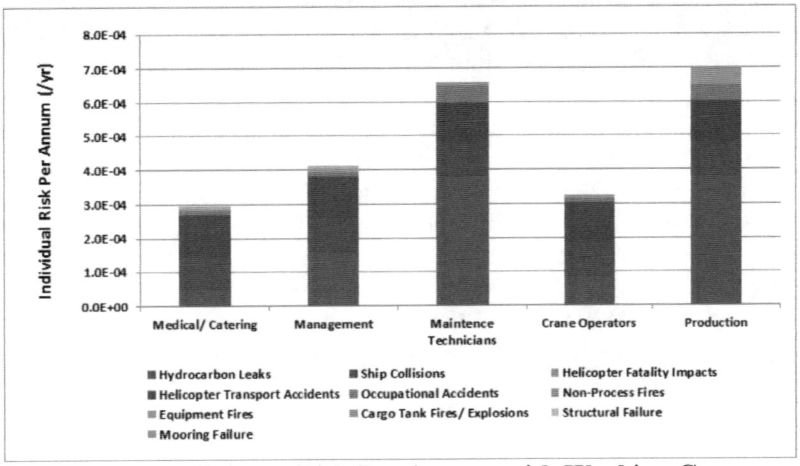

Figure 7. Individual Risk Per Annum with Working Group

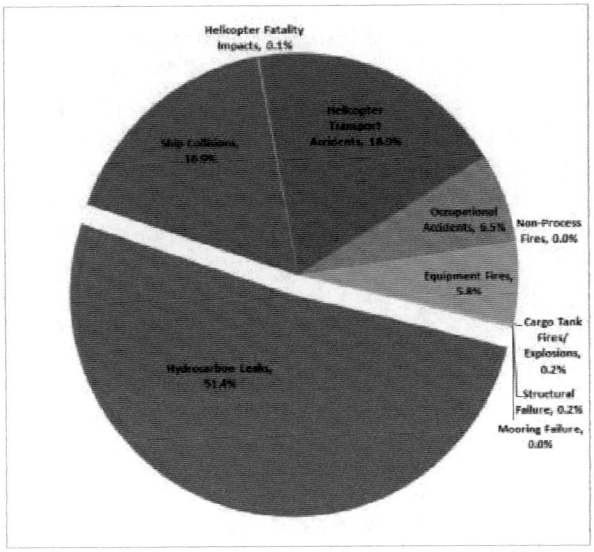

Figure 8. Potential Loss of Life by Hazard Type

Recommendations

The analysis results show the risk level of FLNG against the safety acceptance criteria. Among the hazard type, hydrocarbon leaks have significant contribution to total IPRA. When considering the DAL for fire and explosions, mainly the process modules on the low temperature operation have significant value. The case is similar to personnel risk as production and maintenance group. While safety acceptance criteria for IPRA are satisfied, PLL still is high compared to the existing offshore platform. So, countermeasures should be prescribed to reach the safety engineering within ALARP. Followings are recommendation from the inspection.

- Additional ESDVs: FRA and ERA classifies the most risky modules as P4, P7, S7 and S8, because they have high leak frequency and inventories. One of the ways to reduce risk is to cut off the inventories to the smaller segments by adding ESDVs.
- ESDV shutdown time: Currently, shutdown time of 45s is considered. Reducing shutdown time to 30s can give an opportunity to limit the amount of released flammable gas and hence reduce consequence.
- Manning Policy: Manning distribution is assumed to get assessed for the personnel risk. It is preferred to limit working time in the higher risk ranking area.
- Pipework Minimization: The modules related to LNG production might have at least connection and fully welded pipelines are recommended [4]. The modules of liquefaction and refrigerant compression are based on PFD. Updating the analysis based on the revised data is recommended.
- Others: Hazards from the cryogenic spillage and the turret are not considered. Non-process hazards of ship collision, helicopter transport accident and etc. are based on generic data. It is required to reflect those effects on the safety with an accurate design and data.

Conclusions

Risk based analysis has been applied to the developed FLNG model. Although there is no reference on commercial operation, the risk estimation based on historical experiences broadens the insight on the new installations and risk reduction measures can be drawn at the early design stage. In this aspect, the risk level should be estimated and some risk reduction measures be recommended. More dominant hazard type is hydrocarbon leaks leading to the fire and explosions. These features are induced from more complex process, and so large inventories and part counts. It is expected that additional ESDVs have effect on the risk reduction. Non-process hazard types estimated from the generic data also have significant contribution on the personnel risk. Further update is required with the specific data to reflect the reality. The most important thing is that all of the hazards should be controlled in accordance with ALARP. It is one of the solutions to keep the low inventories and part counts for the installations. Finally, this effort to enhance safety critical items give a deep understanding on FLNG and basis for successful project execution as an EPC contractor.

References

1. Kiil Nam, Jinsang Park, T. J. Kim, Yoon Choon Kim (2012), "Advanced Technology Horizon of FLNG", *AIChE 2012 Spring Meeting,* Huston.

2. Jerome Hocquet, Yann Le Gourrierec, Laurent Paris (2013), "Challenges in Using Risk and Performance Based Design Methods for FLNG Safety Engineering", *17th International Conference and Exhibition on Liquefied Natural Gas (LNG 17),* Huston.
3. NORSOK STANDARD Z-103, "Risk and emergency preparedness analysis".
4. DNV Report (2012), "Failure Frequency Guidance: Process Equipment Leak Frequency Data for Use in QRA".

ADVANCED MODELING FOR CRYOGENIC SPILL RISK ASSESSMENTS OF FLOATING LNG

Filippo Gavelli, GexCon US, Bethesda, MD, USA
Claudio Marsegan, GexCon US, Bethesda, MD, USA

Abstract

The cryogenic spill risk assessment and cryogenic spill protection plan are critical steps in the design of a floating LNG liquefaction facility, given the large quantities of cryogenic materials present and the potential for extensive damage to the facility in the event of cryogenic failure of critical components. However, CSRAs are typically performed using analytical models coupled with somewhat arbitrary assumptions (e.g., artificially capping the hazard distance of a cryogenic jet to take into account the probability of impingement upon a structure). This approach limits the accuracy of the CSRA and does not allow a proper evaluation of cryogenic spill protection and mitigation measures.

This paper will discuss the application of CFD modeling to the analysis of cryogenic spill hazards on FLNGs. The use of the FLACS CFD tool allows an accurate 3D model of the FLNG to be used in the analysis, thus increasing the accuracy of the hazard predictions and removing the need for arbitrary assumptions. Additionally, the effectiveness of cryogenic spill protection and mitigation measures (e.g., curbing critical areas to minimize the pool spreading area, or using shrouded flanges to reduce the momentum of cryogenic jet releases) can be evaluated quantitatively.

Background

In recent years, the LNG industry has been begun the development of facilities to bring to market natural gas from gas fields located in deep waters, particularly in the South Pacific Ocean (i.e., the waters off SouthEast Asia and Australia). Due to the depth and remote location of many of these fields from land, the gas needs to be extracted and liquefied offshore. For these reason, several floating LNG liquefaction facilities (FLNGs) are being designed or built.

The design of an FLNG presents many challenges, associated with installing several process areas (modules) onto the limited footprint of a vessel, which will then be moored at a remote location and be subject to ambient conditions (winds, waves) that can at times be severe. One of the critical elements in the design of an FLNG is the protection of structures and other critical elements from cryogenic damage in the event of an accidental release of LNG or other cryogenic fluid. Given the costs associated with protecting decks and equipment from cryogenic damage, the scope of the cryogenic spill protection plan can have important implications on the overall project cost and construction schedule.

Notwithstanding the above observations, cryogenic spill risk assessments (CSRAs) are often performed using simplified models and assumptions. This can then result in unnecessary costs – if too many conservative assumptions are made – or, even worse, in damage to the facility and extended downtime – if inaccurate assumptions were made or consequence modeling results obtained.

Cryogenic Spill Risk Assessments

The procedure for a CSRA is generally similar to other risk assessments:

1. Equipment and lines are reviewed to identify those relevant to the scope of the CSRA;
2. Leak frequencies, discharge parameters and available inventories are determined;
3. Consequence modeling is performed to determine the extent of the damage area for each release;
4. Risk contours are calculated.

The scope of this paper is to address step 3 in the above list: the assumptions and models used for the consequence modeling.

General assumptions

Several assumptions are typically made in order to define the area affected by a cryogenic release. For example, instantaneous cryogenic damage can be assumed to occur wherever cryogenic liquid (from the spreading of a pool on the deck or from droplets sprayed by a pressurized release) impacts an unprotected structure or piece of equipment. Damage can also occur if a component is exposed to very low temperature vapors for sufficiently long time, as heat transfer to the cold vapors could eventually drop its temperature below critical values (e.g., the threshold for cryogenic damage to carbon steel is typically placed between -30 and -40 °C). A similar set of considerations can be made for impact on facility personnel.

Hazard threshold assumptions are fairly consistent across the industry, although by no means standardized nor generally agreed upon. They are also independent of the consequence modeling tool used.

Modeling assumptions

All other parameters being equal, the consequence model used for the CSRA can have a significant impact on the results of the risk assessment. Frequently, especially at concept or early design stage, analytical models are used to calculate the cryogenic hazard distances for the various release scenarios. These models offer several advantages which make them attractive for studies involving a large number of scenarios, such as: ease of setup, speed of computation, and in some cases coupling with risk calculation packages.

However, the quick setup and calculation come at the price of accuracy: these models are often based on semi-empirical correlations and therefore should be used with caution when the modeled conditions deviate from the range of validation. For example, these correlations do not take into account the presence of obstacles (e.g., piping, equipment, structures and walls) along the path of the release or of the dispersing vapor cloud. Given the high congestion levels in FLNG modules, the assumption of unobstructed jet or cloud dispersion should be evaluated very carefully as it could result in much longer hazard distances than could be possible – especially if, for example, the calculations were to show the hazard extending across a wall.

In order to eliminate some of the non-physical results that can be produced from analytical models, a frequent solution in CSRAs is to set a cap to the hazard distances. Even though this may resolve some specific problem, it is hardly a scientific approach and would have to be evaluated very carefully to ensure that it does not create problems elsewhere in the analysis.

Another example of a severe limitation of certain analytical models is in the treatment of downward releases of cryogenic liquids from pressurized components. The physical phenomenon for this type of scenario is as follows:

- Upon discharge to the atmosphere, a certain fraction of the liquid will flash to vapor, while the rest will be sprayed as an aerosol. The flash fraction will depend upon the degree of superheat of the pressurized liquid in the system;
- As the aerosol travels downward towards the first solid surface (e.g., the main deck), the jet entrains air. Heat transfer between the air and the aerosol leads to progressive evaporation of the liquid droplets, reducing the liquid mass flow;
- For a given release, the amount of liquid the reaches the deck and deposits on its surface (often referred to as "rainout") will depend upon the distance traveled by the spray – that is, the elevation of the release above the deck. If the release location is high enough, or if the discharge has a high degree of superheat, it is possible for all the liquid to evaporate before reaching the deck, resulting in zero rainout.

The PHAST tool, which is often used for this type of studies, is unable to capture this physical phenomenon correctly. In fact, for downward releases of flashing liquids, PHAST neglects the vaporization of the liquid as it travels from the release location to the deck. Instead, PHAST assumes that any liquid that does not immediately flash will rainout and form a pool on the deck. This limitation can result in significant errors in the calculation of the cryogenic hazards from this type of releases – in fact, considering that module decks are often grated, a downward release could travel a significant distance before impacting a solid surface.

These and other limitations of analytical models can be eliminated by using a more detailed modeling approach based on Computational Fluid Dynamics (CFD). CFD models, in fact, do not rely upon correlations to predict the shape and concentration of gas clouds as a function of distance from the release location; instead, CFD models solve the equations of fluid motion throughout the computational domain, which can include the entire FLNG or just a few modules, depending on the characteristics of the release being evaluated. Therefore, the released fluid can interact with the many obstacles and obstructions present in an FLNG, as well as with the complex wind flow patterns that occur inside the modules. This eliminates the need for arbitrary assumptions on the maximum distance traveled by a jet release, or the maximum size of a liquid pool, etc.

The modeling of cryogenic liquid releases requires a CFD tool capable of handling two-phase flows, including the heat and mass transfer between the liquid and vapor phases and the ambient fluid (air). The FLACS Homogeneous Equilibrium Model (HEM) is such example: it combines the ability of the FLACS solver to handle the complex geometries typical of FLNGs (as shown in Figure 1) with the capability of modeling two-phase flows and liquid spills as required for CSRA consequence modeling.

Figure 1. Geometry model of an FLNG.

Case Study

The cryogenic hazard areas predicted by a CFD model (FLACS) and an analytical model (PHAST) are compared for an LNG release from process piping. The release location is shown in Figure 2. The release is assumed to occur from a 10-mm diameter hole approximately 2 m above the deck, and to be directed downward as shown in Figure 2. The LNG is assumed to be at approximately 0.8 barg and -154 °C.

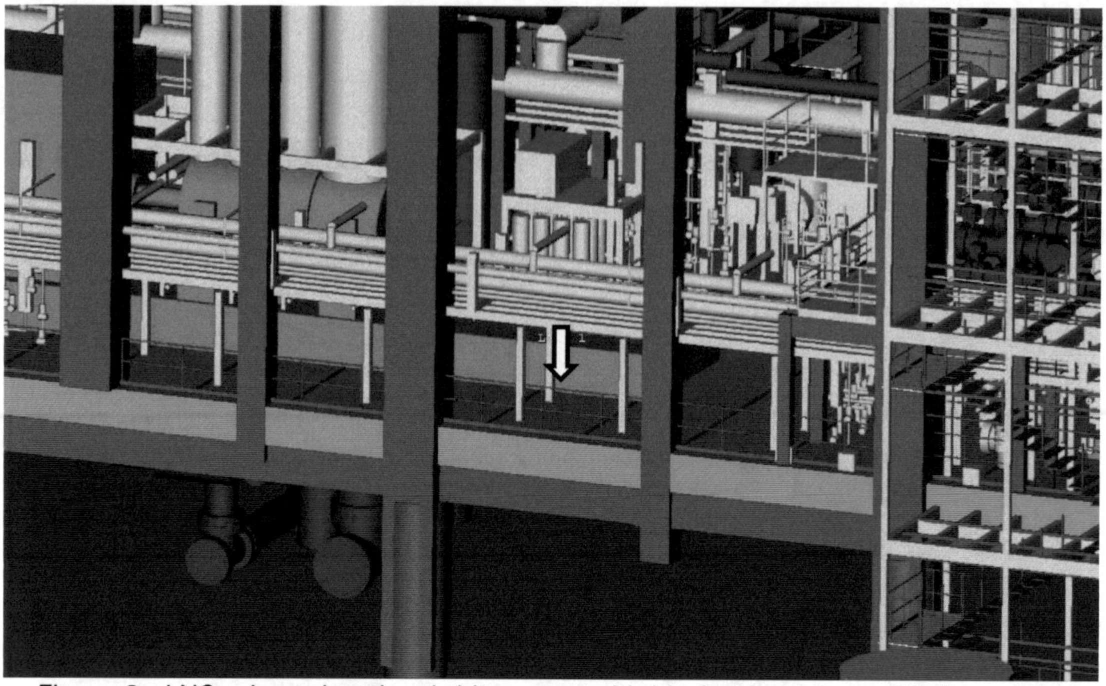

Figure 2. LNG release location (white arrow with black edges) in an FLNG module.

The PHAST tool calculates a rainout fraction of approximately 95% for this release; as discussed above, this result is independent of the elevation of the release above the deck. The FLACS-HEM simulation, instead, shows that a significant amount of the aerosol droplets will evaporate as they travel downwards; the rainout fraction calculated using this method is approximately 53%. For this scenario, therefore, PHAST will predict a much smaller vapor cloud from the pressurized release, but a much larger pool on the deck. Additionally, as shown in Figure 3, the liquid droplets and the cold vapors from this release will interact with the congested geometry of an FLNG module, resulting in complex flow patterns that will likely affect structures and equipment in a manner that cannot be accurately predicted by a simple teardrop shape.

Figure 3. Cryogenic hazard region for pressurized LNG release (defined by the -40 °C temperature threshold).

Recommendations

The results obtained from the Cryogenic Spill Risk Analysis can have a significant impact on the cost and construction schedule for an FLNG, as they determine which areas (including structures and equipment) require cryogenic spill protection. Therefore, the use of simple modeling tools – with their faster computing times but more onerous simplifying assumptions – may not be as cost-effective as commonly assumed, compared with the mode complex but more accurate CFD models.

Fortunately, the two types of models are not exclusive of each other. In fact, the most efficient approach to CSRA consequence modeling can be obtained through a combination of analytical and CFD modeling. A large fraction of the scenarios to be considered in a CSRA tend to be "small", meaning that the hazard areas are limited (e.g., on the order of 1-2 m) and therefore the effect of obstacles and obstructions is likely minimal: for these scenarios, simple modeling tools allow rapid calculation without appreciable loss of accuracy. The CFD models can then be used to evaluate the larger scenarios, which are likely to be affected by the 3D environment.

GexCon's experience with this "hybrid" approach is that it allows extensive studies to be performed in a reasonable time, and that the added accuracy provided by the use of CFD models largely outweighs the increased time and cost of the analysis.

LNG Plant Inlet Facility - Design, Safety & Operability

Haribabu Chittibabu
Bechtel Oil, Gas & Chemicals, Inc.
3000 Post Oak Blvd., Houston, TX
hchittib@bechtel.com

Dipanjan Bhattacharya
Bechtel Oil, Gas & Chemicals, Inc.
3000 Post Oak Blvd., Houston, TX
dbhatta1@bechtel.com

Author Biography

Haribabu Chittibabu is a Senior Process Engineer in Bechtel's Advanced Simulation and Analysis group (Process Engineering). Haribabu's fields of expertise include process design, dynamic simulation analysis, optimization studies and development of Operator Training Simulator for Refinery, Gas plant and LNG plants. Currently Haribabu is responsible for the technical execution of dynamic simulation studies for Bechtel Projects. He has a bachelor's degree in chemical engineering from the University of Madras, India and a Master's in Petroleum Refining and Petrochemicals from Anna University, India.

Dipanjan Bhattacharya is a Senior Engineering Specialist in Bechtel's Advanced Simulation and Analysis group (Process Engineering). Dipanjan's fields of expertise include simulation, LNG system design, refining operations and design, start-up & commissioning and process controls. Prior to Bechtel he has worked with Honeywell and Bharat Petroleum Corporation. Dipanjan holds a Bachelor of Science in Chemical Engineering from Jadavpur University, India and a Master of Science in Chemical Engineering from University of Oklahoma

Key words

LN, Dynamic simulation, Inlet Facility, Onshore, HIPS, engineering and design.

Abstract

In Onshore gas processing Inlet Facility of LNG or Gas Plant the feed gas or mixture of gas & condensate are received by means of high pressure pipelines. The condensate or heavies if any are separated and feed gas is spilt into multiple trains for processing. The capacities of these trains and the number are based on project specific considerations, available gas supply and future sales demand.

In each train the pressure is letdown in single or multiple pressure control valves (PCV), turbo expanders or combination of these depending on the process involved for heavy removal, gas to liquid conversion or export of sales gas. The train through put is controlled by a combination of pressure and flow control. The design and configuration of inlet facility in gas processing is important for operation flexibilities between multiple trains, reducing the relief load and design pressure of the downstream facility which impacts the Capex of LNG plant and its off-sites.

Dynamic simulation of inlet facility helps to solve key engineering problems with safer design and operation flexibility. In recent years, there is an increase in applying dynamic simulation during the engineering phase of a project in addition to its use during plant startup, shutdown, revamp and troubleshooting. There are various aspects in the design of Inlet facility for LNG or gas plant. This paper discusses about an inlet facility configuration of LNG plant, its operating mode and how dynamic simulation is applied in engineering the inlet facility with respect to safety protection, controllability and operation flexibility.

Inlet facility configuration

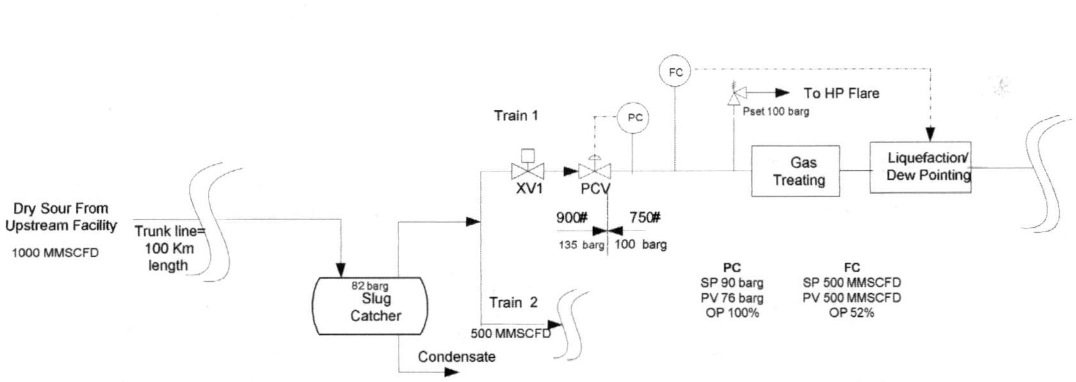

Figure 1: Inlet Facility – Normal Operation

The dry sour gas from upstream facility is delivered to LNG plants via pipelines called as trunklines. The function of the trunkline is to transport gas from the offshore field to the LNG Plant. The trunkline can be operated wet, dry or wet with MEG injection and are few kilometers long. Under normal operations, the gas is dry and free of particulate matter. However, the design of the inlet facilities anticipates liquid slugs and some level of solids which are separated in the slug catcher. The condensate if any are separated and feed gas is spilt into two trains for processing.

The gas separated out of slug catcher split into two identical LNG trains as shown in figure 1 and flows through pressure control station that prevents over pressuring the equipment in the train if a pressure higher than normal delivery feed gas pressure is available from the pipeline. In each LNG train the pressure is letdown in single pressure control valves (PCV) to letdown the pressure and a flow controller (FC) to regulate the flow through the train. FC is provided for the operator to adjust the plant feed rate based on refrigeration compressor capacity and operating mode.

Operating Modes:

- Normal Operating Mode is when both LNG trains running fully utilizing the maximum gas rate. There will be no packing of trunklines line in this mode as the gas utilization rates will be equal to the export rate from the upstream facilities. During normal operating mode when the feed gas pressure at Slug Catcher is 82 barg, the pressure control valve PCV will be fully open to take minimum pressure drop across valves as shown in figure 1.
- Subnormal operating mode or packed line mode is when the downstream gas processing facility has lost one of LNG train. As a consequence there will be packing in the trunkline and the slug catcher as the gas utilization rates are less than the export rate from the upstream facilities. During packed line mode operation when feed gas pressure at Inlet Slug Catcher is higher than 82 barg, pressure control valve will modulate to reduce pressure of feed gas to each LNG train as shown in figure 2.

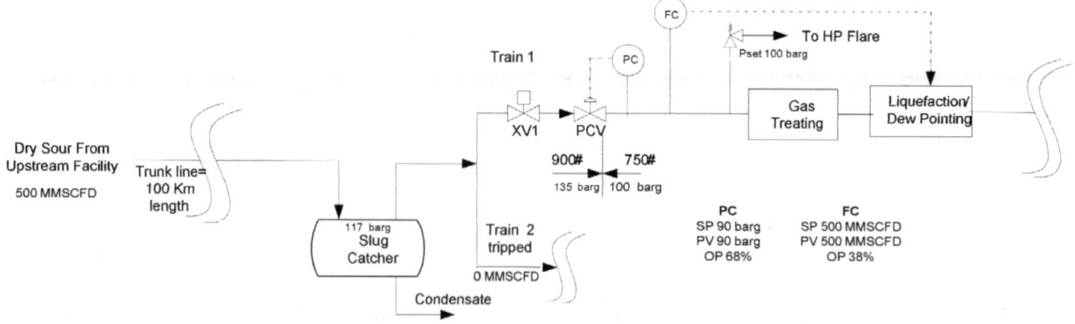

Figure 2: Inlet Facility – Packed line mode of operation

Safety Protection

Generally there is a design pressure rating change downstream of PCV in order to reduce the design pressure of the gas treating unit and associated piping. The PCV is a resistance which modulates to reduce the pressure of feed gas in packed mode operation. Any blocked outlet condition downstream of PCV or when PCV fail open, will pressurize the train and calls for protection. General engineering practice is to provide a relief valve and size for the maximum train capacity. Generally the relief event is dynamic in nature. The relief rates are higher in the beginning and decreases as the system pressure settles out as in the case of PCV fail open scenario. This is not the case in the event of blocked outlet downstream of PCV as the relief event is continuous. These scenarios can be studied by means of dynamic simulation and the safety of the inlet facility can be verified. Figure 3 shows the gas flow rate and the pressure increase through the train when PCV fail opens and when pipeline and slug catcher operating in packed mode operation.

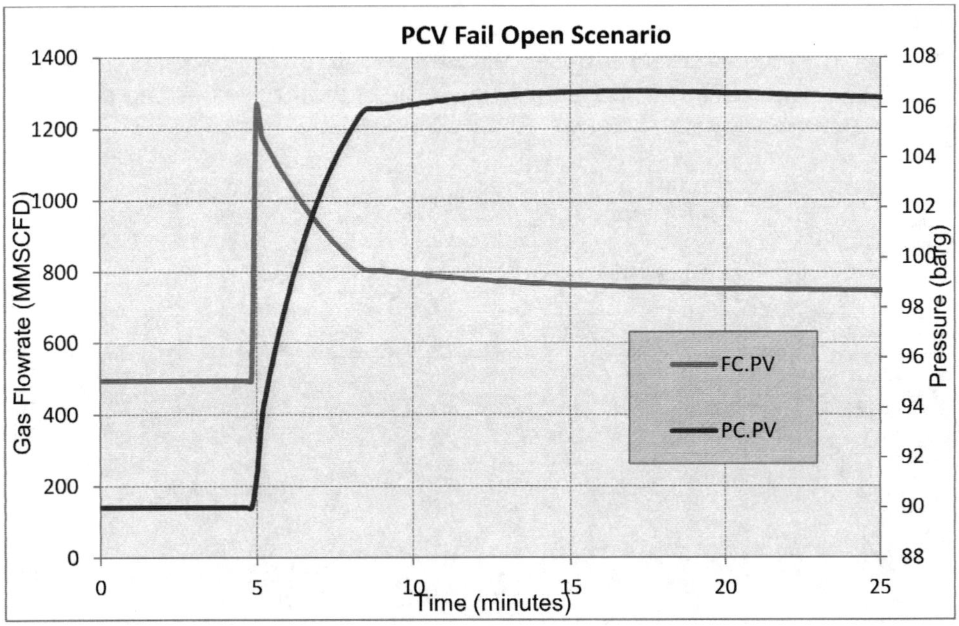

Figure 3: PCV Fail Opens Scenario @ time t = 5 minute

When PCV fail open immediately in the initial few minutes the gas flow through the train is about 1200 MMSCFD. The pressure at the outlet of PCV is 106 barg which is also above the design pressure of Gas Treating unit. The train flow rate has been increased from 500 MMSCFD to 1200 MMSCFD momentarily. Question arises

- Whether Gas treating unit and downstream liquefaction unit can handle this increase in flow rate or will it be tripped?
- Whether to relieve the entire 1200 MMSCFD to flare and its impact on capital cost?
- Whether to relieve the difference 700 MMSCFD to flare assuming downstream can handle only 500 MMSCFD?
- Whether to mitigate the scenario by utilizing Safety Instrumented System (SIS) owing to huge relief load and reducing the flare capacity?

There is various design option to handle this scenario. But it is not preferred to design a flare which could relieve the entire 1200 MMSCFD owing to the capital cost unless you have an existing flare with this capacity. Instead install a relief valve set at design pressure of 100 barg which relieves 700 MMSCFD as the train can handle the capacity of 500 MMSCFD. Even the increase in gas flow rate is momentary for a period of 5 minutes till the pressure equalizes.

Figure 4 shows blocked outlet relief load when any valve downstream of PCV are closed. The relief valve sized for the PCV full open scenario is sufficient for blocked outlet scenario also. In general blocked outlet volumetric rates are lesser than normal flow rates through the train because of the increase in density of gases as the pressure increases and they are continuous.

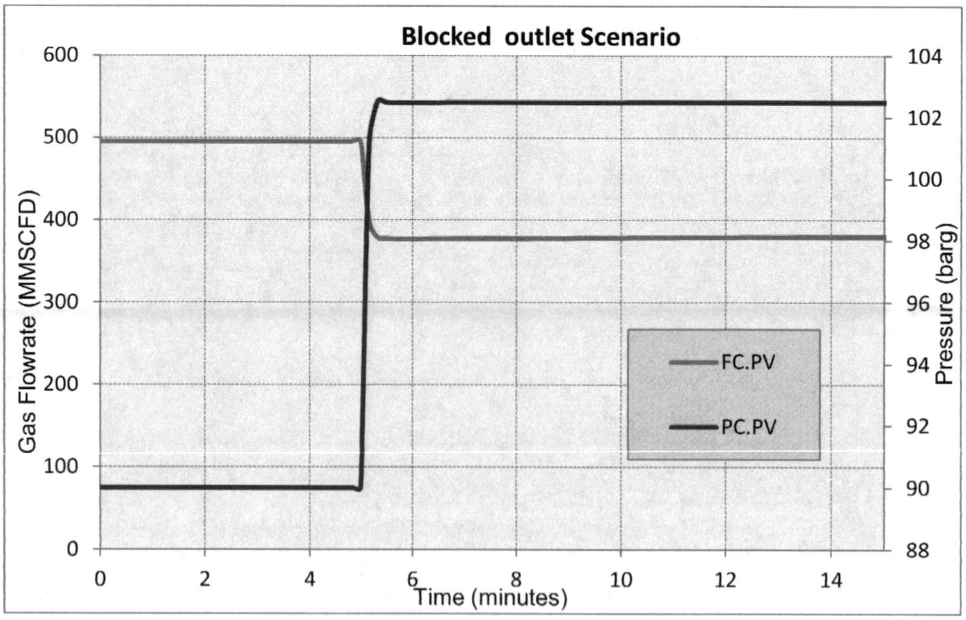

Figure 4: Blocked Outlet Scenario @ time t= 5 minute

One more design option is to mitigate this scenario by installation of HIPS valves.

High Integrity Protection Systems (HIPS):

In conventional designs, pressure relief devices are used as the primary means of pressure protection. The design of each pressure relief device is based on the assessment of overpressure scenarios such as blocked outlets and letdown valve fail open with respect to inlet facilities. In many countries, there is increased pressure from community and regulatory authorities to reduce venting and combustion of gases. In cases it may be unacceptable to flare large volumes of gas. Safety and environmental requirements have resulted in increased focus on using an alternative approach to pressure protection.

API 521 and Code Case 2211 of ASME Section VIII provide an alternative to pressure relief devices – the use of an instrumented system to protect against overpressure. When used, this instrumented system must meet or exceed the protection provided by the pressure relief device. These instrumented systems are safety instrumented systems (SIS), since their failure can result in the release of hazardous chemicals and creation of unsafe working conditions. The risk typically involved with overpressure protection results in the need for high SIS integrity; therefore, these systems are often called High Integrity Protection Systems (HIPS).

In this inlet facility configuration the representation of HIPS is shown in figure 5. The system includes redundancy with three measuring elements (PT A, PT B, and PT C), control back-up, high reliability components and intelligent software. High High pressure downstream of PCV is used to trip HIPS valves (XV1 and XV2). 2003 pressure transmitter downstream of PCV is input to HIPPS.

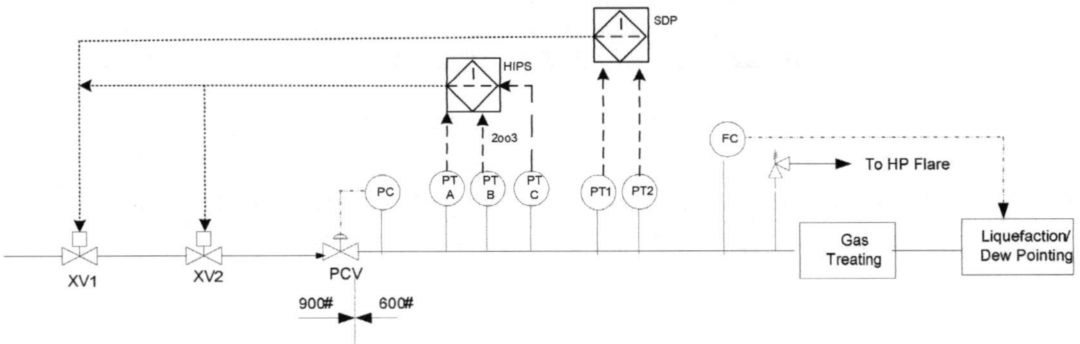

Figure 5: HIPS representation

The design and implementation of HIPS requires special procedure and particular attention during its operational life such as maintenance, testing and inspection. For these reasons, the decision to implement HIPS should be made with a great deal of caution and careful consideration.

Further dynamic model can be extended to finalize High High Pressure trip set point, HIPS valve closing times and their response. In PCV full open scenario when the PT A/B/C reached 97 barg, the HIPS valves (XV1 & XV2) were automatically set to close with a closing time of 5 seconds. Figure 6 shows the pressure at the outlet of PCV after HIPS implementation for full open scenario. The HIPS valve closing time was sufficient as the pressure downstream of PCV did not reach the design pressure.

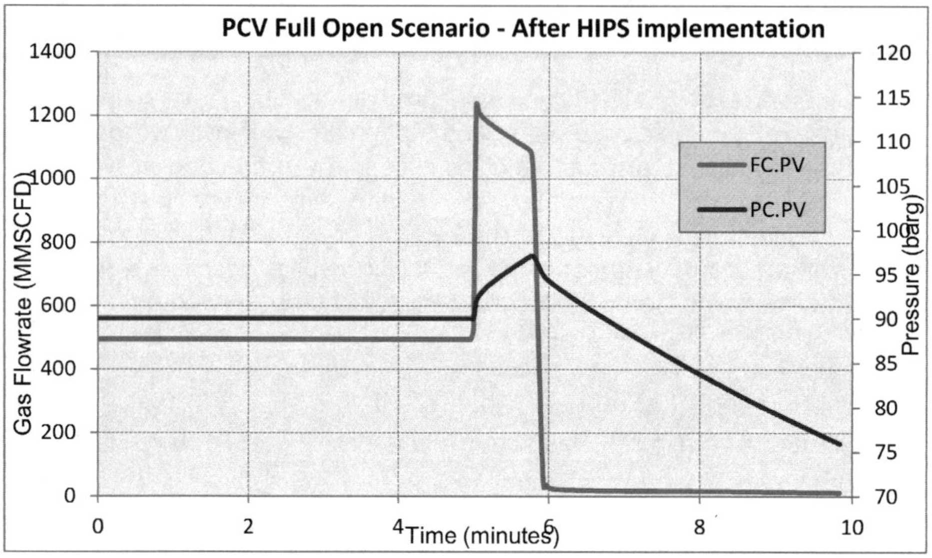

Figure 6: PCV Fail Opens Scenario @ time t = 5 minute – HIPS implementation.

Control System Verification and Operation flexibility

Multiple control configurations arise during design phase of a project and multiple trains make the control system more complex owing to various operating scenarios of the plant. Dynamic behavior and controllability of pressure and flow controllers can be evaluated for various transient scenarios like

- Shut down of one LNG train and its impact on the other train
- Loss of pipeline pressure

Shut down of one LNG train:

In some inlet facilities the feed gas is sourced from a single pipeline or several pipelines which gets dry feed gas from one or more upstream gas treating units (GTU). GTU's removes water from wet gas obtained from one or more wells usually referred as Well

Heads. Any disturbance in inlet facility will affect upstream operation and vice versa. For example shut down of a train will pressurize the pipeline as the gas utilization rate is less than the export rate. Due to the physical or larger volume of the pipeline the rate of pressure increase may be dampened. Dynamic simulation can provide the increase in pipeline pressure profile with respect to time which helps to decide when to trip the Wells or turn down the GTU's.

In this Inlet facility configuration when one train is tripped, gas utilization rate is reduced to half, with export rate from upstream facility continuing at same capacity. The feed gas pipeline and Slug Catcher will get packed due to imbalance between gas export and utilization. Flow through Train 1 increases momentarily and FC throttles to stabilize the Train 1 flow at its set point as shown in figure 7. Simultaneously, PC also throttles to maintain its downstream pressure at 90 barg. Pressure at the Slug Catcher is estimated to increase to 122 barg in about 6 hours which is trip set point of the upstream facility as shown in figure 8. The export rate has to be reduced to match utilization before that time period.

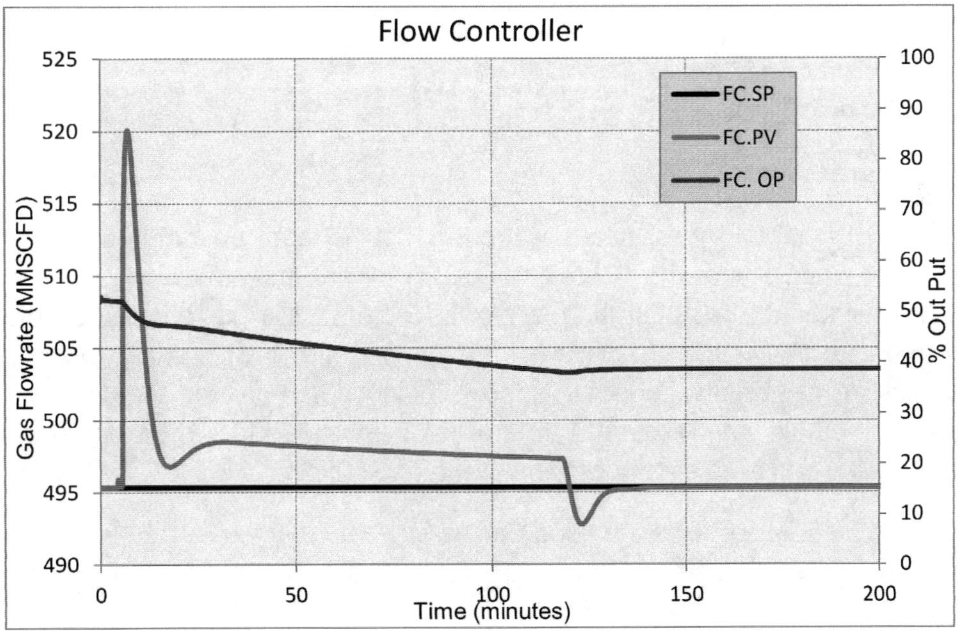

Figure 6: Shut down of one LNG Train @ time t =5 minute – FC response.

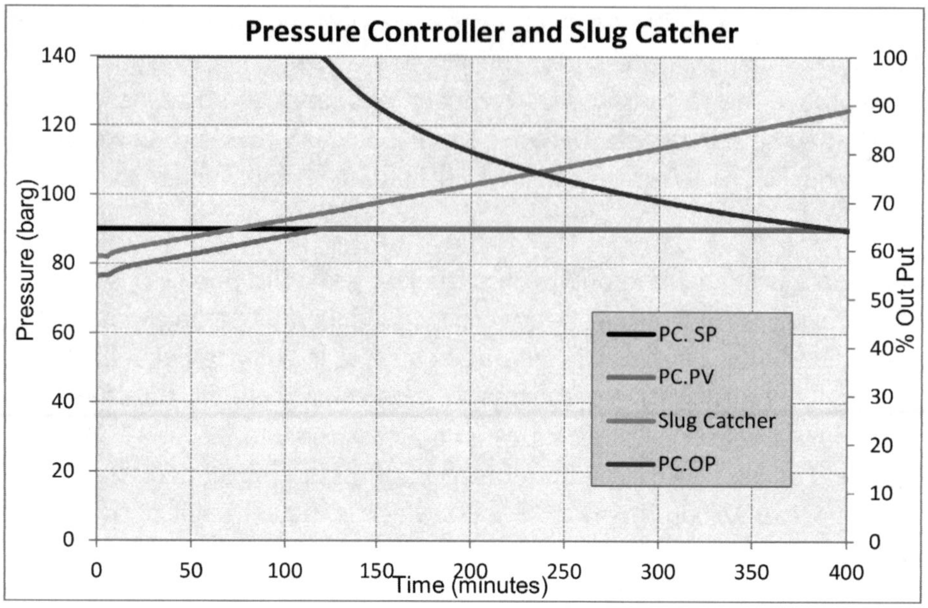

Figure 7: Shut down of one LNG Train @ time t =5 minute – PC response

On the other hand, dynamic analysis can be carried out for sudden loss in pipeline pressure when few wells get tripped.

Loss of pipeline pressure:

Any disturbance in the GTU or well heads will affect downstream operation and vice versa. For example when one of the GTU's is tripped or during turn down the pipeline pressure will go down as the gas utilization rate of LNG trains is more than the export rate from upstream facility. Due to the physical or larger volume of the pipeline the rate of pressure decrease may be dampened. Dynamic simulation can provide the decrease in pipeline pressure profile with respect to time which helps to decide when to reduce the feed rate of LNG trains or to shut down one of the LNG trains.

Figure 8 shows the feed gas pipeline and Slug Catcher pressure decreases due to imbalance between gas export and utilization rate. Figure 9 shows continuous decrease in flow through Train 1. Pressure at the Slug Catcher is estimated to decrease to 70 barg in about 2 hours. The gas utilization rate has to be reduced before that time period.

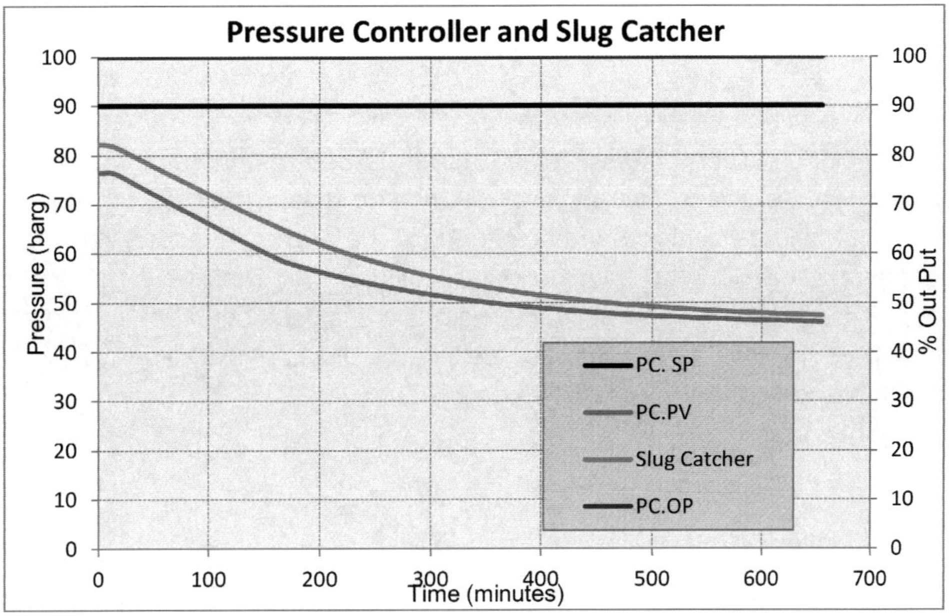

Figure 8: Loss of pipeline pressure @ time t =5 minute – PC response

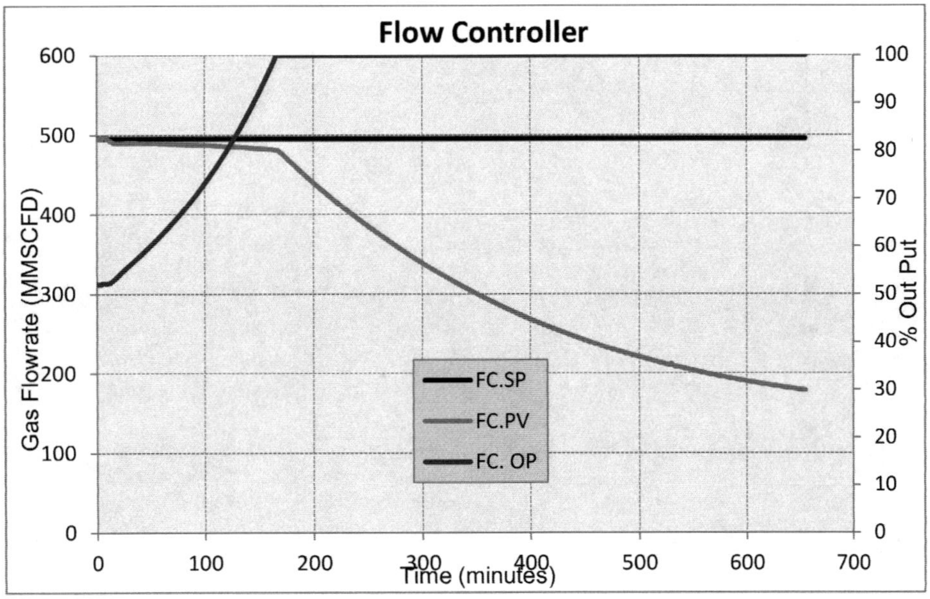

Figure 9: Loss of pipeline pressure @ time t =5 minute – FC response

Conclusion

Inlet facility acts as a sandwich between upstream facility (Well heads & GTU) and downstream facility (Liquefaction or Dew Pointing) and its design and configuration is important for safe and efficient operation of the entire gas processing complex. Dynamic simulation should be deployed in engineering stage of the project for complex Inlet facility configuration which processes multiple trains and each train having multiple letdowns valves. . The versatility of dynamic models and the unique operational insights they offer can help engineers design safe and sustainable processing facilities, as well as provide measurable value in terms of cost savings.

DESIGN CONSIDERATIONS FOR SMALL-SCALE LNG FACILITIES

Jenna M Wilson
CH·IV International
1341A Ashton Road
Hanover, MD 21076
jmwilson@ch-iv.com

Phil J Suter
CH·IV International
1341A Ashton Road
Hanover, MD 21076
pjsuter@ch-iv.com

Filippo Gavelli
GexCon US
4833 Rugby Avenue, Suite 100
Bethesda, MD 20814
fgavelli@gexcon.com

Abstract

With the influx of natural gas production throughout the United States, the small-scale LNG market is growing at a rapid pace. From LNG production and trucking facilities in the central plains to bunkering vessels along the Pacific and Gulf coasts, the jurisdictional requirements for designing and siting such facilities in a safe and practical way vary with each individual operational purpose, infrastructure connection and location.

Proper design of a facility concept will start with the application of good engineering practice and relevant standards followed by preliminary siting and feasibility studies to determine whether a particular parcel of land is appropriate for the desired project. While the applicable design codes for safe design are generally clear, the appropriate requirements and approved models for siting are often less clearly defined and can vary depending on jurisdictional responsibility.

In addition to potential design code differences based on jurisdictional responsibility, the economics and budget for such projects are likely to be considerably different. Therefore, considerations must be included to develop the safest design while providing a project within an appropriate budget.

Introduction

There are many different LNG facilities currently in operation, under construction or still in feasibility stages. LNG facilities providing fuel for trucks and even locomotives is an increasingly appealing industry given the inconsistent crude or diesel pricing indexes associated with production vulnerabilities. With the US becoming a market flooded with natural gas, we are currently in a time where operating vehicles off natural gas is viewed as increasingly smart and popular.

LNG as bunkering fuel for barges, offshore supply vehicles or ships is also becoming increasingly popular for similar reasons as LNG trucking. In addition, recent Environmental Protection Agency (EPA) rulings have enacted more stringent NOx and SOx limits for ships in emission control areas surrounding the continental United States and Hawaii.

Facility Design Codes

During the very early stages of any facility design, the developer must first determine the Authority Having Jurisdiction (AHJ) and the required design codes for the facility. Currently, the applicable jurisdiction for a facility can significantly impact the siting requirements and associated plant footprint to demonstrate control. A large-scale LNG liquefaction and export facility typically could fall under different requirements for permitting than a satellite LNG bunkering facility.

Any facility will fall under the jurisdiction of the Department of Transportation (DOT) Pipeline and Hazardous Material Safety Administration (PHMSA) 49 Code of Federal Regulations (CFR) Part 193 if the facility is connected to a natural gas pipeline that is governed by Part 192. A caveat to the Part 193 governance is if a facility is the end consumer of the LNG which is brought into the facility or if the facility is located in navigable water. For example, if an import terminal was connected only to an on-site co-generation power plant, and no natural gas was sent off-site, the facility would not fall under the jurisdiction of DOT PHMSA.

For a facility which is determined to be non-jurisdictional to DOT PHMSA, a local or state agency will typically be the authority having jurisdiction. Those local agencies may be the state Public Utilities Commission, Fire Marshall or Environmental Department.

The Federal Energy Regulatory Commission (FERC) is responsible for authorizing the siting and construction of onshore and near-shore LNG import or export facilities under Section 3 of the Natural Gas Act whereas the Coast Guard is responsible for off-shore LNG import or export facilities. All of the land-based LNG import and export facilities in the US (current, proposed or under construction) are under both FERC and DOT PHMSA jurisdiction.

When a facility falls under the jurisdiction of DOT PHMSA, the project must be built in compliance with 49 CFR Part 193, the Federal Safety Standards for LNG Facilities. 49 CFR Part 193 incorporates by reference the National Fire Protection Association's 59A- Standard for LNG facilities (2001 edition). It must be noted that the current edition of 59A was published in 2013 and differs from the 2001 edition in several aspects, as it reflects the many technical improvements made in the LNG industry over the past 12 years.

Implications of DOT PHMSA Jurisdiction

In the event that a small-scale facility is determined to be under the jurisdiction of DOT PHMSA, the siting requirements would be significantly different from those of an NFPA 59A-only facility. Among the most notable differences, are the requirements for the determination of vapor dispersion exclusion zones.

As incorporated by reference in 49 CFR Part 193, NFPA 59A (2001) Chapter 2 requires the exclusion zones for flammable vapor dispersion be determined from vapor dispersion associated with LNG tank impoundments and design spills; it also requires the DEGADIS or FEM3A models to be used for this purpose.

Following the publication of the March 2009 report, "LNG Source Term Models for Hazard Analysis: A review of the State-of-the-Art and an Approach to Model Assessment", DOT PHMSA issued a written interpretation requiring that "in addition to analyzing the effects of vapor dispersion from LNG pool spills, the effects of "jetting and flashing" releases must also be considered in calculating the vapor gas dispersion exclusion zone[1]".

Although the DEGADIS model can calculate vapor dispersion distances from LNG pool spills, it cannot calculate dispersion resulting from a jet release. Following petitions by the developers and a thorough review of their validation against the Model Evaluation Protocol, in October 2011, DOT PHMSA approved the use of two new models (PHAST v6.6 and 6.7 developed by Det Norske Veritas - DNV - of Oslo, Norway, and FLACS v9.1 developed by Gexcon AS of Bergen, Norway) to calculate vapor dispersion exclusion zone distances. The new approved models are able to model jet releases as well as liquid spills, effectively replacing DEGADIS as the modeling tool for LNG facility siting.

A combination of the PHAST and FLACS software tools can be used to provide a preliminary indication of whether or not a piece of real estate being considered for an LNG facility covered under 49 CFR 193 will likely be sufficient to accommodate the exclusion zones. Due to the high momentum of the release, jetting and flashing scenarios have become the typical bounding case for modern facility siting, supplanting the thermal radiation hazards associated with a tank impoundment area fire.

For example, the PHAST model can be used as an initial screening tool to determine the unmitigated, unobstructed dispersion distances for the various flashing and jetting scenarios. Since PHAST cannot account for the presence of obstacles or obstruction to the air/vapor flow, these results can be often considered conservative and thus present a likely upper bound to the real estate requirements for the facility.

---

[1] July 16, 2010 letter from DOT PHMSA responding to request for written interpretation on the applicability of 49 CFR 193 to proposed LNG import terminal in Robbinston, Maine.

The results of the PHAST analysis are independent of wind or release direction and therefore the calculated dispersion distances can be easily plotted on a property survey of the facility that illustrates the general location of all major equipment, including marine facilities, LNG and refrigerant storage, regasification, liquefaction and LNG or refrigerant truck loading. The resulting plots will:

- Illustrate potential problem areas that may need to be mitigated using passive mitigation techniques (such as vapor fences or walls) during refined analysis,
- Help in making changes to the preliminary layout of equipment, and
- Illustrate problem areas that may require further consideration with respect to basis of the facility design and additional property requirements.

The computational fluid dynamics (CFD) model FLACS can then be used to solve the problems identified by the PHAST modeling, such as:

- Obtain more accurate results for critical scenarios (e.g., scenarios that extend beyond a property line that can be built upon when using PHAST and/or could be significantly affected by equipment, buildings or other obstructions);
- Evaluate the placement and effectiveness of mitigation measures (e.g., vapor barriers);
- Evaluate the exclusion zones for liquid spills into trenches and troughs (for which PHAST is not approved).

Although not exclusion zones as defined by 49 CFR Part 193, the PHAST and FLACS modeling tools can also evaluate the other hazards associated with an LNG facility, such as:

- Flammable refrigerant dispersion hazards;
- Overpressures associated with ignition of flammable refrigerants; and
- Hazards associated with toxic components.

This combination of a screening tool (PHAST) and a more accurate model (FLACS) has been applied successfully by CH·IV and GexCon to several LNG facilities under PHMSA jurisdiction – both large-scale export or bi-directional projects and smaller-scale LNG facilities.

DOT PHMSA Letters of Interpretation

Letters of code interpretations have been issued by DOT PHMSA describing individual cases which segregate a facility from falling under the jurisdiction of DOT PHMSA. Those cases include two small scale liquefaction facilities. In May 1998, Williams Field Services requested a formal interpretation for the Ignacio LNG production facility. This plant provides compression, natural gas conditioning, liquid recovery, and LNG production, storage and truck out equipment. The facility is fed from a rural gathering system which is not subject to pipeline safety laws and 49 CFR Part 192. Therefore, the facility was determined to be non-jurisdictional by DOT PHMSA.

A second case, issued in April 1998, determined that a facility which received gas from an interstate pipeline for use in an electric generation turbine and liquefaction plant for sale in a trucking station is under the jurisdiction of Part 193 because it receives gas from a Part 192 regulated pipeline.

The background of previous cases helps to provide developers an understanding of jurisdictional responsibility for their concept facilities. Further legal council should be sought for any facility, early in the development stages.

Non-DOT PHMSA Jurisdictional Facilities

For facilities deemed non-DOT PHMSA Jurisdictional, the design and siting of the facility may be based on NFPA 59A 2013.

Chapter 5 of NFPA 59A 2013 provides plant siting and layout standards for non-DOT PHMSA jurisdictional facilities. However, for the installation, design, fabrication, and siting of LNG facilities using containers of 100,000 U.S. gal (379 m^3) water capacity and less constructed in accordance with the ASME *Boiler and Pressure Vessel Code*, and maximum aggregate storage capacity of the facility of 280,000 U.S. gal (1,060 m^3) water capacity, may apply the requirements of NFPA 59A 2013 Chapter 13. This chapter also provides minimum equipment spacing requirements for facilities, which provides the basis for facility siting.

Mitigation Measures for Small-Scale Facilities

The design constrains are often significantly different between a multi-billion dollar world-class export project and a 100,000 gallon per day liquefaction plant. Typically, the smaller bunkering-type facilities need to fit on small properties, in industrial or port areas, and surrounded by other plants or facilities. Therefore, the option to expand with acquiring additional land to accommodate large exclusion zones becomes increasingly complex. For these facilities, budgets also tend to be much more limited and therefore the cost of installing mitigation measures can be prohibitive.

In these cases, an accurate determination of the exclusion zones is critical to determine the viability of a project. The screening calculations using PHAST may not suffice for maintaining control of the property encompasses within the exclusion zones. For example, a flashing and jetting release from a 1-inch hole in an LNG truck loading supply line at 40 psig will travel approximately 510 feet, which may exceed the distance available in many satellite LNG facilities. If the exclusion zones extend offsite, passive mitigation measures can be implemented to further reduce the cloud migration and retain the ½-LFL vapor cloud onsite. The most common mitigation measures, due to their versatility, are vapor barriers which:

- Can be placed in most locations (along the property line, close to the release location, or anywhere in between) as long as they do not interfere with facility operations;
- Can be built in a variety of sizes (height, width) and materials (plastic, metal, masonry), to accommodate site-specific vapor dispersion requirements. For

example, adding slats to a facility's security fence can, in some cases, provide sufficient flow restriction to keep the vapor dispersion exclusion zones onsite; and

- Are relatively inexpensive to build and maintain.

Figure 1 shows the comparison between the jetting and flashing vapor exclusion zones predicted by PHAST and FLACS. The FLACS model includes facility equipment and a vapor barrier for a small-scale LNG liquefaction facility.

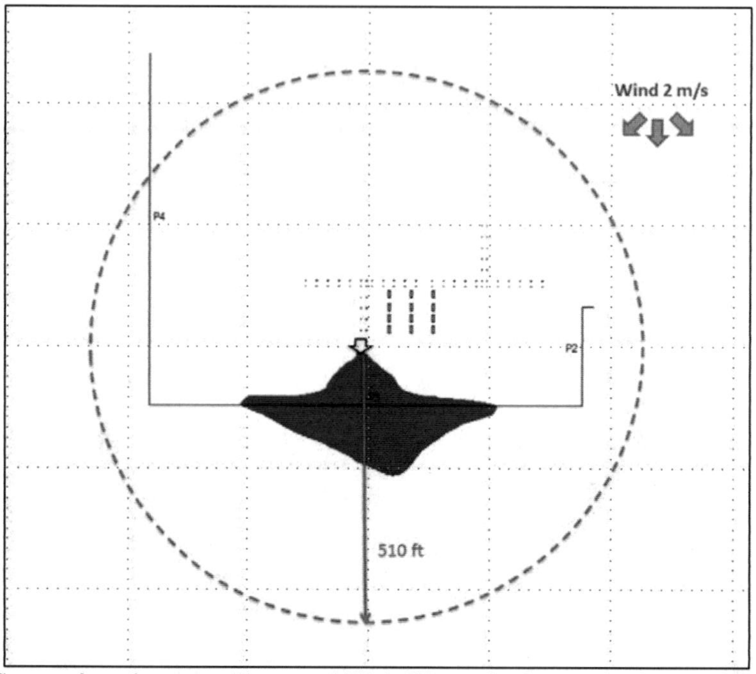

FIGURE 1: Comparison between Phast and FLACS results from a jetting and flashing release.

In addition to jetting and flashing, LNG conveyance spill vapor clouds may be difficult to control at small-scale facilities. However, design recommendations can be incorporated from GexCon and CH·IV's intimate knowledge of facility siting analysis. An advanced layout of the facility's impoundment features (including the location of the sump and routing of the trenches) and detailed FLACS modeling that includes all relevant aspects (including piping, equipment, buildings and topography), can help to significantly reduced vapor cloud production. Figure 2 illustrates a unique design for reducing vaporization rates for large surface areas surrounding LNG storage tanks.

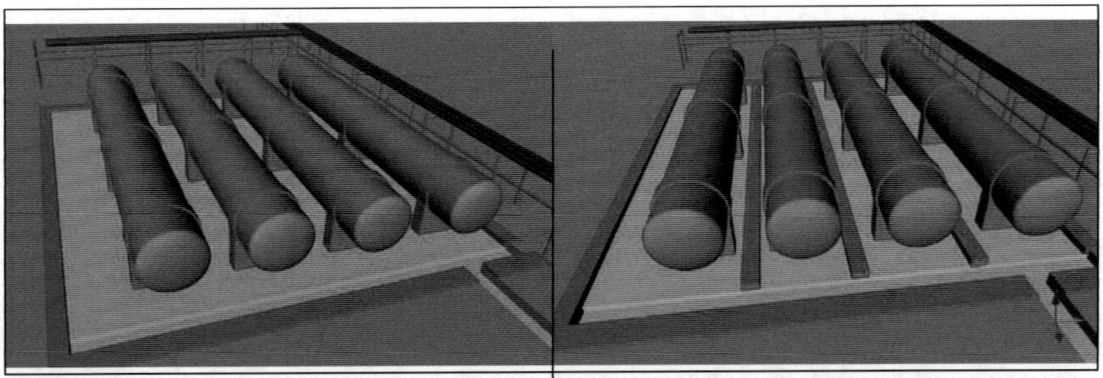

FIGURE 2. LNG storage tank area: (left) as originally designed; (right) mitigated design, with inter-tank curbing to reduce pool spreading.

Figure 3 shows a before and after example of design features applied to a small-scale LNG facility. In this case, modeling the liquid spills using FLACS and including the specialized impoundment features demonstrated that the exclusion zones could be significantly reduced.

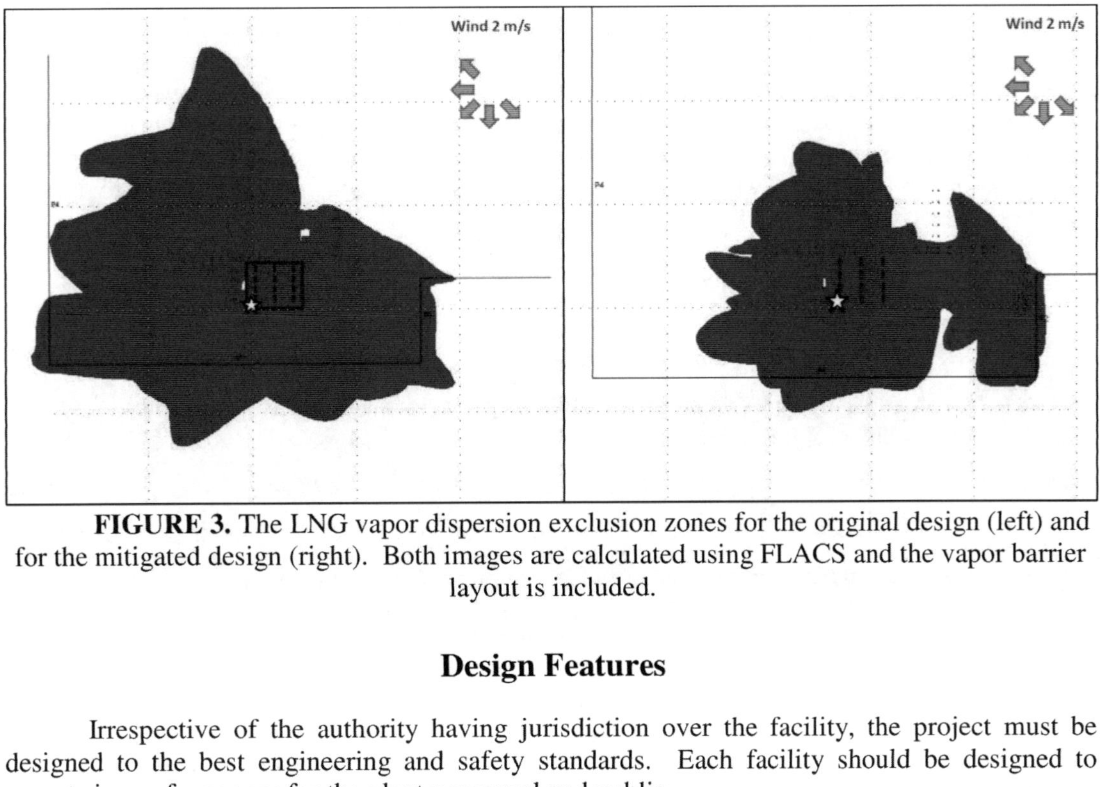

FIGURE 3. The LNG vapor dispersion exclusion zones for the original design (left) and for the mitigated design (right). Both images are calculated using FLACS and the vapor barrier layout is included.

Design Features

Irrespective of the authority having jurisdiction over the facility, the project must be designed to the best engineering and safety standards. Each facility should be designed to operate in a safe manner for the plant personnel and public.

Hazard Detection and Mitigation Design Features

Even though facility siting regulations do not allow credit for the presence of hazard detection systems, such systems should be used in order to reduce the consequences of accidental releases.

Hazard detection system designs focus on the hazards that have severe consequences to the public, to facility personnel, and to terminal equipment. The systems should consist of local detectors, equipment to process data from the detectors, audible and visual notification devices, and system activating devices.

The following hazard detection system features are typically provided for an LNG project:

- Detectors for liquid (LNG or refrigerant) spills, flammable gas (natural gas, or other refrigerant vapor) releases, toxic gas releases,
- Detectors for process and non-process fire conditions,
- Notifications to operators and plant personnel of the hazardous conditions, and
- Automatically initiated mitigating actions/systems to minimize the hazard consequences

The detection features to rapidly detect the presence of vapor formation and / or LNG resulting from LNG leaks from pipelines and equipment are included to rapidly initiate automated emergency shutdown systems that will isolate systems and equipment, thereby rapidly removing system operating pressure. Communications with other systems and additional panels are provided for redundancy, improved response, and activation of other systems (i.e., ESD system). Facilities must have a hazard monitoring philosophy that will define the proper equipment and how it will integrate with the DCS. Fire water and, where appropriate, deluge systems shall be provided to protect personnel, equipment and facilities.

Hazards from potential LNG spills and ignition shall be mitigated by a combination of fire and vapor suppression systems, which may include:

- Dry chemical systems, fixed and mobile
- Dedicated fire water system with water storage, electric and diesel pumps, hydrants and fixed monitors
- Dedicated water deluge and sprinkler applications, if needed
- Dedicated Safe Agent fire suppression systems for electrical and instrument rooms
- High expansion foam systems for LNG impoundment areas

The items listed above are common for large LNG facilities. Small-scale facilities, however, may be located in remote areas and be equipped with minimal equipment, therefore, some of the active mitigation features may be unnecessary, provided sufficient passive features are included in the design. For example, high expansion foam is not required by federal code, but its use is considered good practice for many facilities. Therefore, the additional cost of a high expansion foam system can be avoided in a small facility, if proper spacing is maintained between impoundment areas and critical equipment.

Sound engineering practice should be incorporated in the decision for every facility when determining the extent of mitigation features.

Containment Design Features

Each LNG facility must be designed with a system to contain liquid spills in the event of an equipment leak or break. Recommended practice is that a liquid collection and containment system be provided to convey liquid into a remote impoundment located at a safe distance away from equipment. If a facility includes refrigerants such as propane and ethane, impoundments should be provided for those materials storage and process equipment as well.

NFPA 59A 2013 includes design standards for LNG pipe-in pipe systems. A popular pipe-in-pipe design is a vacuum-jacketed configuration. The vacuum provides insulation properties unrivaled by current mechanical insulation. The outer pipe can also be designed with stainless steel to withstand cryogenic liquid in the event that the inner pipe leaks.

Conclusion

A fundamental requirement for the safe design of LNG facilities in the United States is that the public and public property be protected from credible accidental events within the facility. The current regulatory requirements for facility siting can be significantly different depending on whether a facility is under federal (DOT PHMSA), state or local jurisdiction. Therefore, one of the first steps in the planning of an LNG facility should be to determine the applicable jurisdiction. Engaging technical companies with the proper tools and experience in the various regulatory environments can help steer the early design in a direction that facilitates regulatory compliance. This simple step can significantly reduce the overall duration of the permitting process, by providing the authority having juristiction with the required information (thus minimizing follow-up questions) and by minimizing design changes late in the project.

REFERENCES

1. National Fire Protection Association (NFPA), NFPA 59A: Standard for the Production, Storage and Handling of Liquefied Natural Gas (LNG), NFPA, Qunicy, MA, 2001
2. National Fire Protection Association (NFPA), NFPA 59A: Standard for the Production, Storage and Handling of Liquefied Natural Gas (LNG), NFPA, Qunicy, MA, 2013
3. Title 49 Code of Federal Regulations Part 193(49-CFR-193): Liquefied Natural Gas Facilities: Federal Safety Standards, U.S. Government Printing Office, Washington, DC, 1980.

EFFECTS OF EXPANSION FOAM ON CONTROLLING LNG VAPORIZATION RATE

Bin Zhang[1], Yi Liu[1], Tomasz Olewski[2], Luc Vechot[2] and M. Sam Mannan[1*]

[1]Mary Kay O'Connor Process Safety Center, Artie McFerrin Department of Chemical Engineering, Texas A&M University, College Station, Texas 77843-3122, USA

[2]Chemical Engineering Program, Texas A&M University at Qatar, Doha, Qatar

Abstract

With increasing consumption of natural gas, the safety of liquefied natural gas (LNG) utilization has become a challenge that requires comprehensive studies on risk assessment in the LNG facilities with mitigation measures. The rapid vaporization after an LNG spill generates a vapor cloud, which is flammable at a concentration between roughly 5-15%v/v in the air. The LNG vapor is heavier than air and may migrate at the ground level over a long distance in a downwind direction, which increases the possibility of ignition resulting in fires. High expansion foam has been proved to be effective for mitigating vapor hazard and suppressing a pool fire. With respect to vapor hazard mitigation, high expansion foam may reduce the vaporization rate and increase vapor buoyancy. These effects have not been fully analyzed yet. In this work, a small-scale field experiment was performed to study the effect of high expansion foam on the vaporization rate of liquid nitrogen, used as a safe analog of LNG. It has been found that a foam blanket reduces convective and radiative heat flux to the pool. At the same time, the water drainage from collapsing foam adds additional heat to the pool; however, this was found to be small and negligible with the long lasting pools.

Key words: LNG, high expansion foam, vaporization rate, blanketing effect, mitigation

Introduction

Natural gas is a flammable mixture consisting of up to 99% of methane, and other components include ethane, propane, butane, carbon dioxide and nitrogen. The combustion of natural gas emits little sulfur dioxide, small amounts of nitrogen oxides, no ash, and much less carbon dioxide per unit of energy produced when compared to oil and coal according to the U.S. Energy Information Administration (EIA);[1] therefore, it is favored as an environmentally attractive fuel. Now, natural gas is the world's fastest-growing fossil fuel with a consumption growth projection of 185 trillion cubic feet in 2040 from 113 trillion cubic feet in 2008.[2]

* Corresponding author: M. Sam Mannan, mannan@tamu.edu, +1-979-862-3985

Liquefied Natural Gas (LNG) is the liquid form of natural gas at atmospheric pressure, which facilitates storage and transportation of natural gas, but raises safety concerns in terms of vapor hazard and pool fire. Because of the cryogenic nature of LNG, an accidental release of LNG in massive volume can result in an LNG pool with continuous vaporization followed by the formation of an LNG vapor cloud. The high density of LNG vapor associated with low temperature makes it more likely to migrate in a downwind direction at the ground level, which causes a huge flammable vapor cloud as the vapor hazard. NFPA 59 A requires a thorough evaluation of potential consequences, and requires mitigation measures to reduce the risk to a tolerable level.[3]

High expansion foam has been recommended in NFPA 11 and NFPA 471 for LNG hazard mitigation.[4,5] Expansion foam is produced by aeration of a foam solution. It is classified into three categories by expansion ratio: low expansion foam with expansion ratios up to 20, medium expansion foam with expansion ratios from 20 to 200, and high expansion foam with expansion ratios from 200 to approximately 1,000. Because the water drainage of foam can act as a heat source for the LNG vaporization, high expansion foam possesses better performances than low expansion foam, since it has a smaller water drainage rate. The vapor hazard of an LNG spill is believed to be mitigated by high expansion foam through two mechanisms: (1) reducing the vaporization rate to generate fewer vapors and (2) raising the vapor buoyancy to enhance vapor dispersion. The use of high expansion foam to mitigate the vapor hazard of LNG has been discussed previously, but it is limited to the warming effect of foam on raising LNG vapor buoyancy.[6] On the other hand, high expansion foam provides a large volume of foam blanket above the LNG pool in the impoundment, which can block other heat sources transferred to the LNG pool, *e.g.*, convection and radiation, to reduce vaporization rate, defined as the blockage effect. The boil-off effect is the vaporization caused by the foam water drainage. The blanketing effect of foam on LNG vaporization is a combination of both competing effects discussed above. Since the blockage effect of foam on convection and radiation was not studied, the blanketing effect on the LNG vaporization rate remains unclear.

In 1974, University Engineers Inc. conducted a series of vapor dispersion tests with an LNG spill on land. The effect of expansion foam on vapor dispersion was examined, and the vapor concentration can be reduced up to 80% during an LNG high boil-off period at a distance of one pool diameter of the spill.[7] In 1996, Takeno and his colleagues studied the ability of high expansion foam to raise the temperature of vaporized cryogenic gas; They concluded that approximately 92% of the heat held in the applied foam was transferred into vapor to increase the buoyancy in this specific test.[8] Since 2005, Mary Kay O'Connor Process Safety Center has performed a series of LNG spill tests with high expansion foam at the Brayton Fire Training Field (BFTF) in College Station, TX, USA.[9,10,11] Suardin (2009) confirmed the reduction of LNG vapor concentration at the ground level after high expansion foam application using a hydrocarbon camera;[12] Yun (2011) determined that foam application is effective for reducing the Lower Flammable Limit (LFL) distance by 89%.[10] The effectiveness of high expansion foam has been proven by these studies in terms of reducing downwind LNG vapor concentration or LFL distance. However, more work is required to study the mitigation mechanisms of high expansion foam for vapor hazards of LNG.

With respect to the blanketing effect on vaporization rate, it has been concluded that foam application acted as a heat source for vaporization, but the blockage effect of foam on convection and radiation was not studied. [8]

Experimental setup

This work aims to experimentally study the blanketing effect of high expansion foam on the vaporization rate of an LNG spill on land. All the tests in this work were conducted in a wind tunnel with approximate dimensions of 10m x 2.5m x 2.5m, in which natural interference was prevented, and additional props were installed to apply stable forced convection and radiation. Liquid nitrogen (LN_2) was used due to its similar thermal properties with LNG so that the results are valid to explain LNG behavior. LN_2 is also safer to operate, easier to access, and less expensive. The blockage effect of foam to eliminate convection and radiation was studied through seven well-designed tests, in which various heat transfer mechanisms, *i.e.*, natural convection, forced convection, and radiation, were applied. The boil-off effect on vaporization associated with the water drainage of foam was also examined. Then, the blanketing effect was characterized quantitatively by comparing these two effects.

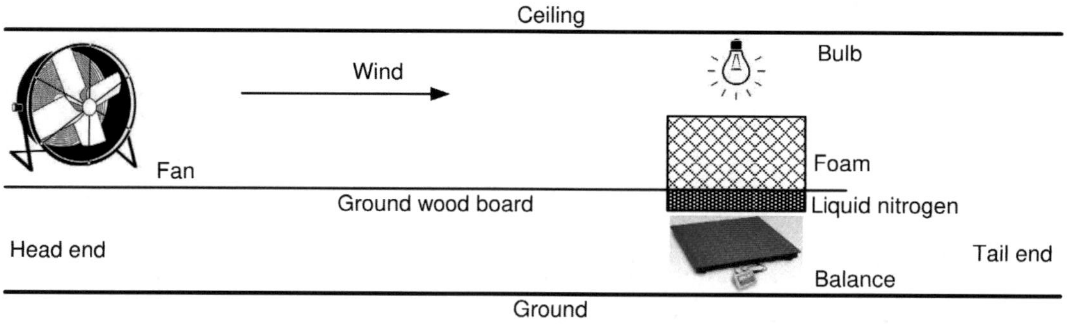

Figure 1 Schematic diagram of the wind tunnel

Test apparatus and materials

To study the vaporization process, a container with a capacity of 0.023 m^3 (0.48 m x 0.48 m x 0.10 m) was used to contain LN_2. It was made by combining several pieces of polystyrene foam board using anti-cryogen paste. The use of polystyrene can minimize the conduction from the container to the LN_2 pool, which is critical to ensure the experimental accuracy of the study on convection and radiation. The container has five heat flux sensors embedded at one depth and eight thermocouples at two different depths within the bottom wall of the container to monitor the temperature profile and heat flux through the bottom wall. Two additional heat flux sensors were attached to the outside surface of the side walls with thermally conductive paste to measure the heat flux through the side walls and the temperature of outside surfaces. An additional 30 centimeters of transparent fence was installed above the polystyrene container in Test 2-Test 7 to be described in detail in the section for experimental procedure, and foam of 30 centimeters high was applied in Test 5-Test 7 ("foam" refers to high expansion foam hereafter unless otherwise specified).

This experiment was conducted in a wind tunnel located at Ras Laffan Industrial City (RLIC) in Qatar with a setup as shown in Figure 1. The inner surface of the wind tunnel was decorated with wood boards to ensure a good wind tunnel effect, and the ground wood boards were fixed at approximately 1m high, the same elevation as the upper edge of the polystyrene container. The test container was placed on a balance (4 load cell mild steel scale, RadWag) with readability of 10 g to monitor the mass dropping rate of LN_2. The scale was placed in the center line of the wind tunnel, approximately 6 m away from the head end of the wind tunnel, where a blower fan was placed to supply the forced convection with a maximum wind speed of 2.2 m/s. The blower fan generated the wind of forced convection, and the wind speed was measured using an ultrasonic anemometer (Model 81000, R.M. Young Company). A panel with 9 bulbs was employed to provide radiation, which was measured using a radiation sensor (LP02, CAMPBELL SCIENTIFIC, INC.). All the measuring equipment, including thermocouples, heat flux sensors, radiation sensor, anemometers, and the balance, were connected to a data acquisition system (DAQ) to record the experimental data for a systematic analysis. A summary of experimental facts is illustrated in Table 1.

Table 1 Summary of Experimental Facts

Air temperature		39.06 ± 1.54 °C
Relative humidity		20.65 ± 9.77
Maximum wind applied		1.82±0.24 m/s
Average radiation to liquid		509±46 W/m^2
Expansion foam	Estimated expansion ratio	800 : 1
	Foam concentrate	Expandol
	Foam generator	Fomax 7

Experimental procedure

To study the blanketing effect of foam on the LN_2 vaporization rate, seven tests were designed and conducted with various conditions through a combination of foam application, convection, and radiation.

Conduction study

An advantage of using polystyrene to build the container is that conduction from the container can reach a steady state in a limited time, and Reid's (1980) work has obtained similar results using other insulating material.[13] In order to characterize the blanketing effect of foam, the constant conduction was determined in Test 1. The head end was covered by two wood boards to prevent air circulation in the wind tunnel. LN_2 was spilled into the container up to 18kg; then, the container was covered by a polystyrene lid to eliminate convection and radiation as shown in Figure 2B. The test continued for approximately one hour to cool down the container, and a constant conduction was achieved at approximately 22 minutes. The constant conduction was determined by the mass measurement using the balance, and validated by a theoretical calculation based on temperature and heat flux measurements at the bottom of the container.

Convection and radiation study without foam

The contribution of various heat sources, including natural convection, forced convection, and radiation, was studied without foam application in Test 2, Test 3, and Test 4, respectively. After Test 1, the lid was removed and the transparent fence was installed to start Test 2 as shown in Figure 2A. In this test, the vaporization rate of LN_2 was determined by conduction and natural convection. After about 30 minutes, the wood boards at the head end were removed, and the fan was turned on to start Test 3. In this test, the vaporization rate of LN_2 was determined by conduction and forced convection. After another 20 minutes, the fan was stopped, and the head end of wind tunnel was covered with wood boards again. The bulbs panel was installed above the container as shown in Figure 2C. Test 4 began by turning on the bulbs. In this test, the vaporization rate of LN_2 was determined by conduction, natural convection, and radiation.

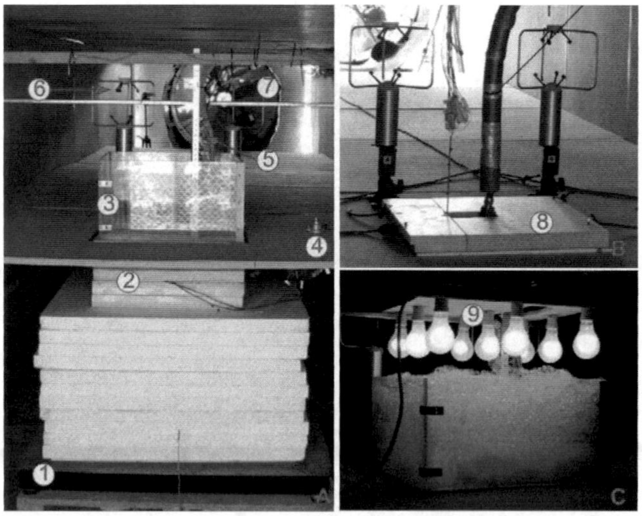

Figure 2 The experimental setup for LN_2 vaporization tests. A: the setup for Test 2-7; B: the setup for Test 1; C: the setup with bulbs panel (1, balance; 2, polystyrene container; 3, fence for foam containment; 4, radiation sensor; 5, aluminum bar with 26 thermocouples; 6, ultrasonic anemometer; 7, fan; 8, polystyrene lid; 9, bulbs panel)

Convection and radiation study with foam

The blanketing effect of foam on the vaporization rate under various conditions was studied through Test 5, Test 6, and Test 7, which were analogous tests of Test 2, Test 3, and Test 4, but with foam present in the transparent fence. Since foam is not a stable medium, it disappears between 10 and 20 minutes depending on the test conditions. In order to obtain reliable data, foam application was repeated for three times in each test.

Data analysis method

A typical mass curve of LN_2 obtained from Test 5, is shown in Figure 3. The vaporization rate of LN_2 is determined by linear regression of the mass curve for each test; therefore, the slope is vaporization rate, and the high coefficient of determination confirms a constant vaporization rate of LN_2 in each test; During each foam application, the vaporization rate is also examined at 100-second intervals, which further confirms a steady control of foam on the vaporization rate. Once the vaporization rate is determined, it will be converted to heat flux into an LN_2 pool.

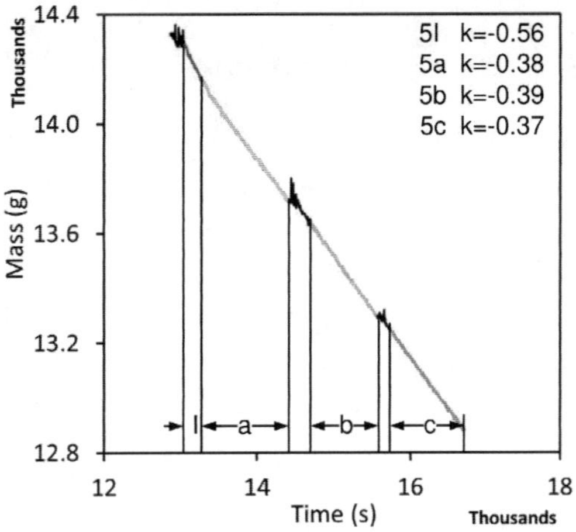

Figure 3 Mass loss rates of Test 5. k is the slope for each test and indicates the vaporization rate; I represents the initial stage of first foam application; a, b, and c represent three repeated tests with foam application

Results and discussion

Conduction

The use of polystyrene for LN_2 container construction benefits the experiment in several aspects. First, it guarantees a steady state of conduction. Conduction is a transient heat source. It supplies an extreme high heat flux at the beginning, and will decrease as the container is cooled by the cryogenic liquid. As discussed by Reid (1980), conduction can reach a steady state in a limited time for a container made of insulating material.[13] Therefore, convection and radiation can be easily obtained by subtracting the constant conduction from the total heat flux. Second, polystyrene greatly reduces the conduction from the container to LN_2. With the container made of polystyrene, the conductive heat flux is only approximately 2% of that in a regular concrete container; therefore, convection and radiation are relatively significant in this experiment, and the effects of foam can be measured more accurately. These two features allow the study of blanketing effects of foam on convection and radiation quantitatively.

The blanketing effect of foam

The blanketing effect is the ability of foam to reduce convection and radiation, and therefore reduce the vaporization rate. By comparing the individual heat flux of each heat transfer mechanism in Tests 5-7 with that in Tests 2-4, natural convection, forced convection, and radiation after foam application are blocked by 74%, 70%, and 62%, respectively. The remaining heat flux are the boil-off effect, which is determined by the intrinsic foam property affecting water drainage rate. It has been noticed that foam provides additional heat flux at the initial stage of first foam application (before the formation of the ice plate) as shown in Figure 3, but it is short, small, and negligible compared to its role in reducing the convective or radiative heat transfer.

Conclusion

This work quantitatively studied the effects of foam on different heat transfer mechanisms for a cryogenic liquid pool, including natural convection, forced convection, and radiation. Experimental conditions were well controlled in a wind tunnel specially built for this experiment. The main findings in this work are that the blanketing effect of foam helps reduce the heat input from convection and radiation to the liquid nitrogen pool, which has been studied quantitatively. Water drainage from breaking foam before the formation of the ice plate adds extra heat to vaporize liquid nitrogen, but it is small and negligible compared to its role in reducing the convective or radiative heat transfer.

Acknowledgement

This research was funded by BP Global Gas SPU and Qatar National Research Foundation (NPRP No. 6-425-2-172). Special thanks to Qatar Petroleum for providing support on experiments conducted at fire station 2 in Ras Laffan Industrial City, Qatar.

References

1. Energy Information Administration, Natural Gas 1998 Issues and Trends, 1999.
2. U.S. Energy Information Administration, International Energy Outlook 2013, Washington, DC, 2013.
3. National Fire Protection Association, NFPA 59A: Standard for the Production, Storage, and Handling Of Liquefied Natural Gas (LNG), (2013).
4. National Fire Protection Association, NFPA 11: Standard for Low-, Medium, and High-Expansion Foam, (2010).
5. National Fire Protection Association, NFPA 471: Recommended Practice for Responding to Hazardous Materials Incidents, (2002).
6. R. Hiltz, Foam blanketing: the use of foam to mitigate the vapor hazard of spilled volatile chemicals, in: Prev. Control Accid. Releases Hazard. Gases, John Wiley & Sons, 1993: pp. 216–231.

7. University Engineers Inc., An Experimental Study on the Mitigation of Flammable Vapor Dispersion and Fire Hazards Immediately Following LNG Spills on Land, The Association, Arlington, Va., 1974.
8. K. Takeno, T. Ichinose, K. Tokuda, R. Ohba, K. Yoshida, K. Ogura, Effects of high expansion foam dispersed onto leaked LNG on the atmospheric diffusion of vaporized gas, J. Loss Prev. Process Ind. 9 (1996) 125–133.
9. J.A. Suardin, Y.J. Wang, M. Willson, M.S. Mannan, Field experiments on high expansion (HEX) foam application for controlling LNG pool fire, J. Hazard. Mater. 165 (2009) 612–622.
10. G. Yun, D. Ng, M.S. Mannan, Key Observations of Liquefied Natural Gas Vapor Dispersion Field Test with Expansion Foam Application, Ind. Eng. Chem. Res. 50 (2011) 1504–1514.
11. G. Yun, D. Ng, M.S. Mannan, Key Findings of Liquefied Natural Gas Pool Fire Outdoor Tests with Expansion Foam Application, Ind. Eng. Chem. Res. 50 (2011) 2359–2372.
12. J.A. Suardin, The application of expansion foam on liquefied natural gas (LNG) to suppress LNG vapor and LNG pool fire thermal radiation, Texas A&M University, 2008.
13. R.C. Reid, Boiling of LNG on Typical Dike Floor Materials, (1980).

AUTHOR INDEX

AUTHOR INDEX

American Institute of Chemical Engineers
3 Park Avenue
New York, New York 10016-5991

CURRAN ASSOCIATES INC
proceedings
.com